# Total Quality in the Construction Supply Chain

## John Oakland
Professor of Business Excellence and Quality Management
Leeds University Business School
Executive Chairman, Oakland Consulting plc

## Marton Marosszeky
Multiplex Professor of Engineering Construction Innovation
School of Civil and Environmental Engineering
University of New South Wales
Managing Director, Marosszeky and Associates Pty Ltd

AMSTERDAM • BOSTON • HEIDELBERG • LONDON • NEW YORK • OXFORD
PARIS • SAN DIEGO • SAN FRANCISCO • SINGAPORE • SYDNEY • TOKYO

Butterworth-Heinemann is an imprint of Elsevier

Butterworth-Heinemann is an imprint of Elsevier
Linacre House, Jordan Hill, Oxford OX2 8DP
30 Corporate Drive, Burlington, MA 01803

First edition 2006

**British Library Cataloguing in Publication Data**
A catalogue record for this book is available from the British Library.

**Library of Congress Cataloguing in Publication Data**
A catalog record for this book is available from the Library of Congress.

ISBN-13: 978-0-7506-6185-0
ISBN-10: 0-7506-6185-2

Typeset by Charon Tec Ltd, Chennai, India
www.charontec.com
Printed and bound in Great Britain

06 07 08 09 10    10 9 8 7 6 5 4 3 2 1

# Contents

# Preface

Relatively few in the construction sector may be familiar with John Oakland's book *Total Quality Management*, first published in 1988 and now in its third edition. The book is widely recognized in other sectors and has now sold over 10 000 copies. When we first spoke about a book for the construction sector, John's reaction was: Is the building and construction industry ready for a book on Total Quality?

But Marton who, for the past fifteen years, has specialized in research, teaching and consulting in building and construction process improvement, in the areas of quality, safety and production improvement, argued that the industry is changing. Pressure from clients and governments as well as commercial competitive pressures are forcing lead companies in the sector to differentiate themselves on the basis of customer focus, overall product and process quality, and especially, service quality.

In response to these pressures, senior management in lead companies worldwide are embracing the basic philosophy of *Total Quality*. Often approaching the overall task from different perspectives, some adopt the framework of performance measurement and benchmarking, others use the goal of continuous improvement while others choose to follow the values and concepts of *Lean* construction. We see all of these as simply different perspectives of the same broad objective, improving performance in all the activities of a business.

Traditionally in conversations about quality, the building and construction sector has had a natural orientation towards product quality. Given the complexity of its organizational relationships and traditional craft based processes, most of the construction quality literature reflects this product focus: either providing a guide to compliance with the ISO 9000 series of international quality standards or providing pragmatic advice on tools for the control of quality. However, lead organizations in every area of the building and construction industry have recognized that the broad focus that TQM brings to all aspects of organizing and managing is as relevant to building and construction as it is to the manufacturing and service sectors. Furthermore, teachers and researchers in building and construction quality are realizing that the traditional product centred paradigm does not provide a sufficiently broad and robust basis for quality improvement within the sector.

This book is designed to address this critical need. As well as providing a broad and robust conceptual platform on which organizations can build their overall process improvement endeavours, the book integrates and places into a unified perspective the many seemingly disparate management innovations of the past twenty years.

Increasing the satisfaction of customers and stakeholders through effective goal deployment, cost reduction, productivity and process improvement has proved to be essential for organizations to stay in operation. It is now widely recognized that quality has developed into the most important competitive weapon that any organization has in its armoury, and that a comprehensive quality focused strategy is the most robust way of managing for the future. TQM is far wider in its application than assuring product or service quality – it is a way of managing organizations and their supply chains so that every aspect of performance, both internally and externally is improved.

Our book provides guidance on how to manage building and construction sector businesses in a total quality way. It is structured in the main around four parts of a new model for TQM – improving *Performance* through better *Planning* and management of *People* and the *Processes* in which they work. The core of the model will always be performance in the eyes of the customer, but must be extended to include performance measures for all the stakeholders. This new core still needs to be surrounded by *commitment* to quality and meeting customer requirements, *communication* of the quality message, and recognition of the need to change the *culture* of most organizations to create total quality. These are the *soft* foundations which must encase the hard management necessities of planning, people and processes.

Under these headings the essential steps for the successful implementation of *Total Quality* are set out in what we believe is a meaningful and practical way. The book guides the reader through the language of *Total Quality* and all associated recent developments and sets down a clear way for an organization to proceed.

At the end of the book there are 14 case studies to support the text. Each of these presents the approach and achievements of a different organization within the building and construction sector. They include private and public sector organizations in a wide range of business areas: quarrying gravel, design and construction, real estate, property management and public sector services and policy. Each case study is linked back to the specific areas of the book that it illustrates.

Many of the new approaches to quality and improving performance appear to present different theories. However in reality they are talking the same language using different dialects. The basic principles of defining quality and taking it into account throughout all the activities of the business are common. Quality has to be managed – it does not just happen. Understanding and commitment by senior management, effective leadership, teamwork and good process management are fundamental parts of the recipe for success. We have tried to use our extensive research and consultancy experience to take what is to many a jigsaw puzzle and to assemble a comprehensive, practical working model for *Total Quality* – the rewards of which are greater efficiencies, lower costs, improved reputation and customer loyalty. Moreover, we have tried to show how holistic TQM now is – embracing the most recent models of excellence, six sigma, and a host of other management methods and teachings.

The book should meet the requirements of both students and practitioners who have or are planning careers within the built environment industry including engineers, architects, building and construction contractors, real estate and facility management professionals. In the operations of any organization within the building and construction sector there is a need to understand the broad implications that the *Total Quality* approach holds for its entire supply chain and internal and external customers.

We hope that this is not seen as a specialist text for specialist practitioners in quality. We see *Total Quality* as providing the fundamental building blocks for the management of any organization, and, hence, people working in every part of each organization need to understand this broad perspective. This book documents a comprehensive approach to the management of any business enterprise – one that has been used successfully by many organizations throughout the world.

We would like to thank our colleagues in the European Centre for Business Excellence, at Oakland Consulting plc and at the Australian Centre for Construction Innovation at the University of New South Wales for the sharing of ideas and help in their development. This book is the result of many years of collaboration in assisting organizations to introduce good methods of management and embrace the concepts of *Total Quality*. We are most grateful to Marti Marosszeky who worked with us throughout this project and helped convert our ideas into fluent text!

Preface

# 1

# Understanding quality

# Quality, competitiveness and customers

Over the past 25 years, professionals in construction have watched industry leaders in other sectors effectively implement quality-based philosophies to transform their enterprises into market leaders. Yet, in this sector, most companies have struggled to fully understand what the quality movement can offer it. Since the mid-1980s a few leading construction industry organizations around the world have demonstrated how to solve this riddle. This book presents their experiences in the form of a series of case studies in the context of a comprehensive framework for total quality management.

The construction industry does have a number of specific issues which create difficulties for the sector. The first of these is the extreme fragmentation of supply chains and the short-term nature of relationships within them. Relationships between head contractors and subcontractors are essentially project based, although in the past decade lead suppliers have started to take a longer-term view of some of their key relationships. A second factor is the long timeframe of procurement. For large projects, conception to completion often takes years and, within this time, many of the individuals involved in the process will change, making continuity a challenge. The final and most profoundly limiting issue is the unilateral definition of quality – the concept is usually regarded almost entirely in product terms. The building process – a collaboration between numerous suppliers from the early design stage to completion of construction – is not *owned* by any one party; rather, it is achieved through the negotiations of the many players whose predominant focus is the product. It is as if the 'process' was incidental and, yet, it is the industry's processes which determine key outcomes. The difficulties that exist within these process are remarkably similar the world over.

- *Safety* – construction is one of the most hazardous industries – its accident record is third behind mining and forestry. New legislation has now made head contractors, designers and project managers more responsible and potentially liable for injuries incurred by workers on a site, regardless of who had employed them.
- *Quality* – research has shown that the cost of rectifying quality errors during and after the contract is of the same order as the profitability of organizations in the sector. Product quality problems are reflected in leaking buildings and premature deterioration of external finishes.[1]
- *Reliability* – on most project sites, only two out of three tasks planned one week out are actually completed according to plan. This reflects that while individual contractors within the supply chain struggle to maintain focus on their own profitability the overall process is highly unreliable.
- *Ineffective decision-making* – design often takes longer than anticipated and early budgets are rarely accurate. This reflects that the gap between design and construction is too wide and that the design decision-making process is not effective, this includes the decision-making role of clients.

Against this backdrop of complexity and poor performance, contractors and consultants both claim that their profit margins are very low. The emergence of very large global organizations in real estate, material supply, general contracting and specialist contracting, in engineering design and, to a lesser extent, in architecture during the past two decades has led to rapid restructuring in some sections of the industry.

Whatever the type of organization, whether it is an engineering or architectural design practice, a developer, a contractor, a building product manufacturer or a project manager – competition is rife: competition for end customers, for employees, for projects and for funds. Any organization basically competes on its *reputation* – for quality,

3

reliability, price and delivery – and most people now recognize that quality is the most important of these competitive weapons. If you doubt that, just look at the way some organizations, even whole industries in certain countries, have used quality to take the heads off their competitors. American, British, French, German, Italian, Japanese, Spanish, Swiss, Swedish organizations, and organizations from other countries, have used quality strategically to win customers, steal business resources or funding, and be competitive. Moreover, this sort of attention to quality improves performance in reliability, delivery, and price.

For any organization, there are several aspects of reputation which are important:

1.  It is built upon the competitive elements of quality, reliability, delivery, and price, of which quality has become strategically the most important.
2.  Once an organization acquires a poor reputation for quality, it takes a very long time to change it.
3.  Reputations, good or bad, can quickly become national reputations.
4.  The management of the competitive weapons, such as quality, can be learned like any other skill, and used to turnaround a poor reputation, in time.

Before anyone will buy the idea that quality is an important consideration, they would have to know what was meant by it.

## What is quality?

'Is this a quality watch?' Pointing to my wrist, I ask this question of a class of students – undergraduates, postgraduates, experienced managers – it matters not who. The answers vary:

- 'No, it's made in Japan.'
- 'No, it's cheap.'
- 'No, the face is scratched.'
- 'How reliable is it?'
- 'I wouldn't wear it.'

My watch has been insulted all over the world – London, New York, Paris, Sydney, Brussels, Amsterdam, Leeds! Clearly, the quality of a watch depends on what the wearer requires from a watch – perhaps a piece of jewellery to give an impression of wealth; a timepiece that gives the required data, including the date, in digital form; or one with the ability to perform at 50 metres under the sea? These requirements determine the quality.

Quality is often used to signify 'excellence' of a product or service – people talk about 'Rolls-Royce quality' and 'top quality'. In building, the different standards of finish specified for plasterwork depend on the location and lighting of the surface. Concrete can be manufactured to have low shrinkage and highly impervious characteristics to ensure that it performs well in a harsh environment. In some manufacturing companies the word may be used to indicate that a piece of material or equipment conforms to certain physical dimensional characteristics often set down in the form of a particularly 'tight' specification. In a hospital it might be used to indicate some sort of 'professionalism'. If we are to define quality in a way that is useful in its *management*, then we must recognize the need to include in the assessment of quality the true requirements of the 'customer' – the needs and expectations.

*Qual*  s simply *meeting the customer requirements*, and this has been expressed in mar  by other authors:

- 'Fitnes. or purpose or use' – Juran, an early doyen of quality management.
- 'The totality of features and characteristics of a product or service that bear on its ability to satisfy stated or implied needs' – BS 4778: 1987 (ISO 8402, 1986) *Quality Vocabulary*: Part 1, *International Terms*.
- 'Quality should be aimed at the needs of the consumer, present and future' – Deming, another early doyen of quality management.
- 'The total composite product and service characteristics of marketing, engineering, manufacture and maintenance through which the product and service in use will meet the expectation by the customer' – Feigenbaum, the first man to publicize a book with 'Total Quality' in the title.
- 'Conformance to requirements' – Crosby, an American consultant famous in the 1980s.
- 'Degree to which a set of inherent characteristics fulfils requirements' – ISO (EN) 9000:2000 *Quality Management Systems – fundamentals and vocabulary*.

In the case of constructed products, the work of the architect, engineers, product manufacturers, general and specialist contractors plays a critical role in meeting the expectations of the end user, the final customer for the product. A failure in the work of any one of them can jeopardize the work of the entire team.

Another word that we should define properly is *reliability*. 'Why do you buy a BMW car?' 'Quality and reliability' comes back the answer. The two are used synonymously, often in a totally confused way. Clearly, part of the acceptability of a product or service will depend on its ability to function satisfactorily *over a period of time*, and it is this aspect of performance that is given the name *reliability*. It is the ability of the product or service to *continue* to meet the customer requirements. Reliability ranks with quality in importance, since it is a key factor in many purchasing decisions where alternatives are being considered. Many of the general management issues related to achieving product or service quality are also applicable to reliability.

It is important to realize that the 'meeting the customer requirements' definition of quality is not restrictive to the functional characteristics of products or services. Anyone with children knows that the quality of some of the products they purchase is more associated with *satisfaction in ownership* than some functional property. This is also true of many items, from antiques to certain items of clothing. The requirements for status symbols account for the sale of some executive cars, certain bank accounts and charge cards, and even hospital beds! The requirements are of paramount importance in the assessment of the quality of any product or service.

By *consistently* meeting customer requirements, we can move to a different plane of satisfaction – *delighting the customer*. There is no doubt that many organizations have so well ordered their capability to meet their customers' requirements, time and time again, that this has created a reputation for 'excellence'. A development of this thinking regarding customers and their satisfaction is *customer loyalty*, an important variable in an organization's success. Research shows that focus on customer loyalty can provide several commercial advantages:

- Customers cost less to retain than acquire.
- The longer the relationship with the customer, the higher the profitability.
- A loyal customer will commit more spend to its chosen supplier.
- About half of new customers come through referrals from existing clients (indirectly reducing acquisition costs).

Companies like 3M use measures of customer loyalty to identify customers who are 'completely satisfied', would 'definitely recommend' and would 'definitely repurchase'.

Refer to the Sekisui Heim case study (p. 469). Here is a company that is seeking to establish a truly long-term (more than 30-year) relationship with their purchasers as a central plank of their long-term business strategy. Not only are they selling expertise in whole-of-life costs and guiding purchasers to make informed investment decisions, but they also want to maintain and refurbish the homes that they sell as a key strategy for the growth of their business. This clearly signals to everyone in the company that they are committed to satisfying their customers for life.

Total Quality in the Construction Supply Chain

# Understanding and building the quality chains

The ability to meet the customer requirements is vital, not only between two separate organizations, but within the same organization.

When the air stewardess pulled back the curtain across the aisle and set off with a trolley full of breakfasts to feed the early morning travellers on the short domestic flight into an international airport, she was not thinking of quality problems. Having stopped at the row of seats marked 1ABC, she passed the first tray onto the lap of the man sitting by the window. By the time the second tray had reached the lady beside him, the first tray was on its way back to the hostess with a complaint that the bread roll and jam were missing. She calmly replaced it in her trolley and reached for another – which also had no roll and jam.

The calm exterior of the girl began to evaporate as she discovered two more trays without a complete breakfast. Then she found a good one and, thankfully, passed it over. This search for complete breakfast trays continued down the aeroplane, causing inevitable delays, so much so that several passengers did not receive their breakfasts until the plane had begun its descent. At the rear of the plane could be heard the mutterings of discontent. 'Aren't they slow with breakfast this morning?' 'What is she doing with those trays?' 'We will have indigestion by the time we've landed.'

The problem was perceived by many on the aeroplane to be one of delivery or service. They could smell food but they weren't getting any of it, and they were getting really wound up! The air hostess, who had suffered the embarrassment of being the purveyor of defective product and service, was quite wound up and flushed herself, as she returned to the curtain and almost ripped it from the hooks in her haste to hide. She was heard to say through clenched teeth, 'What a bloody mess!'

A problem of quality? Yes, of course; requirements not being met, but where? The passengers or customers suffered from it on the aircraft, but in another part of the organization there was a man whose job it was to assemble the breakfast trays. On this day the system had broken down – perhaps he ran out of bread rolls, perhaps he was called away to refuel the aircraft (it was a small airport!), perhaps he didn't know or understand, perhaps he didn't care.

Three hundred miles away on a building site … 'What the hell is Quality Control doing? We will have to demolish the entire wall, not only do we lose the tiles, the waterproof membrane and the sheeting, the electrician and plumber will also have to come back to relocate the electrical and hydraulic services so that we can move the wall frame into its correct location. Where is the idiot surveyor who set the thing out?' This was accompanied by a barrage of verbal abuse (which will not be repeated here) aimed at the shrinking quality control manager who tried to slink into his office, as the red-faced project manager advanced menacingly.

Was it the surveyor? Or was it the design detail of the architect or was it the tradesman who built the wall? What about the three subsequent trades, who all could have looked at the relationship between the position of the wall and the doorframe and seen that the finished wall would not line up. Each did his or her work well in isolation; however, a fundamental flaw at the outset compromised the entire job.

Do you recognize these two situations? Do they not happen every day of the week – possibly every minute somewhere in manufacturing or the service industries? Is it any different in banking, insurance, and the health service? The inquisition of checkers and testers is the last bastion of desperate systems trying in vain to catch mistakes, stop defectives, hold lousy materials, before they reach the external customer – and woe betide the idiot who lets them pass through!

Two everyday incidents, but why are events like these so common? The answer is the acceptance of one thing – *failure*. Not doing it right the first time at every stage of the process.

Why do we accept failure in the production of artefacts, the provision of a service, or even the transfer of information? In many walks of life we do not accept it. We do not say 'Well, the nurse is bound to drop the odd baby in a thousand – it's just going to happen.' We do not accept that!

In each department, each office, even each household, there are a series of suppliers and customers. The PA is a supplier to the boss. Are the requirements being met? Does the boss receive error-free information set out as it is wanted, when it is wanted? If so, then we have a quality PA service. Does the air steward receive from the supplier in the airline the correct food trays in the right quantity, at the right time?

Throughout and beyond all organizations, whether they be manufacturing concerns, banks, retail stores, universities, hospitals or hotels, there is a series of *quality chains* of customers and suppliers (Figure 1.1) that may be broken at any point by one person or one piece of equipment not meeting the requirements of the customer, internal or external. The interesting point is that this failure usually finds its way to the interface between the organization and its outside customers, and the people who operate at that interface – like the air steward – usually experience the ramifications. The concept of internal and external customers/suppliers forms the *core* of total quality management.

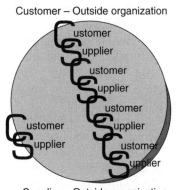

**Figure 1.1**
The quality chains

A great deal is written and spoken about employee motivations as a separate issue. In fact the key to motivation *and* quality is for everyone in the organization to have well-defined customers – an extension of the word beyond the outsider that actually purchases or uses the ultimate product or service to anyone to whom an individual gives a part, a service, information – in other words the results of his or her work.

In construction these relationships are often implicit rather than explicit. Throughout the process there is a very long chain of events: in planning and design, information is built on information; and in construction, material is built upon material. At every step of the process each person has a customer, though, often, because the contractual relationship is with a third party, the client or the head contractor, the process relationship is not perceived in this way. The plasterer is the customer of the bricklayer, and the plasterboard sheet-fixer is the customer of the carpenter who has framed the wall, even though the head contractor pays both of them.

Quality has to be managed – it will not just happen. Clearly it must involve everyone in the process and be applied throughout the organization. Many people in the support functions of organizations never see, experience, or touch the products or services that their organizations buy or provide, but they do handle or produce things like purchase orders or invoices. If every fourth invoice carries at least one error, what image of quality is transmitted?

Failure to meet the requirements in any part of a quality chain has a way of multiplying and a failure in one part of the system creates problems elsewhere, leading to yet more failure, more problems and so on. The price of quality is the continual examination of the requirements and our ability to meet them. This alone will lead to a 'continuing improvement' philosophy. The benefits of making sure the requirements are met at every stage, every time, are truly enormous in terms of increased competitiveness and market share, reduced costs, improved productivity and delivery performance, and the elimination of waste.

## Meeting the requirements

If quality is meeting the customer requirements, then this has wide implications. The requirements may include availability, delivery, reliability, maintainability and cost-effectiveness, among many other features. On construction sites accuracy of work, cleanliness of the work area and the completion of the work in a manner that will allow the following trade to continue in an optimal way are critical issues. The first item on the list of things to do is find out what the requirements are. If we are dealing with customer/supplier relationship crossing two organizations, then the supplier must establish a 'marketing' activity or process charged with this task.

The marketing process must of course understand not only the needs of the customer but also the ability of its own organization to meet them. If my customer places a requirement on me to run 1500 metres in 4 minutes, then I know I am unable to meet this demand, unless something is done to improve my running performance. Of course I may never be able to achieve this requirement.

Real customer needs are rarely discussed on-site. The formworker, who is paid by the square metre, is busy laying as much flat slab sheeting as possible so that his payment is maximized. He rarely stops to think that the steel-fixer, coming right behind him, is slowed down by the absence of edge boards. In the design process the architect and client often continue to make decisions well after the design engineers have started their design work, in the belief that the design decisions are locked in. Throughout the construction supply chain we see examples of parties optimizing their own positions at the expense of others, acting independently rather than interdependently.

To achieve quality throughout an organization, each person in the quality chain must interrogate every interface as follows:

## Customers

- Who are my immediate customers?
- What are their true requirements?
- How do or can I find out what the requirements are?
- How can I measure my ability to meet the requirements?
- Do I have the necessary capability to meet the requirements?
  (If not, then what must change to improve the capability?)
- Do I consistently meet the requirements?
  (If not, then what prevents this from happening, when the capability exists?)
- How do I monitor changes in the requirements?

## Suppliers

- Who are my immediate suppliers?
- What are my true requirements?
- How do I communicate my requirements?
- How do I, or they, measure their ability to meet the requirements?
- Do my suppliers have the capability to meet the requirements?
- Do my suppliers continually meet the requirements?
- How do I inform them of changes in the requirements?

The measurement of capability is extremely important if the quality chains are to be formed within and without an organization. Each person in the organization must also realize that the supplier's needs and expectations must be respected if the requirements are to be fully satisfied.

To understand how quality may be built into a product or service, at any stage, it is necessary to examine the two distinct, but interrelated aspects of quality:

- Quality of design.
- Quality of conformance to design.

# Quality of design

We are all familiar with the old story of the tree swing (Figure 1.2), but in how many places in how many organizations is this chain of activities taking place? To discuss the quality of, say, a chair it is necessary to describe its purpose. What it is to be used for? If it is to be used for watching TV for 3 hours at a stretch, then the typical office chair will not meet this requirement. The difference between the quality of the TV chair and the office chair is not a function of how it was manufactured, but its *design*.

Quality of design is a measure of how well the product or service is designed to achieve the agreed requirements. The beautifully presented gourmet meal will not necessarily please the recipient if he or she is travelling on the motorway and has stopped for a quick bite to eat. The most important feature of the design, with regard to achieving quality, is the specification. Specifications must also exist at the internal supplier/customer interfaces if one is to achieve a total quality performance. For example, the

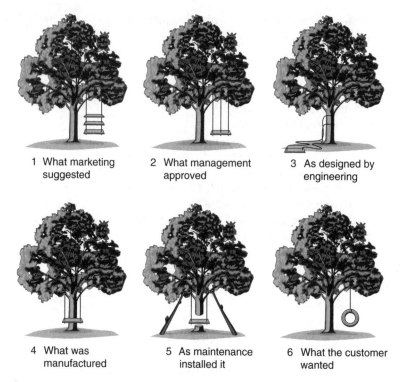

1  What marketing
   suggested

2  What management
   approved

3  As designed by
   engineering

4  What was
   manufactured

5  As maintenance
   installed it

6  What the customer
   wanted

**Figure 1.2**
Quality of design

company lawyer asked to draw up a purchasing contract for a major item from an overseas supplier by the project manager requires a specification as to its content:

1. What is the type of contract to be used, fixed price or schedule of rates?
2. Who are the contracting parties?
3. In which countries are the parties located?
4. What are the products involved?
5. What are the performance requirements?
6. What is the volume?
7. What are the shipping details? Timing of deliveries?
8. What are the financial aspects, e.g. price escalation, currency risk?

The financial controller must issue a specification of the information he or she needs, and when, to ensure that foreign exchange fluctuations do not put the project finances at risk. The business of sitting down and agreeing on a specification at every interface will clarify the true requirements and capabilities. It is the vital first stage for a successful total quality effort.

There must be a corporate understanding of the organization's quality position in the marketplace. It is not sufficient that marketing specifies the product or service 'because that is what the customer wants'. There must be an agreement that the operating departments can achieve that requirement. Should they be incapable of doing so, then one of two things must happen: either the organization finds a different position in the marketplace or substantially changes the operational facilities.

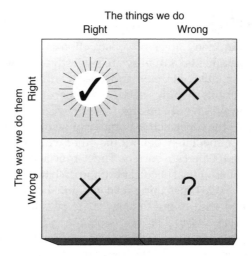

**Figure 1.3**
How much time is spent doing the right things right?

A specific challenge in construction is that every design has to meet multiple requirements. It is not enough that the client likes the look of the building or structure: it also must be able to be built safely, to the required standard of quality and economy. The design process is an exercise in team-based optimization against multiple criteria – there is almost always conflict between some of the objectives.

## Quality of conformance to design

This is the extent to which the product or service achieves the quality of design. What the customer actually receives should conform to the design, and operating costs are tied firmly to the level of conformance achieved. Quality cannot be inspected into products or services; the customer satisfaction must be designed into the whole system. The conformance check then makes sure that things go according to plan.

An important issue, on construction sites, is the avoidance of design generated, unnecessary risks. This requires experienced management. Designers are often unaware of the construction implications of their detailing; architects routinely alter concealed details unnecessarily, without thinking through the construction consequences; and structural engineering reinforcement details are at times complex and unbuildable. Especially in high risk areas, it makes sense to standardize construction details so that tradesmen on site get to know their work and instinctively do it right.

A high level of inspection or checking at the end is often indicative of attempts to inspect in quality. This may well result in spiralling costs and decreasing viability. The area of conformance to design is concerned largely with the quality performance of the actual operations. It may be salutary for organizations to use the simple matrix of Figure 1.3 to assess how much time they spend doing the right things right. A lot of people, often through no fault of their own, spend a good proportion of the available time doing the right things wrong. There are people (and organizations) who spend time doing the wrong things very well, and even those who occupy themselves doing the wrong things wrong, which can be very confusing!

# Managing quality

The quality of finishes in many areas of construction is hard to judge, standards are loosely defined and assessments of quality are subjective. These areas cause more conflict among the architect, subcontractor and head contractor than almost any other issue, often leading to serious disputes between the parties.

On many construction sites we see an argument between the quality manager and the construction supervisor – the construction supervisor wants to see progress and wants to see one trade follow on immediately behind the other while the quality manager wants the preceding work to be 100 percent correct. They argue and debate the evidence before them, the rights and wrongs of the specification, and each tries to convince the other of the validity of his argument. Sometimes things can get quite heated.

This ritual is associated with trying to answer the question '*Have we done the job correctly?*', or put another way '*What is the risk associated with proceeding correctly?*' and *risk* depending on the interpretation given to the specification on that particular day. This is not quality *control*, it is *detection* – wasteful detection of questionable product when it is already too late. There is still a belief in some quarters that to achieve quality we must check, test, inspect or measure – the ritual pouring on of quality at the end of the process, when in fact the checks have to be built into the process, they must be an integral part of it.

In the office one finds staff checking other people's work before it goes out, validating computer data, checking invoices, word processing, etc. There is also quite a lot of looking for things, chasing why things are late, apologizing to customers for lateness, and so on. Waste, waste, waste!

To get away from the natural tendency to rush into the detection mode, it is necessary to ask different questions in the first place. We should not ask whether the job has been done correctly, we should ask first '*Are we capable of doing the job correctly?*' This question has wide implications, and this book is devoted largely to the various activities necessary to ensure that the answer is yes. However, we should realize straight away that such an answer will only be obtained by means of satisfactory methods, right through the supply chain, appropriate design and documentation, materials and equipment, suitable skills, instruction and leadership, and a satisfactory overall 'process'.

# Quality and processes

As we have seen, quality chains can be traced right through the business or service processes used by any organization. A process is the transformation of a set of inputs into outputs that satisfy customer needs and expectations, in the form of products, information or services. Everything we do is a process; so in each area or function of an organization there will be many processes taking place. For example, a finance department may be engaged in budgeting processes, accounting processes, salary and wage processes, costing processes, etc. Each process in each department or area can be analysed by an examination of the inputs and outputs. This will determine some of the actions necessary to improve quality. There are also cross-functional processes.

The output from a process is that which is transferred to somewhere or to someone – the *customer*. Clearly to produce an output that meets the requirements of the customer, it is necessary to define, monitor and control the inputs to the process, which in turn may be supplied as output from an earlier process. At every supplier/customer interface then there resides a transformation process (Figure 1.4), and every single task throughout an organization must be viewed as a process in this way.

Once we have established that our process is capable of meeting the requirements, we can address the next question '*Do we continue to do the job correctly?*', which brings a

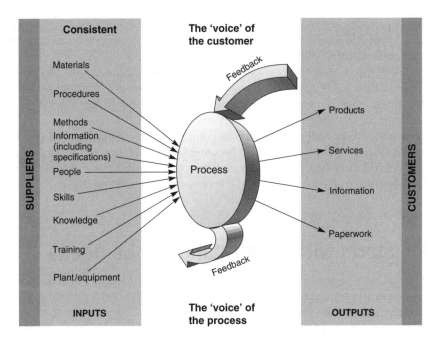

**Figure 1.4**
A process

requirement to monitor the process and the controls on it. If we now re-examine the first question 'Have we done the job correctly?', we can see that, if we have been able to answer the other two questions with a yes, we *must* have done the job correctly. Any other outcome would be illogical. By asking the questions in the right order, we have moved the need to ask the 'inspection' question and replaced a strategy of *detection* with one of *prevention*. This concentrates all the attention on the front end of any process – the inputs – and changes the emphasis to making sure the inputs are capable of meeting the requirements of the process. This is a managerial responsibility and is discharged by efficiently organizing the inputs and its resources and controlling the processes.

These ideas apply to every transformation process; they all must be subject to the same scrutiny of the methods, the people, skills, equipment and so on to make sure they are correct for the job. A person giving a lecture whose audio/visual equipment will not focus correctly, or whose teaching materials are not appropriate, will soon discover how difficult it is to provide a lecture that meets the requirements of the audience.

In every organization there are some very large processes – groups of smaller processes often called *core business processes*. These are activities the organization must carry out especially well if its mission and objectives are to be achieved. This area will be dealt with in some detail in Chapter 4. These activities are crucial if the management of quality is to be integrated into the strategy of the organization.

The *control* of quality can only take place at the point of operation or production – where the letter is word-processed, the design is documented, the physical work is done, or the building product manufactured. The act of *inspection is not quality control.* When the answer to 'Have we done the job correctly?' is given indirectly by answering the questions of capability and control, then we have *assured* quality, and the activity of checking becomes one of *quality assurance* – making sure that the product or service represents the output from an effective *system* to ensure capability and control. It is frequently found that organizational barriers between functional or departmental empires encouraged the development of testing and checking of services or products in a vacuum, without interaction with other departments.

*Quality control* then is essentially the activities and techniques employed to achieve and maintain the quality of a product, process, or service. It includes a monitoring activity, but is also concerned with finding and eliminating causes of quality problems so that the requirements of the customer are continually met.

*Quality assurance* is broadly the prevention of quality problems through planned and systematic activities (including quality documentation). These will include the establishment of a good quality management system and the assessment of its adequacy, the audit of the operation of the system, and the review of the system itself.

# Quality starts with understanding the needs

The marketing processes of an organization must take the lead in establishing the true requirements for the product or service. Having determined the need, the organization should define the market sector and demand, to determine such product or service features as grade, price, quality, timing, etc. For example, a major hotel chain thinking of opening a new hotel or refurbishing an old one will need to consider its location and accessibility before deciding whether it will be predominantly a budget, first-class, business or family hotel. This will determine its capital budget for building or refurbishment.

The organization will also need to establish customer requirements by reviewing the market needs, particularly in terms of unclear or unstated expectations or preconceived ideas held by immediate and end-user customers. It is central to identify the key characteristics that determine the suitability of the product or service in the eyes of the customer. Depending on the situation, this may, of course, call for the use of market research techniques, data-gathering, and analysis of customer complaints. If necessary, quasi-quantitative methods may be employed, giving proxy variables that can be used to grade the characteristics in importance, and decide in which areas superiority over competitors exists. It is often useful to compare these findings with internal perceptions. However, in many instances within the construction supply chain, where the customer is the following trade contractor or designer, meetings to tease out perceptions, competing issues and needs have to take place face to face through productive dialogue, early at the start of a new project. Even here, support from second party research may be of assistance.

Excellent communication between customers and suppliers is the key to a total quality performance; it will eradicate the 'demanding nuisance/idiot' view of customers, which still pervades some organizations. Poor communications often occur in the supply chain between organizations, when neither party realizes how poor they are. Feedback from both customers and suppliers needs to be improved where dissatisfied customers and suppliers do not communicate their problems. In such cases, non-conformance of purchased products or services is often due to customers' inability to communicate their requirements clearly. If these ideas are also used within an organization, then the internal supplier/customer interfaces will operate much more smoothly.

All the efforts devoted to finding the nature and timing of the demand will be pointless if there are failures in communicating the requirements throughout the organization promptly, clearly, and accurately. The marketing processes should be capable of producing a formal statement or outline of the requirements for each product or service.

This constitutes a preliminary set of *specifications*, which can be used as the basis for service or product design. The information requirements include:

1. Characteristics of performance and reliability – these must make reference to the conditions of use and any environmental factors that may be important.
2. Aesthetic characteristics, such as style, colour, smell, task, feel, etc.
3. Any obligatory regulations or standards governing the nature of the product or service.

The organization must also establish systems for feedback of customer information and reaction, and these systems should be designed on a continuous monitoring basis. Any information pertinent to the product or service should be collected and collated, interpreted, analysed, and communicated, to improve the response to customer experience and expectations. These same principles must also be applied inside the organization if continuous improvement at every transformation process interface is to be achieved. If one function or department in a company has problems recruiting the correct sort of staff, for example, and HR has not established mechanisms for gathering, analysing, and responding to information on new employees, then frustration and conflict will replace communication and co-operation.

One aspect of the analysis of market demand that extends back into the organization is the review of market readiness of a new product or service. Items that require some attention include assessment of:

1. The suitability of the distribution and customer-service processes.
2. Training of personnel in the 'field'.
3. Availability of 'spare parts' or support staff.
4. Evidence that the organization is capable of meeting customer requirements.

All organizations receive a wide range of information from customers through invoices, payments, requests for information, letters of complaint, responses to advertisements and promotion, etc. An essential component of a system for the analysis of demand is that this data is channelled quickly into the appropriate areas for action and, if necessary, response.

There are various techniques of research, which are outside the scope of this book, but have been well documented elsewhere. It is worth listing some of the most common and useful general methods that should be considered for use, both externally and internally:

- Surveys – questionnaires, etc.
- Panel or focus group techniques.
- In-depth interviews.
- Brainstorming and discussions.
- Role rehearsal and reversal.
- Interrogation of trade associations.

The number of methods and techniques for researching the market is limited only by imagination and funds. The important point to stress is that the supplier, whether the internal individual or the external organization, keeps very close to the customer. Good research, coupled with analysis of complaints data, is an essential part of finding out what the requirements are, and breaking out from the obsession with inward scrutiny that bedevils quality.

Have a close look at the Graniterock case study (p. 375) and in particular at the way in which the company uses regular processes to collect market information and guide the development of strategies for improvement. Product development and process improvement are both driven by a comprehensive market research process.

# Quality in all functions

For an organization to be truly effective, each component of it must work properly together. Each part, each activity, each person in the organization affects and is in turn affected by others. Errors have a way of multiplying, and failure to meet the requirements in one part or area creates problems elsewhere, leading to yet more errors, yet more problems, and so on. The benefits of getting it right first time everywhere are enormous. The challenges for achieving this in a construction project setting (where the mix of organizational goals, styles and cultures are very diverse) are considerable and require excellent leadership.

Everyone experiences – almost accepts – problems in working life. This causes people to spend a large part of their time on useless activities – correcting errors, looking for things, finding out why things are late, checking suspect information, rectifying and reworking, apologizing to customers for mistakes, poor quality and lateness. The list is endless, and it is estimated that about one-third of our efforts are still wasted in this way. In the service sector it can be much higher.

Quality, the way we have defined it as meeting the customer requirements, gives people in different functions of an organization a common language for improvement. It enables all the people, with different abilities and priorities, to communicate readily with one another, in pursuit of a common goal. When business and industry were local, the craftsman could manage more or less on his own. Business is now so complex and employs so many different specialist skills that everyone has to rely on the activities of others in doing their jobs.

Some of the most exciting applications of TQM have materialized from groups of people that could see little relevance when first introduced to its concepts. Following training, many different parts of organizations can show the usefulness of the techniques. Sales staff can monitor and increase successful sales calls, office staff have used TQM methods to prevent errors in word-processing and improve inputting to computers, customer-service people have monitored and reduced complaints, distribution has controlled lateness and disruption in deliveries.

It is worthy of mention that the first points of contact for some outside customers are the telephone operator, the security people at the gate, or the person in reception. Equally the e-business, paperwork and support services associated with the product, such as websites, invoices and sales literature and their handlers, must match the needs of the customer. Clearly TQM cannot be restricted to the 'production' or 'operations' areas without losing great opportunities to gain maximum benefit.

Managements that rely heavily on exhortation of the workforce to 'do the right job right the first time', or 'accept that quality is your responsibility', will not only fail to achieve quality but may create division and conflict. These calls for improvement infer that faults are caused only by the workforce and that problems are departmental or functional when, in fact, the opposite is true – most problems are interdepartmental. The commitment of all members of an organization is a requirement of 'organization-wide quality improvement'. Everyone must work together at every interface to achieve improved performance and that can only happen if the top management is really committed.

# Reference

1. Thomas, R., Marosszeky, M., Karim, K., Davis, S. and McGeorge, D. (2003) *Enhancing Project Completion*, Australian Centre for Construction Innovation, ISBN 0733420613, pp. 28.

## Chapter highlights

### Quality, competitiveness and customers

- Understand the construction setting and its challenges, only then can you develop strategies for improvement that will be sound.
- When looking at the problems faced by the sector it is important to understand the linkages between safety, quality and production.
- The reputation enjoyed by an organization is built by quality, reliability, delivery and price. Quality is the most important of these competitive weapons.
- Reputations for poor quality last for a long time, and good or bad reputations can become national or international. The management of quality can be learned and used to improve reputation.
- Quality is meeting the customer requirements, and this is not restricted to the functional characteristics of the product or service.
- Reliability is the ability of the product or service to continue to meet the customer requirements over time.
- Organizations 'delight' the customer by consistently meeting customer requirements, and then achieve a reputation of 'excellence' and customer loyalty.

### Understanding and building the quality chains

- Throughout all organizations there are a series of internal suppliers and customers. These form the so-called 'quality chains', the core of 'company-wide quality improvement'.
- The internal customer/supplier relationships must be managed by interrogation, i.e. using a set of questions at every interface. Measurement of capability is vital.
- Similar quality chains exist across the whole of the construction procurement process between subcontract suppliers who work jointly to produce the end product.
- There are two distinct but interrelated aspects of quality, design and conformance to design. *Quality of design* is a measure of how well the product or service is designed to achieve the agreed requirements. *Quality of conformance to design* is the extent to which the product or service achieves the design. Organizations should assess how much time they spend doing the right things right.

### Managing quality

- Asking the question 'Have we done the job correctly?' should be replaced by asking 'Are we capable of doing the job correctly?' and 'Do we continue to do the job correctly?'
- Asking the questions in the right order replaces a strategy of *detection* with one of *prevention*.
- Everything we do is a process, which is the transformation of a set of inputs into the desired outputs.
- In every organization there are some core business processes that must be performed especially well if the mission and objectives are to be achieved.

- Inspection is not *quality control*. The latter is the employment of activities and techniques to achieve and maintain the quality of a product, process or service.
- *Quality assurance* is the prevention of quality problems through planned and systematic activities.

### Quality starts with understanding the needs

- Marketing processes establish the true requirements for the product or service. These must be communicated properly throughout the organization in the form of specifications.
- Excellent communications between customers and suppliers is the key to a total quality performance – the organization must establish feedback systems to gather customer information.
- Appropriate research techniques should be used to understand the 'market' and keep close to customers and maintain the external perspective.

### Quality in all functions

- All members of an organization need to work together on organization-wide quality improvement. The co-operation of everyone at every interface is necessary to achieve improvements in performance, which can only happen if the top management is really committed.

# 2

# Models and frameworks for total quality management

# Early TQM frameworks

In the early 1980s manufacturers in the US and Europe found that Japanese industry had become much more competitive. In some instances, Japanese products of similar or better quality were available in the market at less than the cost price of making them in the West. This created an immediate interest in Japanese manufacturing processes; and the most obvious point of difference was the adoption of quality management practices. Initially, industry leaders were confused because they saw the practice of quality management as being inextricably linked with Japanese culture. It took some years to untangle the two and develop a clearer understanding of the core management practices that had created the differences in process productivity. At this stage there were many attempts to construct lists and frameworks to help this process.

Famous American 'gurus' of quality management, such as W. Edwards Deming, Joseph M. Juran and Philip B. Crosby, started to try to make sense of the labyrinth of issues involved, including the tremendous competitive performance of Japan's manufacturing industry. Deming and Juran had contributed to building Japan's success in the 1950s and 1960s and it was appropriate that they should set down their ideas for how organizations could achieve success.

Deming had 14 points to help management as follows:

1. Create constancy of purpose towards improvement of product and service.
2. Adopt the new philosophy. We can no longer live with commonly accepted levels of delays, mistakes, defective workmanship.
3. Cease dependence on mass inspection. Require, instead, statistical evidence that quality is built in.
4. End the practice of awarding business on the basis of price tag.
5. Find problems. It is management's job to work continually on the system.
6. Institute modern methods of training on the job.
7. Institute modern methods of supervision of production workers. The responsibility of foremen must be changed from numbers to quality.
8. Drive out fear, so that everyone may work effectively for the company.
9. Break down barriers between departments.
10. Eliminate numerical goals, posters, and slogans for the workforce asking for new levels of productivity without providing methods.
11. Eliminate work standards that prescribe numerical quotas.
12. Remove barriers that stand between the hourly worker and his right to pride of workmanship.
13. Institute a vigorous programme of education and retraining.
14. Create a structure in top management that will push every day on the above 13 points.

Juran's ten steps to quality improvement are:

1. Build awareness of the need and opportunity for improvement.
2. Set goals for improvement.
3. Organize to reach the goals (establish a quality council, identify problems, select projects, appoint teams, designate facilitators).
4. Provide training.
5. Carry out projects to solve problems.
6. Report progress.
7. Give recognition.

8. Communicate results.
9. Keep score.
10. Maintain momentum by making annual improvement part of the regular systems and processes of the company.

Crosby, who spent time as Quality Director of ITT, had four absolutes:

1. Definition – conformance to requirements.
2. System – prevention.
3. Performance standard – zero defects.
4. Measurement – price of non-conformance.

He also offered management 14 steps to improvement:

1. Make it clear that management is committed to quality.
2. Form quality improvement teams with representatives from each department.
3. Determine where current and potential quality problems lie.
4. Evaluate the cost of quality and explain its use as a management tool.
5. Raise the quality awareness and personal concern of all employees.
6. Take actions to correct problems identified through previous steps.
7. Establish a committee for the zero defects programme.
8. Train supervisors to actively carry out their part of the quality improvement programme.
9. Hold a 'zero defects day' to let all employees realize that there has been a change.
10. Encourage individuals to establish improvement goals for themselves and their groups.
11. Encourage employees to communicate to management the obstacles they face in attaining their improvement goals.
12. Recognize and appreciate those who participate.
13. Establish quality councils to communicate on a regular basis.
14. Do it all over again to emphasize that the quality improvement programme never ends.

## A comparison

One way to compare directly the various approaches of the three American gurus is in Table 2.1 which shows the differences and similarities clarified under 12 different factors.

Our understanding of 'total quality management' developed through the 1980s and in earlier editions of this author's books on TQM, a broad perspective was given, linking the TQM approaches to the direction, policies and strategies of the business or organization. These ideas were captured in a basic framework – the TQM model (Figure 2.1) which was widely promoted in the UK through the activities of the Department of Trade and Industry (DTI) 'Quality Campaign' and 'Managing into the 90s' programmes. These approaches brought together a number of components of the quality approach, including quality circles (teams), problem solving and statistical process control (tools) and quality systems, such as BS 5750 and later ISO 9000 (systems). It was recognized that culture played an enormous role in whether organizations were successful or not with their TQM approaches. Good *communications*, of course, were seen to be vital to success but the most important of all was *commitment*, not only from the senior management but from everyone in the organization, particularly those operating directly at the customer interface. The customer/supplier or 'quality chains' were the core of this TQM model.

**Table 2.1** The American quality gurus compared

|  | Crosby | Deming | Juran |
|---|---|---|---|
| Definition of quality | Conformance to requirements | A predictable degree of uniformity and dependability at low cost and suited to the market | Fitness for use |
| Degree of senior-management responsibility | Responsible for quality | Responsible for 94% of quality problems | Less than 20% of quality problems are due to workers |
| Performance standard/motivation | Zero defects | Quality has many scales. Use statistics to measure performance in all areas. Critical of zero defects | Avoid campaigns to do perfect work |
| General approach | Prevention, not inspection | Reduce variability by continuous improvement. Cease mass inspection | General management approach to quality – especially 'human' elements |
| Structure | Fourteen steps to quality improvement | Fourteen points for management | Ten steps to quality improvement |
| Statistical process control (SPC) | Rejects statistically acceptable levels of quality | Statistical methods of quality control must be used | Recommends SPLC but warns that it can lead to too-driven approach |
| Improvement basis | A 'process', not a programme. Improvement goals | Continuous to reduce variation. Eliminate goals without methods | Project-by-project team approach. Set goals |
| Teamwork | Quality improvement teams. Quality councils | Employee participation in decision-making. Break down barriers between departments | Team and quality circle approach |
| Costs of quality | Cost of non-conformance. Quality is free | No optimum – continuous improvement | Quality is not free – there is an optimum |
| Purchasing and goods received | State requirements. Supplier is extension of business. Most faults due to purchasers themselves | Inspection too late – allows defects to enter system through AQLs. Statistical evidence and control charts required | Problems are complex. Carry out formal surveys |
| Vendor rating | Yes *and* buyers. Quality audits useless | No – critical of most systems | Yes, but help supplier improve |
| Single sources of supply |  | Yes | No – can neglect to sharpen competitive edge |

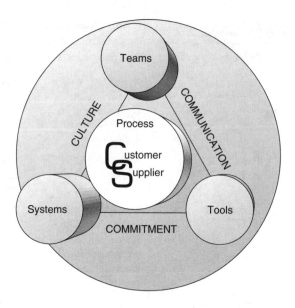

**Figure 2.1**
TQM model

Many companies and organizations in the public sector found this simple framework useful and it helped groups of senior managers throughout the world get started with TQM. The key was to integrate the TQM activities, based on the framework, into the business or organization strategy and this has always been a key component of the author's approach.

Today, the construction sector worldwide is faced with a similar dilemma to that confronted by manufacturers in the 1980s. Construction executives can see the benefits that other industry sectors have won through the adoption of quality management practices and they would like to obtain similar benefits for themselves. However, they cannot see how those ideas can be applied in construction because they see construction as being essentially different. In a sense they have the same disconnect as Western industry leaders had in trying to untangle management practices from Japanese culture. Construction is a one-of-a-kind production: its products are highly complex; its supply chain is extremely fragmented; it has a complex engagement with its clients over a long period of time and, as a result, its processes are somewhat different. The challenge for the sector is to untangle the differences between industries and their management practices. This book presents the theoretical underpinnings for that analysis and, through the case studies of companies that have bridged the gap, demonstrates how the concepts of quality management can be leveraged to benefit construction organizations.

## Quality award models

Starting in Japan with the Deming Prize, companies started to get interested in quality frameworks that could be used essentially in three ways:

1. As the basis for awards.
2. As the basis for a form of 'self-assessment'.
3. As a descriptive 'what-needs-to-be-in-place' model.

The earliest approach to a *total* quality audit process is that established in the Japanese-based 'Deming Prize', which is based on a highly demanding and intrusive process. The categories of this award were established in 1950 when the Union of Japanese Scientists and Engineers (JUSE) instituted the prize(s) for 'contributions to quality and dependability of product'.

The emphasis of the Deming Prize now is on finding out how effectively the applicant is implementing TQM by focusing on the quality of its products and services. The examiners are looking to see if 'TQM has been implemented properly to achieve business objectives and strategies', and that outstanding results have been obtained.

The Deming 'examination viewpoints' now include:

1. Top management leadership and organizational vision and strategies.
2. TQM frameworks:
   - organizational structure and its operations;
   - daily management;
   - policy management;
   - relationship to ISO 9000 and ISO 14000;
   - relationship to other management improvement programmes;
   - TQM promotion and operation.
3. Quality Assurance System:
   - QA system;
   - new product and new technology development;
   - process control;
   - test, quality evaluation and quality audits;
   - activities covering the whole life cycle;
   - purchasing, subcontracting, and distribution management.
4. Management systems for business elements:
   - cross-functional management and its operations;
   - quantity/delivery management;
   - cost management;
   - environmental management;
   - safety, hygiene and work environment management.
5. Human resources development:
   - positioning of people in management;
   - education and training;
   - respect for people's dignity.
6. Effective utilization of information:
   - positioning of information in management;
   - information systems;
   - support for analysis and decision-making;
   - standardization and configuration management.
7. TQM concepts and values:
   - quality;
   - maintenance and improvement;
   - respect for humanity.
8. Scientific methods:
   - understanding and utilization of methods;
   - understanding and utilization of problem-solving methods.
9. Organizational powers:
   - core technology;
   - speed;
   - vitality.

10. Contribution to realization of corporate objectives:
   - customer relations;
   - employee relations;
   - social relations;
   - supplier relations;
   - shareholder relations;
   - realization of corporate mission;
   - continuously securing profits.

There is a general 'TQM Features (Shining Examples)' piece at the end which looks for the company promoting TQM activities that are unique and suitable to its own conditions, and contributing to the development of new TQM concepts, methodologies and technologies in anticipation of its future needs.

The recognition that total quality management is a broad culture change vehicle with internal and external focus embracing behavioural and service issues, as well as quality assurance and process control, prompted the United States to develop in the late 1980s one of the most famous and now widely used frameworks, the Malcolm Baldrige National Quality Award (MBNQA). The award itself, which is composed of two solid crystal prisms 14 inches high, is presented annually to recognize companies in the USA that have excelled in quality management and quality achievement. But it is not the award itself, or even the fact that it is presented each year by the President of the USA which has attracted the attention of most organizations, it is the excellent framework for TQM and organizational self-assessments.

The Baldrige National Quality Program Criteria for Performance Excellence, as it is now known, aim to:

- help improve organizational performance practices, capabilities and results;
- facilitate communication and sharing of best practices information;
- serve as a working tool for understanding and managing performance and for guiding, planning and opportunities for learning.

The award criteria are built upon a set of interrelated core values and concepts:

- visionary leadership;
- customer-driven excellence;
- organizational and personal learning;
- valuing employees and partners;
- agility;
- focus on the future;
- managing for innovation;
- management by fact;
- public responsibility and citizenship;
- focus on results and creating value;
- systems developments.

These are embodied in a framework of seven categories which are used to assess organizations:

1. Leadership:
   - organizational leadership;
   - public responsibility and citizenship.

2. Strategic planning:
   - strategy development;
   - strategy deployment.
3. Customer and market focus:
   - customer and market knowledge;
   - customer relationships and satisfaction.
4. Information and analysis:
   - measurement and analysis of organizational performance;
   - information management.
5. Human resource focus:
   - work systems;
   - employee education training and development;
   - employee well-being and satisfaction.
6. Process management:
   - product and service processes;
   - business processes;
   - support processes.
7. Business results:
   - customer-focused results;
   - financial and market results;
   - human resource results;
   - organizational effectiveness results.

Figure 2.2 shows how the framework's system connects and integrates the categories. This has three basic elements: organizational profile, system, and information and analysis. The main driver is the senior executive leadership which creates the values, goals and systems, and guides the sustained pursuit of quality and performance objectives. The system includes a set of well-defined and well-designed processes for meeting the organization's direction and performance requirements. Measures of progress provide a results-oriented basis for channelling actions to deliver ever-improving customer values and organization performance. The overall goal is the delivery of customer satisfaction and market success leading, in turn, to excellent business results. The seven criteria categories are further divided into items and areas to address. These are described

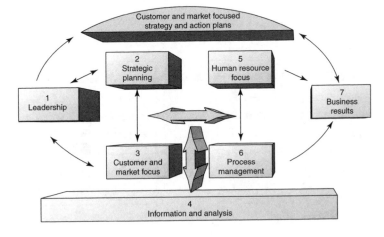

**Figure 2.2**
Baldrige criteria

in some detail in the 'Criteria for Performance Excellence' available from the US National Institute of Standards and Technology (NIST) in Gaithesburg, USA.

The Baldrige Award led to a huge interest around the world in quality award frameworks that could be used to carry out self-assessment and to build an organization-wide approach to quality, which was truly integrated into the business strategy. It was followed in Europe in the early 1990s by the launch of the European Quality Award by the European Foundation for Quality Management (EFQM). This framework was the first one to include 'Business Results' and to really represent the whole business model.

Like the Baldrige, the EFQM Model recognized that processes are the means by which an organization harnesses and releases the talents of its people to produce results/performance. Moreover, improvement in performance can be achieved only by improving the processes by involving the people. This simple model is shown in Figure 2.3.

Figure 2.4 displays graphically the 'non-prescriptive' principles of the full Excellence Model. Essentially customer results, employee results and favourable society results are achieved through leadership driving policy and strategy, people partnerships, resources and processes, which lead ultimately to excellence in key performance results – the

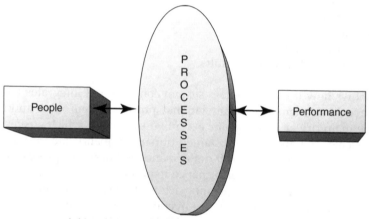

Achieve better performance through involvement of all employees (people) in continuous improvement of their processes

**Figure 2.3**
The simple model for improved performance

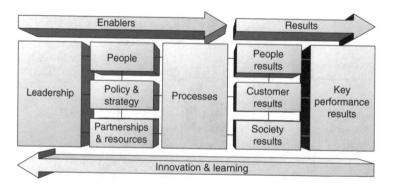

**Figure 2.4**
The EFQM Excellence Model

enablers deliver the results which in turn drive innovation and learning. The EFQM have provided a weighting for each of the criteria which may be used in scoring self-assessments and making awards (see Chapter 8).

Through usage and research, the Baldrige and EFQM Excellence Models continued to grow in stature throughout the 1990s. They were recognized as descriptive holistic business models, rather than just quality models and mutated into frameworks for (Business) Excellence.

The NIST and EFQM have worked together well over recent years to learn from each other's experience in administering awards and supporting programmes, and from organizations which have used their frameworks 'in anger'.

The EFQM publication for the new millennium of the so-called 'Excellence Model' captures much of this learning and provides a framework which organizations can use to follow ten new steps:

1. Set direction through leadership.
2. Establish the results they want to achieve.
3. Establish and drive policy and strategy.
4. Set up and manage appropriately their approach to processes, people, partnerships and resources.
5. Deploy the approaches to ensure achievement of the policies, strategies and thereby the results.
6. Assess the 'business' performance, in terms of customers, their own people and society results.
7. Assess the achievements of key performance results.
8. Review performance for strengths and areas for improvement.
9. Innovate to deliver performance improvements.
10. Learn more about the effects of the enablers on the results.

# The four Ps and three Cs of TQM – a new model for TQM

We have seen in Chapter 1 how *processes* are the key to delivering quality of products and services to customers. It is clear from Figure 2.4 that processes are a key linkage between the enablers of *planning* (leadership driving policy and strategy partnership and resources), through *people* into the *performance* of people, society, customers, and key outcomes.

These 'four Ps' form the basis of a simple model for TQM and provide the 'hard management necessities' to take organizations successfully into the twenty-first century. These form the structure of the remainder of this book.

From the early TQM frameworks, however, we must not underestimate the importance of the three Cs – *culture, communication* and *commitment*. The new TQM Model is complete when these 'soft outcomes' are integrated into the four Ps' framework to move organizations successfully forward (Figure 2.5).

This new TQM Model, based on all the excellent work done during the last century, provides a simple framework for excellent performance, covering all angles and aspects of an organization and its operation.

Performance is achieved using a business excellence approach, and by planning the involvement of people in the improvement of processes. This has to include:

■ *Planning* – the development and deployment of policies and strategies; setting up appropriate partnerships and resources; and designing in quality.

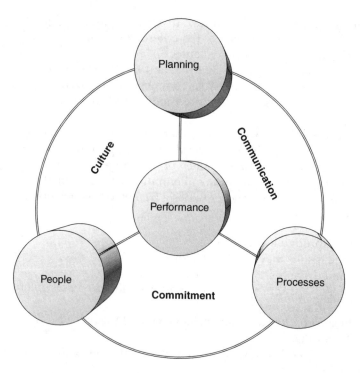

**Figure 2.5**
The new framework for total quality management

- *Performance* – establishing a performance measure framework – a 'balanced scorecard' for the organization; carrying out self-assessment, audits, reviews and benchmarking.
- *Processes* – understanding, management, design and redesign; quality management systems; continuous improvement.
- *People* – managing the human resources; culture change; teamwork; communications; innovation and learning.

Wrapping around all this to ensure successful implementation is, of course, effective leadership and commitment, the subject of the next chapter.

## Chapter highlights

### Early TQM frameworks
- In the early 1980s when US and European business leaders first tried to understand how to apply quality business practices that they saw gave their Japanese competitors an advantage, they had a challenge to untangle the business ideas from the cultural differences.
- There have been many attempts to construct lists and frameworks to help organizations understand how to implement good quality management.
- The 'quality gurus' in America, Deming, Juran and Crosby, offered management 14 points, ten steps and four absolutes (plus 14 steps) respectively. These similar but different approaches may be compared using a number of factors, including definition of quality, degree of senior management responsibility and general approach.

- The understanding of quality developed and, in Europe and other parts of the world, the author's early TQM Model, based on a customer/supplier chain core surrounded by systems, tools and teams, linked through culture, communications and commitment, gained wide usage.
- Today managers in the construction sector have a similar challenge to that faced by general business leaders in the 1980s. They have to untangle the differences between the business setting of manufacturing and construction and identify those practices that are essential for the achievement of quality outcomes in their operations.

## Quality award models

- Quality frameworks may be used as the basis for awards for a form of 'self-assessment' or as a description of what should be in place.
- The Deming Prize in Japan was the first formal quality award framework established by JUSE in 1950. The examination viewpoints include: top management leadership and strategies; TQM frameworks, concepts and values; QA and management systems; human resources; utilization of information; scientific methods; organizational powers; realization of corporate objectives.
- The USA Baldrige Award aims to promote performance excellence and improvement in competitiveness through a framework of seven categories which are used to assess organizations: leadership; strategic planning; customer and market focus; information and analysis; human resource focus; process management; business results.
- The European (EFQM) Excellence Model operates through a simple framework of performance improvement through involvement of people in improving processes.
- The full Excellence Model is a non-prescriptive framework for achieving good results – customers, people, society, key performance – through the enablers – leadership, policy and strategy, people, processes, partnerships and resources. The framework includes proposed weightings for assessment.

## The four Ps and three Cs – a new model for TQM

- *Planning, people* and *processes* are the keys to delivering quality products and services to customers and generally improving overall *performance*. These four Ps form a structure of 'hard management necessities' for a new simple TQM Model which forms the structure of this book.
- The three Cs of *culture, communication,* and *commitment* provide the glue or 'soft outcomes' of the model which will take organizations successfully into the twenty-first century.

# 3

# Leadership and commitment

# The total quality management (TQM) approach

'What is quality management?' Something that is best left to the experts is often the answer to this question. But this is avoiding the issue, because it allows executives and managers to opt out. Quality is too important to leave to the so-called 'quality professionals'; it cannot be achieved on a company-wide basis if it is left to the experts. Equally dangerous, however, are the uninformed who try to follow their natural instincts because they 'know what quality is when they see it'. Usually this belies a narrow product quality focus and can only yield limited benefits. This type of intuitive approach can lead to serious attitude problems, which do no more than reflect the understanding and knowledge of quality that are present in an organization.

The organization which believes that the traditional quality control techniques and the way they have always been used will resolve their quality problems may be misguided. Employing more inspectors, tightening up standards, developing correction, repair and rework teams do not improve quality. Traditionally, quality has been regarded as the responsibility of the QA or QC department, and still it has not yet been recognized in some organizations that many quality problems originate in the commercial, service or administrative areas.

TQM is far more than shifting the responsibility of *detection* of problems from the customer to the producer. It requires a comprehensive approach that must first be recognized and then implemented if the rewards are to be realized. Today's business environment is such that managers must plan strategically to maintain a hold on market share, let alone increase it. In building construction quality problems have been found to seriously erode margins due to the cost of rectifying defective work both during the contract period and afterwards. We have known for years that consumers place a higher value on quality than on loyalty to suppliers, and price is often not the major determining factor in consumer choice. Price has been replaced by quality, and this is true in industrial, service, hospitality, and many other markets. This perception was somewhat distorted in building construction because of the substantial capital growth of property values in major cities the world over, this created markets where there was almost always a speculator prepared to buy even substandard product. However, many of these markets have matured, consumer protection laws and class legal action against poor quality builders have become more common and customers are becoming more demanding of quality.

The case studies in this book describe companies that have been able to obtain a market advantage from the provision of high quality products and services in the construction sector. In Singapore CDL (Case study 2, p. 389), one of the country's leading developers is prepared to pay a premium to builders who can guarantee and deliver a higher standard of product and service quality. In Australia the medium density housing developer and builder Mirvac (Case study 5, p. 415) has managed to differentiate itself on the basis of product quality and reliability, it is common knowledge in the marketplace that their dwelling units hold their prices better than those of other suppliers. In the US quarry and concrete supplier Graniterock (Case study 1, p. 375) has demonstrated that purchasers of concrete are prepared to pay a premium for higher standards of service and a better quality product.

TQM is an approach to improving the competitiveness, effectiveness and flexibility of a whole organization. It is essentially a way of planning, organizing and understanding each activity, and depends on each individual at each level. For an organization to be truly effective, each part of it must work properly together towards the same goals, recognizing that each person and each activity affect and in turn are affected by others. TQM is also a way of ridding people's lives of wasted effort by bringing everyone into the processes of improvement, so that results are achieved in less time. The methods and techniques used in TQM can be applied throughout any organization. They are equally useful in the manufacturing, public service, health care, education and hospitality industries.

The impact of TQM on an organization is, first, to ensure that the management adopts a strategic overview of quality. The approach must focus on developing a *problem-prevention* mentality; but it is easy to underestimate the effort that is required to change attitudes and approaches. Many people will need to undergo a complete change of 'mindset' to unscramble their intuition, which rushes into the detection/inspection mode to solve quality problems: 'We have a quality problem, we had better double check every single item' – whether it is service, waterproofing or finishes related. Managers who worked on the new Australian Parliament House in the late 1980s, where there were three tiers of quality inspection to make sure that everything was right, reported that no one was responsible, the worker relied on the first inspector, the first on the second and so on, no one accepted responsibility.

The correct mindset may be achieved by looking at the sort of barriers that exist in key areas. Staff may need to be trained and shown how to reallocate their time and energy to studying their processes in teams, searching for causes of problems, and correcting the causes, not the symptoms, once and for all. This often requires of management a positive, thrusting initiative to promote the right-first-time approach to work situations. Through *quality or processor performance improvement teams*, these actions will reduce the inspection/rejection syndrome in due course. If things are done correctly first time round, the usual problems that create the need for inspection for failure should disappear.

The managements of many firms may think that their scale of operation is not sufficiently large, that their resources are too slim, or that the need for action is not important enough to justify implementing TQM. Before arriving at such a conclusion, however, they should examine their existing performance by asking the following questions:

1. Is any attempt made to assess the costs arising from errors, defects, waste, customer complaints, lost sales, etc.? If so, are these costs minimal or insignificant?
2. Is the standard of management adequate and are attempts being made to ensure that quality is given proper consideration at the design stage?
3. Are the organization's quality management systems – documentation, processes, operations, etc. – in good order?
4. Have people been trained in how to prevent errors and problems? Do they anticipate and correct potential causes of problems, or do they find and reject?
5. Are subcontract suppliers being selected on the basis of the quality of their people and services as well as price?
6. Do job instructions contain the necessary quality elements, are they kept up to date, and are employers doing their work in accordance with them?
7. What is being done to motivate and train employees to do work right first time?
8. How many errors and defects, and how much wastage occurred last year? Is this more or less than the previous year?

If satisfactory answers can be given to most of these questions, an organization can be reassured that it is already well on the way to using adequate quality management. Even so, it may find that the introduction of TQM causes it to reappraise activities throughout. If answers to the above questions indicate problem areas, it will be beneficial to review the top management's attitude to quality. Time and money spent on quality-related activities are *not* limitations of profitability; they make significant contributions towards greater efficiency and enhanced profits.

# Commitment and policy

To be successful in promoting business efficiency and effectiveness, TQM must be truly organization-wide, it must include the supply chain and it must start at the top with the chief executive or equivalent. The most senior directors and management must all demonstrate that they are serious about quality. The middle management have a particularly important role to play, since they must not only grasp the principles of TQM, they must go on to explain them to the people for whom they are responsible, and ensure that their own commitment is communicated. Only then will TQM spread effectively throughout the organization. This level of management also needs to ensure that the efforts and achievements of their subordinates obtain the recognition, attention and reward that they deserve. Project managers in the construction sector have a critical role as they often have responsibility for selecting the subcontractors and have the challenge of creating a cohesive quality focused team on the project. They have to explain the quality strategy to all suppliers, their on-site supervisors and workers and ensure that all parts of the team are committed to the shared values and goals.

The chief executive of an organization should accept the responsibility for and commitment to a quality policy in which he/she must really believe. This commitment is part of a broad approach extending well beyond the accepted formalities of the quality assurance function. It creates responsibilities for a chain of quality interactions between the marketing, design, production/operations, purchasing, distribution and service functions. Within each and every department of the organization at all levels, starting at the top, basic changes of attitude may be required to implement TQM approaches. If the owners or directors of the organization do not recognize and accept their responsibilities for the initiation and operation of TQM, then these changes will not happen. Controls, systems and techniques are very important in TQM, but they are not the primary requirement. It is more an attitude of mind, based on pride in the job and teamwork, and it requires from the management total commitment, which must then be extended to all employees at all levels and in all departments.

Senior management commitment should be obsessional, not lip service. It is possible to detect real commitment; it shows on the construction site, in the head offices, in the design offices – where the work is being done. Going into organizations sporting poster campaigns of quality instead of belief, one is quickly able to detect the falseness. The people are told not to worry if problems arise, 'just do the best you can', 'the customer will never notice'. The opposite is an organization where total quality means something can be seen, heard, felt. Things happen at this operating interface as a result of *real* commitment. Material problems are corrected with suppliers, equipment difficulties are put right by improved maintenance programmes or replacement, people are trained, change takes place, partnerships are built, continuous improvement is achieved.

At the project level, similar leadership is called for. In the building construction sector in Australia, the UK and USA subcontractors, who both prefabricate and assemble materials and products on site, do more than 80 percent of the actual work. This means that the bulk of the operational workforce on site is from the subcontract supply chain. One of the major challenges on such projects is the need to influence the values and behaviours of the entire delivery team. To be successful the project manager needs to select suppliers on the basis of the fit in values as well as on their technical competence and price. It is the effectiveness of the interorganizational team, working side by side, that will determine the success or failure of the project. In such a setting, the challenge of leadership is to create cohesion in values and behaviours across the entire project.

# The quality policy

A sound quality policy, together with the organization and facilities to put it into effect, is a fundamental requirement, if an organization is to fully implement TQM. Every organization should develop and state its policy on quality, together with arrangements for its implementation. The content of the policy should be made known to all employees. The preparation and implementation of a properly thought out quality policy, together with continuous monitoring, make for smoother production or service operation, minimize errors and reduce waste.

Management should be dedicated to the regular improvement of quality, not simply a one-step improvement to an acceptable plateau. These ideas can be set out in a *quality policy* that requires top management to:

1. Identify the end customer's needs (including perception).
2. Assess the ability of the organization to meet these needs economically.
3. Ensure that bought-in materials' reliability meets the required standards of performance and efficiency.
4. Ensure that subcontract suppliers working on site share your values and process goals.
5. Concentrate on the prevention rather than detection philosophy.
6. Educate and train for quality improvement and ensure that your subcontractors do so as well.
7. Measure customer satisfaction at all levels, the end-customer as well as customer satisfaction between the links of the supply chain.
8. Review the quality management systems to maintain progress.

The quality policy should be the concern of all employees, and the principles and objectives communicated as widely as possible so that it is understood at all levels of the organization and within the subcontract supply chain on construction projects. Practical assistance and training should be given, where necessary, to ensure the relevant knowledge and experience are acquired for successful implementation of the policy throughout the supply chain.

> As an exercise, look at the way in which the case study companies have enunciated their quality policies; these demonstrate the strategies of some of the best companies in the sector worldwide.

# Creating or changing the culture

The culture within an organization is formed by a number of components:

1. Behaviours based on people interactions.
2. Norms resulting from working groups.
3. Dominant values adopted by the organization.
4. Rules of the game for 'getting on'.
5. The climate.

Culture in any 'business' may be defined as how business is conducted, and how employees behave and are treated. Any organization needs a *vision framework* that includes its *guiding philosophy, core values and beliefs* and a *purpose*. These should be combined into a *mission*, which provides a vivid description of what things will be like when it has been achieved (Figure 3.1).

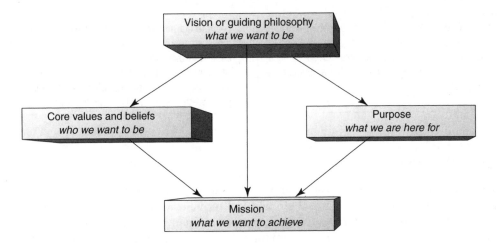

**Figure 3.1**
Vision framework for an organization

The *guiding philosophy* drives the organization and is shaped by the leaders through their thoughts and actions. It should reflect the vision of an organization rather than the vision of a single leader, and should evolve with time, although organizations must hold on to the *core* elements.

The *core values and beliefs* represent the organization's basic principles about what is important in business, its conduct, its social responsibility and its response to changes in the environment. They should act as a guiding force, with clear and authentic values, which are focused on employees, suppliers, customers, society at large, safety, shareholders, and generally stakeholders.

The *purpose* of the organization should be a development from the vision and core values and beliefs and should quickly and clearly convey how the organization is to fulfil its role.

The *mission* will translate the abstractness of philosophy into tangible goals that will move the organization forward and make it perform to its optimum. It should not be limited by the constraints of strategic analysis, and should be proactive not reactive. Strategy is subservient to mission, the strategic analysis being done after, not during, the mission setting process.

Two examples of how leaders of organizations – one in the private sector and one in the public sector – develop their vision, mission and values and are role models of a culture of total quality excellence are given in the insets below:

### Private sector

To enable the company to set direction and achieve its vision, the senior management team address priorities for improvement. These are driven by a business improvement process, which consists of: articulate a vision, determine the actions to realize the vision, define measures and set targets, then implement a rigorous review mechanism.

Each member of the team takes responsibility for one of the Excellence Model criteria. They develop improvement plans and personally ensure that these

### Public sector

The purpose and direction of the organization – the mission – is developed by a task team. Senior, middle and junior managers review and update the mission, vision and values annually to ensure it supports policy and strategy.

Leaders invite input from stakeholders via the employee involvement initiative, monthly update meetings and customer service seminars. The values have been placed on help cards for every employee and are continually re-emphasized at monthly update meetings.

are properly resourced and implemented, and that progress is monitored. Improvements identified at local level are prioritized and resourced by local management against the organization's annual business plan.

Leaders act as role models and have a list of role model standards to follow, which they are measured against in their performance management system. All managers include TQM objectives in their performance agreements and personal development plan, which are reviewed through the review.

## Control

The effectiveness of an organization and its people depends on the extent to which each person and department perform their role and move towards the common goals and objectives. Control is the process by which information or feedback is provided so as to keep all functions on track. It is the sum total of the activities that increase the probability of the planned results being achieved. Control mechanisms fall into three categories, depending upon their position in the managerial process:

| Before the fact | Operational | After the fact |
|---|---|---|
| Strategic plan | Observation | Annual reports |
| Action plans | Inspection and correction | Variance reports |
| Budgets | Progress review | Audits |
| Job descriptions | Staff meetings | Surveys |
| Individual performance objectives | Internal information and data systems | Performance review |
| Training and development | Training programmes | Evaluation of training |

Many organizations use after-the-fact controls, causing managers to take a reactive rather than a proactive position. Such 'crisis orientation' needs to be replaced by a more anticipative one in which the focus is on preventive or before-the-fact controls.

Attempting to control performance through systems, procedures, or techniques *external* to the individual is not an effective approach, since it relies on 'controlling' others; individuals should be responsible for their own actions. An externally based control system can result in a high degree of concentrated effort in a specific area if the system is overly structured, but it can also cause negative consequences to surface:

1. Since all rewards are based on external measures, which are imposed, the 'team members' often focus all their effort on the measure itself, e.g. to have it set lower (or higher) than possible, to manipulate the information which serves to monitor it, or to dismiss it as someone else's goal not theirs. In the budgeting process, for example, distorted figures are often submitted by those who have learned that their 'honest projections' will be automatically altered anyway.
2. When the rewards are dependent on only one or two limited targets, all efforts are directed at those, even at the expense of others. If short-term profitability is the sole criterion for bonus distribution or promotion, it is likely that investment for longer-term growth areas will be substantially reduced. Similarly, strong emphasis and reward for output or production may result in lowered quality.

3. The fear of not being rewarded, or even being criticized, for performance that is less than desirable may cause some to withhold information that is unfavourable but nevertheless should be flowing into the system.
4. When reward and punishment are used to motivate performance, the degree of risk-taking may lessen and be replaced by a more cautious and conservative approach. In essence, the fear of failure replaces the desire to achieve.

The following problem situations have been observed by the authors and their colleagues, within companies that have taken part in research and consultancy:

■ The goals imposed are seen or known to be unrealistic. If the goals perceived by the subordinate are in fact accomplished, then the subordinate has proved himself wrong. This clearly has a negative effect on the effort expended, since few people are motivated to prove themselves wrong!
■ Where individuals are stimulated to commit themselves to a goal, and where their personal pride and self-esteem are at stake, then the level of motivation is at a peak. For most people the toughest critic and the hardest taskmaster they confront is not their immediate boss but themselves.
■ Directors and managers are often afraid of allowing subordinates to set the goals for fear of them being set too low, or loss of control over subordinate behaviour. It is also true that many do not wish to set their own targets, but prefer to be told what is to be accomplished.
■ Where external project managers are recruited to run projects and a reward package is negotiated on the basis of a bonus package reflecting time and cost performance, all too often the company is left with the legacy of quality defects long after the project manager has finished his assignment and pocketed his/her bonuses.
■ Some public sector client organizations, in moving towards the delivery of infrastructure projects through alliances, have developed very complex performance frameworks to attempt to drive project outcomes in non-cost areas such as safety, quality, community and legacy. The complexity of these can be such that the performance measures become an end in themselves and get in the way of management initiative.

TQM is concerned with moving the focus of control from outside the individual to within, the objective being to make everyone accountable for their own performance, and to get them committed to attaining quality in a highly motivated fashion. The assumptions a director or manager must make in order to move in this direction are simply that people do not need to be coerced to perform well, and that people want to achieve, accomplish, influence activity, and challenge their abilities. If there is belief in this, then only the techniques remain to be discussed.

TQM is user-driven – it cannot be imposed from outside the organization, as perhaps can a quality management standard or statistical process control. This means that the ideas for improvement must come from those with knowledge and experience of the processes, activities and tasks; this has massive implications for training and follow-up. TQM is not a cost-cutting or productivity improvement device in the traditional sense, and it must not be used as such. Although the effects of a successful programme will certainly reduce costs and improve productivity, TQM is concerned chiefly with changing attitudes and skills so that culture of the organization becomes one of preventing failure – doing the right things, right first time, every time.

Most construction organizations rely on external control and it is commonplace to believe that workers don't care about anything other than their pay and their beer, that they are relatively unskilled and do not have the potential to change. It is important to realize that management's attitudes can easily become self-fulfilling, and the task of changing workers' behaviour and attitudes is purely a management challenge. Without a change in the attitudes and strategies of leadership, workers' behaviours will not change.

# Effective leadership

Some management teams have broken away from the traditional style of management; they have made a 'managerial breakthrough'. Their approach puts their organization head and shoulders above others in the fight for sales, profits, resources, funding and jobs. Many public service organizations are beginning to move in the same way, and the successful quality-based strategy they are adopting depends very much on effective leadership.

Effective leadership starts with the chief executive's and his top team's vision, capitalizing on market or service opportunities, continues through a strategy that will give the organization competitive or other advantage, and leads to business or service success. It goes on to embrace all the beliefs and values held, the decisions taken and the plans made by anyone anywhere in the organization, and the focusing of them into effective, value-adding action.

Together, effective leadership and total quality management result in the company or organization doing the right things, right first time.

The five requirements for effective leadership are the following:

## 1. Developing and publishing clear documented corporate beliefs and purpose – a mission statement

Executives should express values and beliefs through a clear vision of what they want their company to be and its purpose – what they specifically want to achieve in line with the basic beliefs. Together, they define what the company or organization is all about. The senior management team will need to spend some time away from the 'coal face' to do this and develop their programme for implementation.

Clearly defined and properly communicated beliefs and objectives, which can be summarized in the form of vision and mission statements, are essential if the directors, managers and other employees are to work together as a winning team. The beliefs and objectives should address:

- The definition of the business, e.g. the needs that are satisfied or the benefits provided.
- A commitment to effective leadership and quality.
- Target sectors and relationships with customers, and market or service position.
- The role or contribution of the company, organization, or unit, e.g. example, profit-generator, service department, opportunity-seeker.
- The distinctive competence – a brief statement which applies only to that organization, company or unit.
- Indications for future direction – a brief statement of the principal plans that would be considered.
- Commitment to monitoring performance against customers' needs and expectations, and continuous improvement.

The vision, mission statement and the broad beliefs and objectives may then be used to communicate an inspiring vision of the organization's future. The top management must then show *TOTAL COMMITMENT* to it.

## 2. Develop clear and effective strategies and supporting plans for achieving the mission

The achievement of the company or service vision and mission requires the development of business or service strategies, including the strategic positioning in the 'marketplace'. Plans can then be developed for implementing the strategies. Such strategies and plans can be developed by senior managers alone, but there is likely to be more commitment to them if employee participation in their development and implementation is encouraged.

## 3. Identify the critical success factors and critical processes (Figure 3.2)

The next step is the identification of the *critical success factors* (CSFs), a term used to mean the most important subgoals of a business or organization. CSFs are what must be accomplished for the mission to be achieved. The CSFs are followed by the key, *core business processes* for the organization – the activities that must be done particularly well for the CSFs to be achieved. This process is described in more detail in later chapters. It is also critical for each of these to develop some performance measures that indicate progress towards each goal that has been set.

## 4. Review the management structure

Defining the corporate vision, mission, strategies, CSFs and core processes might make it necessary to review the organizational structure. Directors, managers and other employees can be fully effective only if an effective structure based on process management exists. This includes both the definition of responsibilities for the organization's management and the operational procedures they will use. These must be the agreed best ways of carrying out the core processes.

The review of the management structure should also include the establishment of a process improvement team structure throughout the organization.

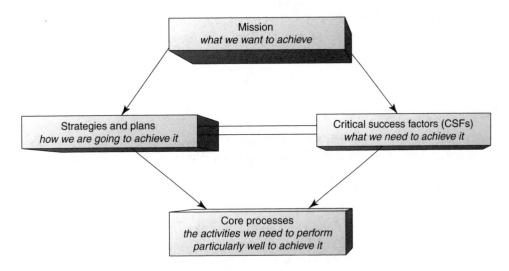

**Figure 3.2**
Mission into action through CSFs and core processes

# 5. Empowerment – encouraging effective employee participation

For effective leadership it is necessary for management to get very close to the employees. They must develop effective communications – up, down and across the organization – and take action on what is communicated; and they should encourage good communications between all suppliers and customers. Particular attention must be paid to the following.

## Attitudes

The key attitude for managing any winning company or organization may be expressed as follows: 'I will personally understand who my customers are and what are their needs and expectations and I will take whatever action is necessary to satisfy them fully. I will also understand and communicate my requirements to my suppliers, inform them of changes and provide feedback on their performance.' This attitude should start at the top – with the chairman or chief executive. It must then percolate down, to be adopted by each and every employee. That will happen only if managers lead by example. Words are cheap and will be meaningless if employees see from managers' actions that they do not actually believe or intend what they say.

## Abilities

Every employee should be able to do what is needed and expected of him or her, but it is first necessary to decide what is really needed and expected. If it is not clear what the employees or subcontractors are required to do and what standards of performance are expected, how can managers expect them to do it? A good example of such confusion has been created on construction projects over the past few decades. Management has repeatedly said that we want the job done fast, but rarely have they stressed that it should be correct the first time. This has resulted in a culture of 'we can fix it later, let's just get on with the job' and numerous defects have been incorporated into buildings, defects that are more expensive to rectify later.

Train, train, train and train again. Training is very important, but it can be expensive if the money is not spent wisely. The training should be related to needs, expectations, and process improvement. It must be planned and *always* its effectiveness reviewed.

## Participation

If all employees are to participate in making the company or organization successful (directors and managers included), then they must also be trained in the basics of disciplined management.

They must be trained to:

- Evaluate – the situation and define their *objectives*.
- Plan – to achieve those objectives fully.
- Do – implement the plans.
- Check – that the objectives are being achieved.
- Amend – take corrective action if they are not.

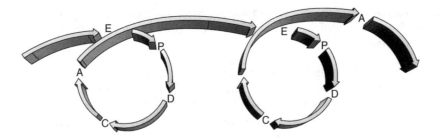

**Figure 3.3**
The helix of never-ending improvement

The word 'disciplined' applied to people at all levels means that they will do what they say they will do. It also means that in whatever they do they will go through the full process of Evaluate, Plan, Do, Check and Amend, rather than the more traditional and easier option of starting by doing rather than evaluating. This will lead to a never-ending improvement helix (Figure 3.3).

This basic approach needs to be backed up with good project management, planning techniques and problem-solving methods, which can be taught to anyone in a relatively short period of time. The project management enables changes to be made successfully and the people to remove the obstacles in their way. Directors and managers need this training as much as other employees.

In construction projects, the project manager has a very important and difficult leadership challenge. He or she has to create a cohesive team at the project level, often with more than 80 percent of the operational staff employed by subcontractors and most of the supervisors employed by other organizations as well. It is up to the project manager to galvanize the entire project team towards a set of shared goals. It is generally true that project managers who do this well do so instinctively; it is not an area that has been identified and taught with any focus. Furthermore on construction projects, because most problems occur at the interfaces between suppliers, the challenge is that process innovation and problem solving has to be achieved by teams of workers from the collaborating organizations who are working side by side in any task. Hence while the TQM principles in construction are the same as in any enterprise, the challenge is to achieve shared goals and common action across the supply chain.

# Excellence in leadership

The vehicle for achieving excellence in leadership is TQM. We have seen that its framework covers the entire organization, all the people and all the functions, including external organization and suppliers. In the first two chapters, several facets of TQM have been reviewed, including:

- Recognizing customers and discovering their needs, this refers to immediate and end-user customers equally.
- Setting standards that are consistent with internal and end-user customer requirements.
- Controlling processes, including systems, and improving their capability.
- Management's responsibility for setting the guiding philosophy, quality policy, etc., and providing motivation through leadership and equipping people to achieve quality.

■ Selecting the right employees and supply chain partners and empowering people at all levels in the organization and across the supply chain to act for quality improvement.

The task of implementing TQM can be daunting, and the chief executive and directors faced with it may become confused and irritated by the proliferation of theories and packages. A simplification is required. The *core* of TQM is the customer/supplier interfaces, both internally and externally, and the fact that at each interface there are processes to convert inputs to outputs. Clearly, there must be commitment to building in quality through management of the inputs and processes.

How can senior managers and directors be helped in their understanding of what needs to be done to become committed to quality and implement the vision? The American and Japanese quality 'gurus' each set down a number of points or absolutes – words of wisdom in management and leadership – and many organizations have used these to establish a policy based on quality.

Similarly, the EFQM have defined the criterion of leadership and its subcriteria as part of their model of Excellence. A fundamental principle behind all these approaches is that the behaviours of the leaders in an organization need to create clarity and constancy of purpose. This may be achieved through development of the vision, values, purpose and mission needed for longer-term performance success.

Using as a construct of the new 'Oakland TQM model', the four Ps and three Cs plus a fourth C – *customers* (which resides in *performance*), the main items for attention to deliver excellence in leadership are given below:

# Planning

■ Develop the vision and mission needed for constancy of purpose and for long-term success.
■ Develop, deploy and update policy and strategy.
■ Align organizational structure to support delivery of policy and strategy.

# Performance

■ Identify critical areas of performance.
■ Develop measures to indicate levels of current performance.
■ Set goals and measure progress towards their achievement.
■ Provide feedback to people at all levels regarding their performance against agreed goals.

# Processes

■ Ensure a system for managing processes is developed and implemented.
■ Ensure through personal involvement that the management system is developed, implemented and continuously improved.
■ Prioritize improvement activities and ensure they are planned on an organization-wide basis.

# People

■ Train managers and team leaders at all levels in leadership skills and problem solving.

- Stimulate empowerment ('experts') and teamwork to encourage creativity and innovation.
- Encourage, support and act on results of training, education and learning activities.
- Motivate, support and recognize the organization's people – both individually and in teams.
- Help and support people to achieve plans, goals, objectives and targets.
- Respond to people and encourage them to participate in improvement activities.

## Customers

- Be involved with customers and other stakeholders.
- Ensure customer (external and internal) needs are understood and responded to.
- Establish and participate in partnerships – as a customer demands continuous improvement in everything.

## Commitment

- Be personally and actively involved in quality and improvement activities.
- Review and improve effectiveness of own leadership.

## Culture

- Develop the values and ethics to support the creation of a total quality culture across the entire supply.
- Implement the values and ethics through actions and behaviours.
- Ensure creativity, innovation and learning activities are developed and implemented.

## Communication

- Stimulate and encourage communication and collaboration.
- Personally communicate the vision, values, mission, policies and strategies.
- Be accessible and actively listen.

TQM should not be regarded as a woolly-minded approach to running an organization. It requires strong leadership with clear direction and a carefully planned and fully integrated strategy derived from the vision. One of the greatest tangible benefits of excellence in leadership is the improved overall performance of the organization. The evidence for this can be seen in some of the major consumer and industrial markets of the world. Moreover, effective leadership leads to improvements and superior quality which can be converted into premium prices. Research now shows that leadership and quality clearly correlate with profit but the less tangible benefit of greater employee participation is equally, if not more, important in the longer term. The pursuit of continual improvement must become a way of life for everyone in an organization if it is to succeed in today's competitive environment.

## Chapter highlights

### The total quality management approach

- TQM is a comprehensive approach to improving competitiveness, effectiveness and flexibility through planning, organizing and understanding each activity, and involving each individual at each level. It is useful in all types of organization.
- TQM ensures that management adopts a strategic overview of quality and focuses on prevention, not detection, of problems.
- It often requires a mindset change to break down existing barriers. Managements that doubt the applicability of TQM should ask questions about the operation's costs, errors, wastes, standards, systems, training and job instructions.

### Commitment and policy

- TQM starts at the top, where serious obsessional commitment to quality and leadership must be demonstrated. Middle management also has a key role to play in communicating the message.
- Every chief executive must accept the responsibility for commitment to a quality policy that deals with the organization for quality, the customer needs, the ability of the organization, supplied materials and services, education and training, and review of the management systems for never-ending improvement.

### Creating or changing the culture

- The culture of an organization is formed by the beliefs, behaviours, norms, dominant values, rules and climate in the organization.
- Any organization needs a vision framework, comprising its guiding philosophy, core values and beliefs, purpose, and mission.
- The effectiveness of an organization depends on the extent to which people perform their roles and move towards the common goals and objectives.
- TQM is concerned with moving the focus of control from the outside to the inside of individuals, so that everyone is accountable for his/her own performance.

### Effective leadership

- Effective leadership starts with the chief executive's vision and develops into a strategy for implementation.
- Top management should develop the following for effective leadership: clear beliefs and objectives in the form of a mission statement; clear and effective strategies and supporting plans; the critical success factors and core processes; the appropriate management structure; employee participation through empowerment, and the EPDCA helix.

### Excellence in leadership

- The vehicle for achieving excellence in leadership is TQM. Using the construct of the new Oakland TQM model, the four Ps and four Cs provide a framework for this: *planning, performance, processes, people, customers, commitment, culture, communication.*

# 4

# Policy, strategy and goal deployment

# Integrating TQM into the policy and strategy

In the previous chapter on leadership the main message was that leaders should have a clear sense of direction and purpose, which they communicate effectively throughout the organization and to their supply chain partners. This involves the development of the vision, values and mission which are clearly aspects of policy and strategy. Included in the EFQM Excellence Model, the criterion policy and strategy is concerned with:

> How the organization implements its mission and vision via a clear stakeholder-focused strategy, supported by relevant policies, plans, objectives, targets and processes.

For this to happen the vision and mission and their deployment must be based on the needs and expectations of the organization's stakeholders – present and future. This in turn requires information from research and learning activities and even more importantly performance measurement, on which to base the policies and strategies. Of course, time and the world around us do not stand still, so the policy and strategies must be reviewed, updated and generally developed to meet the changing needs of the organization.

Many companies have difficulty starting a continuous improvement programme; they do not know how to state their vision and mission, simply because management is not used to thinking in these terms. A good place to start is through research into external customer satisfaction. Feedback can then provide the foundation definition of corporate vision, and as the process matures it will involve feedback from all stakeholders.

Relatively few construction organizations would consider that their supply chain is a stakeholder in their organization; yet, subcontractors supply most of the labour, materials and equipment on construction sites. Because *design and construction* as a method of procurement has becomes more popular, the supply chain now includes the designers as well. Subcontract employees constitute upwards of 80 percent of workers on most building construction sites, fewer on civil projects. Hence, a very important part of a construction organization's strategy must consider such issues as the values, skills, knowledge and attitudes of subcontractor management and employees.

There are six basic steps for achieving this and providing a good foundation for the implementation of TQM.

## Step 1 Develop a shared vision and mission for the business/organization

Once the top team is reasonably clear about the direction the organization should be taking it can develop vision and mission statements that will help to define process-alignment, roles and responsibilities. This will lead to a co-ordinated flow of analysis of processes that cross the traditional functional areas at all levels of the organization, without changing formal structures, titles, and systems which can create resistance. The vision framework was introduced in Chapter 3 (Figure 3.1) and has been extended in Figure 4.1.

The mission statement gives a purpose to the organization or unit. It should answer the questions 'what are we here for?' or 'what is our basic purpose?' and 'what have we got to achieve?' It therefore defines the boundaries of the business in which the organization operates. This will help to focus on the 'distinctive competence' of the organization, and to orient everyone in the same direction of what has to be done. The mission must be documented, agreed by the top management team, sufficiently explicit to enable

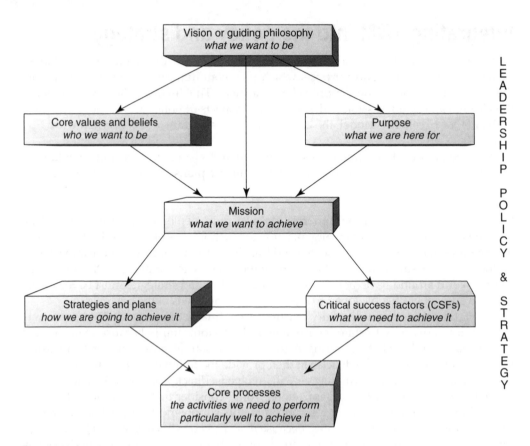

**Figure 4.1**
Vision framework for an organization

its eventual accomplishment to be verified, and ideally be no more than four sentences. The statement must be understandable, communicable, believable, and usable.

The mission statement is:

- an expression of the aspiration of the organization;
- the touchstone against which all actions or proposed actions can be judged;
- usually long term;
- short term if the mission is survival.

Typical content includes a statement of:

- the role or contribution of the business or unit – for example, profit generator, service department, opportunity seeker;
- the definition of the business – for example, the needs you satisfy or the benefits you provide. Do not be too specific or too general;
- your distinctive competence – this should be a brief statement that applies only to your specific unit. A statement, which could apply equally to any organization, is unsatisfactory;
- indications for future direction – a brief statement of the principal things you would give serious consideration to.

Some questions that may be asked of a mission statement are, does it:

- define the organization's role?
- contain the need to be fulfilled:
  - is it worthwhile/admirable?
  - will employees identify with it?
  - how will it be viewed externally?
- take a long-term view, leading to, for example, commitment to a new product or service development, or training of personnel?
- take into account all the 'stakeholders' of the organization?
- ensure the purpose remains constant despite changes in top management?

It is important to establish in some organizations whether or not the mission is survival. This does not preclude a longer-term mission, but the short-term survival mission must be expressed, if it is relevant. The management team can then decide whether they wish to continue long-term strategic thinking. If survival is a real issue it is inadvisable to concentrate on long-term planning initially.

There must be open and spontaneous discussion during generation of the mission, but there must in the end be convergence on one statement. If the mission statement is wrong, everything that follows will be wrong too, so a clear understanding is vital.

## Step 2 Develop the 'mission' into its critical success factors (CSFs) to coerce and move it forward

The development of the mission is clearly not enough to ensure its implementation. This is the 'danger gap' which many organizations fall into because they do not foster the skills needed to translate the mission through its CSFs into the core processes. Hence, they have 'goals without methods' and change is not integrated properly into the business.

Once the top managers begin to list the CSFs they will gain some understanding of what the mission or the change requires. The first step in going from mission to CSFs is to brainstorm all the possible impacts on the mission. In this way 30 to 50 items ranging from politics to costs, from national cultures to regional market peculiarities may be derived.

The CSFs may now be defined – *what* the organization must accomplish to achieve the mission, by examination and categorization of the impacts. This should lead to a balanced set of deliverables for the organization in terms of:

- financial and non-financial performance;
- customer/market satisfaction;
- people/internal organization satisfaction;
- supply chain satisfaction;
- environmental/societal satisfaction.

There should be no more than eight CSFs, and no more than four if the mission is survival. They are the building blocks of the mission – minimum key factors or subgoals that the organization must have or needs and which together will achieve the mission. They are the *whats* not the *hows*, and are not directly manageable – they may be in some case statements of hope or fear. But they provide direction and the success criteria, and are the end product of applying the processes. In CSF determination, a management team should follow the rule that each CSF is necessary and together they are sufficient for the mission to be achieved.

| | | Corporate CSFs / KPOs | | | | | | | | |
|---|---|---|---|---|---|---|---|---|---|---|
| Divisional or functional CSFs | CSF No. | 1 | 2 | 3 | 4 | 5 | 6 | 7 | 8 | KPIs |
| | 1 | | | | | | | | | |
| | 2 | | | | | | | | | |
| | 3 | | | | | | | | | |
| | 4 | | | | | | | | | |
| | 5 | | | | | | | | | |
| | 6 | | | | | | | | | |
| | 7 | | | | | | | | | |
| | 8 | | | | | | | | | |

**Figure 4.2**
Interaction of corporate and divisional CSFs

Some examples of CSFs may clarify their understanding:

- We must have right-first-time suppliers.
- We must have motivated, skilled workers.
- Suppliers working on our sites must have motivated, skilled workers.
- We need new products that satisfy market needs.
- Our designs must be well documented.
- We need new business opportunities.
- We must have best-in-the-field product quality.

The list of CSFs should be an agreed balance of strategic and tactical issues, each of which deals with a 'pure' factor, the use of 'and' being forbidden. It will be important to know when the CSFs have been achieved, but an equally important step is to use the CSFs to enable the identification of the processes.

Senior managers in large complex organizations may find it necessary or useful to show the interaction of divisional CSFs with the corporate CSFs in an impact matrix (see Figure 4.2 and discussion under Step 6).

## Step 3 Define the key performance indicators as being the quantifiable indicators of success in terms of the mission and CSFs

The mission and CSFs provide the what of the organization, but they must be supported by measurable key performance indicators (KPIs) that are tightly and inarguably linked. These will help to translate the directional and sometimes 'loose' statements of the mission into clear targets, and in turn to simplify management's thinking. The KPIs will be

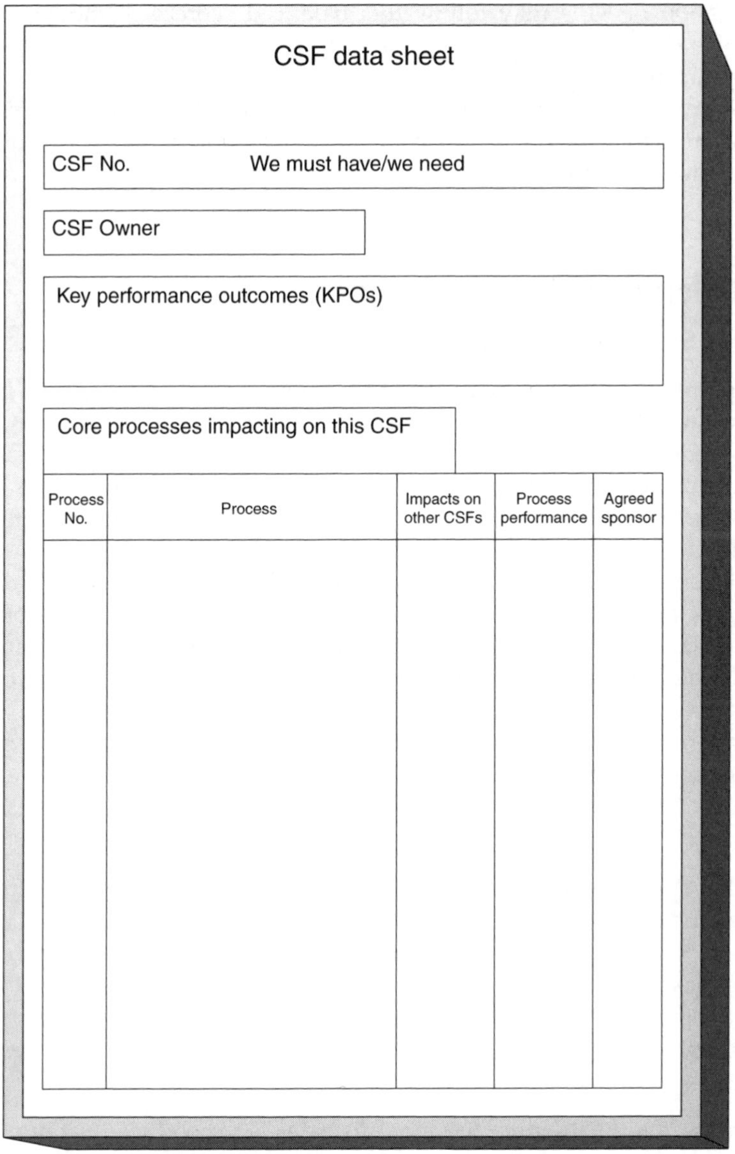

CSF data sheet

CSF No. We must have/we need

CSF Owner

Key performance outcomes (KPOs)

Core processes impacting on this CSF

| Process No. | Process | Impacts on other CSFs | Process performance | Agreed sponsor |
|---|---|---|---|---|
| | | | | |

**Figure 4.3**
CSF data sheet

used to monitor progress and as evidence of success for the organization, in every direction, internally and externally.

Each CSF should have an 'owner' who is a member of the management team that agreed the mission and CSFs. The task of an owner is to:

- define and agree the KPIs and associated *targets;*
- ensure that appropriate data is collected and recorded;
- monitor and report progress towards achieving the CSF (KPIs and targets) on a regular basis;
- review and modify the KPIs and targets where appropriate.

A typical CSF data sheet for completion by owners is shown in Figure 4.3.

The derivation of KPIs may follow the 'balanced scorecard' model, proposed by Kaplan, which divides measures into financial, customer, internal business and innovation and learning perspectives (see Chapter 7).

Some CSFs may involve gathering feedback from supply chain partners. It is important in these areas to bring the supply chain partners along in the process of developing and implementing those CSFs.

## Step 4 Understand the core processes and gain process sponsorship

This is the point when the top management team have to consider how to institutionalize the mission in the form of processes that will continue to be in place, until major changes are required.

The core business processes describe what actually is or needs to be done so that the organization meets its CSFs. As with the CSFs and the mission, each process which is necessary for a given CSF must be identified, and together the processes listed must be sufficient for all the CSFs to be accomplished. To ensure that processes are listed, they should be in the form of verb plus object, such as research the market, recruit competent staff, or manage supplier performance. The core processes identified frequently run across 'departments' or functions, yet they must be measurable.

Each core process should have a sponsor, preferably a member of the management team that agreed the CSFs.

The task of a sponsor is to:

- ensure that appropriate resources are made available to map, investigate and improve the process;
- assist in selecting the process improvement team leader and members;
- remove blocks to the team's progress;
- ensure that supply chain collaborations are in place, where necessary;
- report progress to the senior management team.

The first stage in understanding the core processes is to produce a set of processes of a common order of magnitude. Some smaller processes identified may combine into core processes, others may be already at the appropriate level. This will ensure that the change becomes entrenched, the core processes are identified and that the right people are in place to sponsor or take responsibility for them. This will be the start of getting the process team organization up and running.

The questions will now come thick and fast; is the process currently carried out? By whom? When? How frequently? With what performance and how well compared with competitors? The answering of these will force process ownership into the business. The process sponsor may form a process team which takes quality improvement into the next steps. Some form of prioritization using process performance measures is necessary at this stage to enable effort to be focused on the key areas for improvement. This may be carried out by a form of impact matrix analysis (see Figure 4.4). The outcome should be a set of 'most critical processes' (MCPs) which receive priority attention for improvement, based on the number of CSFs impacted by each process and its performance on a scale A to E.

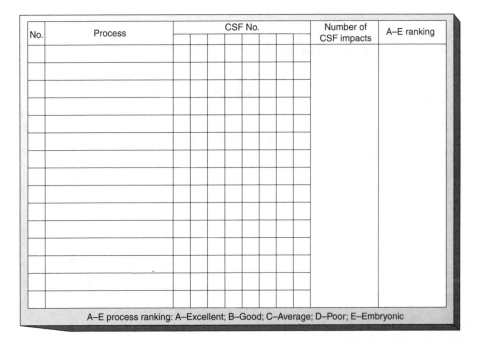

| No. | Process | CSF No. | | | | | | | | Number of CSF impacts | A–E ranking |
|-----|---------|--|--|--|--|--|--|--|--|-----------------------|-------------|
| | | | | | | | | | | | |
| | | | | | | | | | | | |
| | | | | | | | | | | | |
| | | | | | | | | | | | |
| | | | | | | | | | | | |
| | | | | | | | | | | | |
| | | | | | | | | | | | |
| | | | | | | | | | | | |
| | | | | | | | | | | | |
| | | | | | | | | | | | |
| | | | | | | | | | | | |
| | | | | | | | | | | | |

A–E process ranking: A–Excellent; B–Good; C–Average; D–Poor; E–Embryonic

**Figure 4.4**
A–E process ranking

## Step 5 Break down the core processes into subprocesses, activities and tasks and form improvement teams around these

Once an organization has defined and mapped out the core processes, people need to develop the skills to understand how the new process structure will be analysed and made to work. The very existence of new process teams with new goals and responsibilities will force the organization into a learning phase. The changes should foster new attitudes and behaviours.

In the construction procurement process, whether in the design stage, say during design documentation, or in the construction stage, for say site process quality improvement, some of these teams will necessarily involve supply chain partners. Although it may be difficult to establish such cross-organizational process teams, it is important to identify and sell the benefits of the new process structures to all parties involved.

An illustration of the breakdown from mission through CSFs and core processes, to individual tasks may assist in understanding the process required (Figure 4.5).

> *Mission*

Two of the statements in a well-known management consultancy's mission statement are:

> Gain and maintain a position as Europe's foremost management consultancy in the development of organizations through management of change.

> Provide the consultancy, training and facilitation necessary to assist with making continuous improvement an integral part of our customers' business strategy.

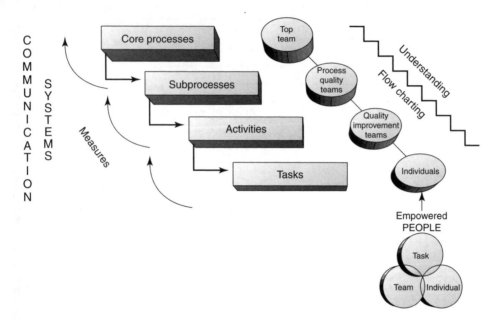

**Figure 4.5**
Breakdown of core processes

*Critical success factor*

One of the CSFs which clearly relates to this is:

We need a high level of awareness.

↓

*Core process*

One of the core processes which clearly must be done particularly well to achieve this CSF is to:

Promote, advertise, and communicate the company's business capability.

↓

*Subprocess*

One of the subprocesses which results from a breakdown of this core process is:

Prepare the company's information pack.

↓

*Activity*

One of the activities which contributes to this subprocess is:

Prepare one of the subject booklets, i.e. 'Business Excellence and Self-Assessment'.

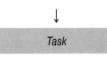

$\downarrow$

Task

One of the tasks which contributes to this is:

> Write the detailed leaflet for a particular seminar, e.g. one or three day seminars on self-assessment.

### Individuals, tasks and teams

Having broken down the processes into subprocesses, activities and tasks in this way, it is now possible to link this with the Adair model of action-centred leadership and team-work (see Chapter 15).

The tasks are clearly performed, at least initially, by individuals. For example, some-body has to sit down and draft out the first version of a seminar leaflet. There has to be an understanding by the individual of the task and its position in the hierarchy of processes. Once the initial task has been performed, the results must be checked against the activity of co-ordinating the promotional booklet – say for TQM. This clearly brings in the team, and there must be interfaces between the needs of the *tasks*, the *individuals* who performed them and the *team* concerned with the *activities*.

### Performance measurement and metrics

Once the processes have been analysed in this way, it should be possible to develop metrics for measuring the performance of the processes, subprocesses, activities, and tasks. These must be meaningful in terms of the *inputs* and *outputs* of the processes, and in terms of the *customers* and of *suppliers* to the processes (see Figure 4.5).

At first thought, this form of measurement can seem difficult for processes such as preparing a sales brochure or writing leaflets advertising seminars, but, if we think care-fully about the *customers* for the leaflet-writing tasks, these will include the *internal* ones, i.e. the consultants, and we can ask whether the output meets their requirements. Does it really say what the seminar is about, what its objectives are and what the programme will be? Clearly, one of the 'measures' of the seminar leaflet-writing task could be the number of typing errors in it, but is this a *key* measure of the performance of the process? Only in the context of office management is this an important measure. Elsewhere it is not.

The same goes for the *activity* of preparing the subject booklet. Does it tell the 'cus-tomer' what TQM or SPC is and how the consultancy can help? For the *subprocess* of preparing the company brochure, does it inform people about the company and does it bring in enquiries from which customers can be developed? Clearly, some of these meas-ures require *external market research*, and some of them *internal research*. The main point is that metrics must be developed and used to reflect the *true performance* of the processes, subprocesses, activities and tasks. These must involve good contact with external and internal customers of the processes. The metrics may be quoted as *ratios*, e.g. numbers of customers derived per number of brochures mailed out. Good data collection, record keeping, and analysis are clearly required.

It is hoped that this illustration will help the reader to:

■ understand the breakdown of processes into subprocesses, activities and tasks;

- understand the links between the process breakdowns and the task, individual and team concepts;
- link the hierarchy of processes with the hierarchy of quality teams;
- begin to assemble a cascade of flowcharts representing the process breakdowns, which can form the basis of the quality management system and communicate what is going on throughout the business;
- understand the way in which metrics must be developed to measure the true performance of the process, and their links with the customers, suppliers, inputs and outputs of the processes.

The changed patterns of co-ordination, driven by the process maps, should increase collaboration and information sharing. Clearly the senior and middle managers need to provide the right support. Once employees, at all levels, identify what kinds of new skill are needed, they will ask for the formal training programmes in order to develop those skills further. This is a key area, because teamwork around the processes will ask more of employees, so they will need increasing support from their managers.

This has been called 'just-in-time' training, which describes very well the nature of the training process required. This contrasts with the blanket or carpet bombing training associated with many unsuccessful change programmes, which targets competencies or skills, but does not change the organization's patterns of collaboration and co-ordination.

## Step 6 Ensure process and people alignment through a policy deployment or goal translation process

One of the keys to integrating excellence into the business strategy is a formal 'goal translation' or 'policy deployment' process. If the mission and measurable goals have been analysed in terms of critical success factors and core processes, then the organization has begun to understand how to achieve the mission. Goal translation ensures that the 'whats' are converted into 'hows', passing this right down through the organization, using a quality function deployment (QFD) type process, Figure 4.6 (see Chapter 6). The method is best described by an example.

At the top of an organization in the chemical process industries, five measurable goals have been identified. These are listed under the heading 'What' in Figure 4.7. The top team listens to the 'voice of the customer' and tries to understand *how* these

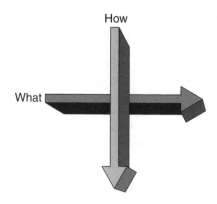

**Figure 4.6**
The goal translation process

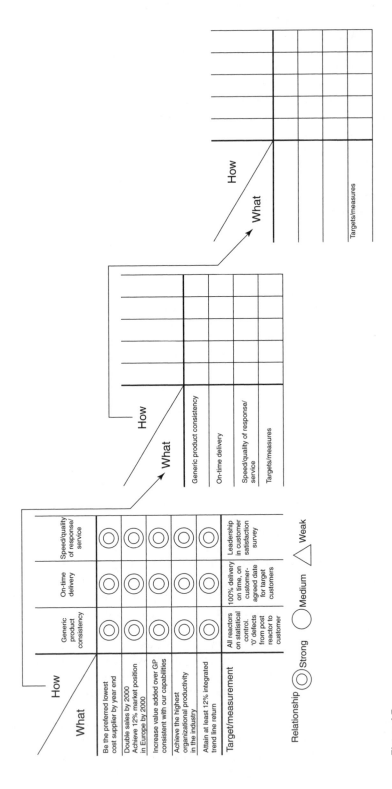

**Figure 4.7**
The goal translation process in practice

business goals will be achieved. They realize that product consistency, on-time delivery, and speed or quality of response are the keys. These CSFs are placed along the first row of the matrix and the relationships between the *what* and the *how* estimated as strong, medium or weak. A measurement target for the hows is then specified.

The *how* becomes the *what* for the next layer of management. The top team share their goals with their immediate reports and ask them to determine their *hows*, indicate the relationship and set measurement targets. This continues down the organization through a 'catch-ball' process until the senior management goals have been translated through the *what/how* → *what/how* → *what/how* matrices to the individual tasks within the organization. This provides a good discipline to support the breakdown and understanding of the business process mapping described in Chapter 9.

A successful approach to policy/goal deployment and strategic planning in an organization with several business units or division, is that mission, CSFs with KPIs and targets, and core processes, are determined at the corporate level, typically by the board. While there needs to be some flexibility about exactly how this is translated into the business units, typically it would be expected that the process is repeated with the senior team in each business unit or division. Each business unit head should be part of the top team that did the work at the corporate level, and each of them would develop a version of the same process with which they feel comfortable.

Each business unit would then follow a similar series of steps to develop their own mission (perhaps) and certainly their own CSFs and KPIs with targets. A matrix for each business unit showing the impact of achieving the business unit CSFs on the corporate CSFs would be developed. In other words, the first deployment of the corporate 'whats' CSFs is into the 'hows' – the business unit CSFs (see Figure 4.2).

If each business unit follows the same pattern, the business unit teams will each identify unit CSFs, KPIs with targets and core processes, which are interlinked with the ones at corporate level. Indeed the core processes at corporate and business unit level may be the same, with any specific additional processes identified at business unit level to catch the flavour and business needs of the unit. It cannot be overemphasized how much ownership there needs to be at the business unit management level for this to work properly. For example, to ensure that planned outcomes are achieved on a construction project, it is the project director's task to ensure that the corporate CSFs and KPIs are effectively addressed while the project manager's task is to ensure that any necessary training at the project level is implemented.

With regard to core processes, each business unit or function will begin to map these at the top level. This will lead to an understanding of the purpose, scope, inputs, control, and resources for each process and provide an understanding of how the subprocesses are linked together. Flowcharting showing connections with procedures will then allow specific areas for improvement to be identified so that the continuous improvement, 'bottom-up' activities can be deployed, and benefit derived from the process improvement training to be provided (Figure 4.8).

It is important to get clarity at the corporate and business unit management levels about the whats/hows relationships, but the ethos of the whole process is one of involvement and participation in goal/target setting, based on good understanding of processes – so that it is known and agreed what can be achieved and what needs measuring and targeting at the business unit level.

Senior management may find it useful to monitor performance against the CSFs, KPIs and targets, and to keep track of process using a reporting matrix, perhaps at their monthly meetings. A simplified version of this developed for use in a small company is shown in Figure 4.9. The frequency of reporting for each CSF, KPI, and process can be determined in a business planning calendar.

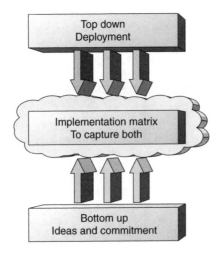

**Figure 4.8**
Implementation top down and bottom up

As previously described, in a larger organization, this approach may be used to deploy the goals from the corporate level through divisions to site/departmental level, Figure 4.10. This form of implementation should ensure the top-down *and* bottom-up approach to the deployment of policies and goals.

## Deliverables

The deliverables after one planning cycle of this process in a business will be:

1. An agreed framework for policy/goal deployment through the business.
2. Agreed mission statement for the business and, if required, for the business units/division.
3. Agreed critical success factors (CSFs) with ownership at top team level for the business and business units/divisions.
4. Agreed key performance indicators (KPIs) with targets throughout the business.
5. Agreed core business processes, with sponsorship at top team level.
6. A corporate CSF/business unit CSF matrix showing the impacts and the first 'whats/hows' deployment.
7. At what/how (CSF/process) matrix approach for deploying the goals into the organization through process definition, understanding, and measured improvement at the business unit level.
8. Focused business improvement, linked back to the CSFs, with prioritized action plans and involvement of employees.

## Strategic and operational planning

Changing the culture of an organization to incorporate a sustainable ethos of continuous improvement and responsive business planning will come about only as the result of a carefully planned and managed process. Clearly many factors are involved, including:

- identifying strategic issues to be considered by the senior management team;
- balancing the present needs of the business against the vital needs of the future;

| Core processes | CSFs: We must have | | | | | Measures | Year targets | Target CSF owner |
|---|---|---|---|---|---|---|---|---|
| | Satisfactory financial and non-financial performance | A growing base of satisfied customers | A sufficient number of committed and competent people | Research projects properly completed and published | * = Priority for improvement | | | |
| | | | | | | Sales volume. Profit. Costs versus plan. Shareholder return Associate/employee utilization figures | Turnover £2m. Profit £200k. Return for shareholders. Days/month per person | |
| | | | | | | Sales/customer Complaints/recommendations Customer satisfaction | >£200k = 1 client. £100k–£200k = 5 clients. £50k–£100k = 6 clients <£50k = 12 clients | |
| | | | | | | No. of employed staff/associates Gaps in competency matrix. Appraisal results Perceptions of associates and staff | 15 employed staff 10 associates including 6 new by end of year | |
| | | | | | | Proportion completed on time, in budget with customers satisfied. Number of publications per project | 3 completed on time, in budget with satisfied customers | |
| | | | | | | Process owner | | |
| | | | | | | Process performance | | |
| | | | | | | Measures and targets | | |
| Manage people | X | X | X | X | ** | | | |
| Develop products | | X | | | | | | |
| Develop new business | X | X | | | ** | | | |
| Manage our accounts | X | X | | X | ** | | | |
| Manage financials | X | | X | | | | | |
| Manage int. systems | X | X | X | | | | | |
| Conduct research | | X | | X | | | | |

**Figure 4.9**
CSF/core process matrix

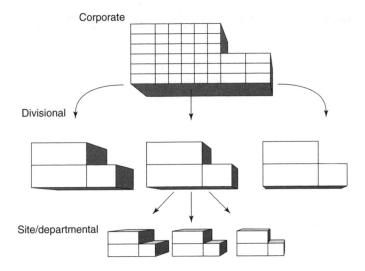

**Figure 4.10**
Deployment – what/how

- concentrating finite resources on important things;
- providing awareness of impending changes in the business environment in order to adapt more rapidly, and more appropriately.

Strategic planning is the continuous process by which any organization will describe its destination, assess barriers standing in the way of reaching that destination, and select approaches for dealing with those barriers and moving forward. Of course the real contributors to a successful strategic plan are the participants.

The strategic and operational planning process described in this chapter will:

- Provide the senior management team with the means to manage the organization and to identify strengths and weaknesses through the change process.
- Allow the senior management team members to have a clear understanding and to achieve agreement on the strategic direction, including vision and mission.
- Identify and document those factors critical to success (CSFs) in achieving the strategic direction and the means by which success will be measured (KPIs) and targeted.
- Identify, document and encourage ownership of the core processes that drive the business.
- Reach agreement on the priority processes for action by process improvement teams, incorporating current initiatives into an overall, cohesive framework.
- Provide a framework for successfully deploying all goals and objectives through all organizational levels through a two-way 'catch-ball' process.
- Provide a mechanism by which goals and objectives are monitored, reviewed, and appropriate actions taken, at appropriate frequencies throughout the operational year.
- Transfer the skills and knowledge necessary to sustain the process.

The components outlined above will provide a means of effectively deploying a common vision and strategy throughout the organization. They will also allow for the incorporation of all change projects, as well as 'business as usual' activities, into a common framework which will form the basis of detailed operating plans.

# The development of policies and strategies

Let us assume that a management team are to develop the policies and strategies based on stakeholder needs and the organization's capabilities, and that it wants to ensure these are communicated, implemented, reviewed and updated. Clearly a detailed review is required of the major stakeholders' needs, the performance of competitors, the state of the market and industry/sector conditions. This can then form the basis of top level goals, planning activities and setting of objectives and targets.

How individual organizations do this varies greatly, of course, and some of this variation can be seen in the case studies at the end of the book. However, some common themes emerge under six headings:

## Customer/market

- Data collected, analysed and understood in terms of where the organization will operate.
- Customers' needs and expectations understood, now and in the future.
- Developments anticipated and understood, including those of competitors and their performance.
- The organization's performance in the marketplace known.
- Benchmarking against best in class organizations.

## Shareholders/major stakeholders

- Shareholders'/major stakeholders' needs and ideas understood.
- Appropriate economic trends/indicators and their impact analysed and understood.
- Policies and strategies appropriate to shareholder/stakeholder needs and expectations developed.
- Needs and expectations balanced.
- Various scenarios and plans to manage risks developed.

## People

- The needs and expectations of the employees are understood.
- The needs and expectations of the subcontractor's employees are understood.
- Data collected, analysed and understood in terms of the internal performance of the organization.
- Output from learning activities understood.
- Everyone appropriately informed about the policies and strategies.

## Processes

- A key process framework to deliver the policies and strategies designed, understood and implemented.
- Key process owners identified.
- Each key process and its major stakeholders defined.
- Key process framework reviewed periodically in terms of its suitability to deliver to organization's requirements.

## Partners/resources

- Appropriate technology understood.
- Impact of new technologies analysed.
- Needs and expectations of partners understood.
- Policies and strategies aligned with those of partners.
- Financial strategies developed.
- Appropriate buildings, equipment and materials identified/sourced.

# Society

- Social, legal and environmental issues understood.
- Environment and corporate responsibility policies developed.

The whole field of business policy, strategy development and planning is huge and there are many excellent texts on the subject. It is outside the scope of this book to cover this area in detail, of course, but one of the most widely used and comprehensive texts is *Exploring Corporate Strategy – text and cases*, 6th edition by Gerry Johnson and Kevan Scholes. This covers strategic positioning and choices, and strategy implementation at all levels.

## *Chapter highlights*

### Integrating TQM into the policy and strategy
- Policy and strategy is concerned with how the organization implements its mission and vision in a clear stakeholder-focused strategy supported by relevant policies, plans, objectives, targets and processes.
- Senior management may begin the task of alignment through six steps:
    - develop a shared vision and mission;
    - develop the critical success factors;
    - define the key performance indicators (balanced scorecard);
    - understanding the core process and gain ownership;
    - break down the core processes into subprocesses, activities and tasks;
    - ensure process and people alignment through a policy deployment or goal translation process.
- The deliverables after one planning cycle will include: an agreed policy/goal deployment framework; agreed mission statements; agreed CSFs and owners; agreed KPIs and targets; agreed core processes and sponsors; whats/hows deployment matrices; focused business improvement plans.

### The development of policies and strategies
- The development of policies and strategies requires a detailed review of the major stakeholders' needs, the performance of competitors, the market/industry/sector conditions to form the basis of top level goals, planning activities and setting of objectives and targets.
- The common themes for planning strategies may be considered under the headings of customer/market, shareholders/major stakeholders, people, processes, partners/resources and society.
- The field of policy and strategy development is huge and the text by Johnson and Scholes is recommended reading.

Policy, strategy and goal deployment

**67**

# 5

# Partnerships and resources

# Partnering

In recent years business, technologies and economies have developed in such a way that organizations recognize the increasing needs to establish mutually beneficial relationships with other organizations, often called 'partners'. The philosophies behind the various TQM and Excellence models support the establishment of partnerships and lay down principles and guidelines for them. This is particularly important in construction where because of the cost focused mindset, organizations within the supply chain are often out of alignment. This creates situations in which parties that rely on each other for services have inconsistent objectives.

Because in the late 1980s and early 1990s *partnering* was promoted and adopted as a formal procurement process, many people in the construction sector use the word *partnering* to describe a form of contracting. We, however, are *not* using the word in this sense. In this book when we use the term *partnering* we simply mean an effective and close collaboration between two or more parties working together towards shared objectives.

How companies in the private sector plan and manage their partnerships can mean the difference between success and failure for it is extremely rare to find companies which can sustain a credible business operation now without a network of co-operation between individuals and organizations or parts of them. This extends the internal customer/ supplier relationship ideas into the supply chain of a company making sure that all the necessary materials, services, equipment, information skills and experience are available in totality to deliver the right products or services to the end customer. Gone are the days, hopefully, of conflict and dispute between customers and their suppliers. An efficient supply chain process, built on strong confident partnerships, will create high levels of people satisfaction, customer satisfaction and support and, in turn, good business results.

This is especially true in construction where much of the work is outsourced and where on construction sites as many as 85 percent of workers can be in the employ of subcontractors. Construction contractors are increasingly finding that their risks, whether they are safety, quality, production or environment related, lie in the hands of their subcontractors' employees and supervisors. They are the people at the workface who must get it *right-first-time* if the process is to be truly efficient. It is a critical challenge for general contractors to work out how they can build long-term relationships with their subcontractors, even when their joint work is discontinuous and project based. They need to work out how to create viable and stable relationships, which will create advantages they can leverage to their mutual benefit, when working together.

How an organization plans and manages the external partnerships must be in line with its overall policies and strategies, being designed and developed to support the effective operation of its processes (Figure 5.1). A key part of this, of course, is identifying with whom those key strategic partnerships will be formed. Whether it is working with key suppliers to deliver materials or components to the required quality, plan (lead times) and costs, or the supply of information technology, transport, broadcasting or design consultancy services, the quality of partnerships has been recognized throughout the world as a key success criterion.

There are many essential ingredients to ensuring that partnership processes work well for an organization. Key aspects include a clear definition of goals, roles, responsibilities, processes and performance measures backed up by good communications and exchange of information. This supports learning between two organizations and often leads to innovative solutions to problems that have remained unsolved in the separate organizations, prior to their close collaboration. To ensure that objectives are achieved, activities should be supported by quality management processes that might include the use of system audits and reviews, certificates of competence or performance reviews and joint action plans.

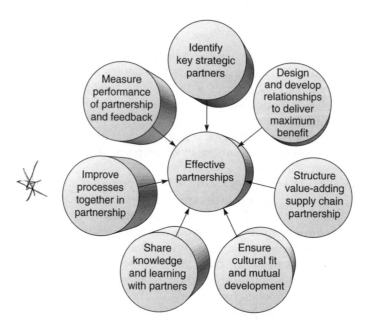

**Figure 5.1**
The contributors to effective partnerships

Although in construction the focus for some years has been on fixed price contracts, some examples of genuine partnerships do exist. For example, in recent years, long-term, stable and beneficial partnering arrangements have been created in the maintenance of buildings, roads and rail infrastructure, and many of these have yielded very significant benefits to the service providers and clients alike.

In construction, examples are fewer; however, an outstanding example is of a piling contractor and its parent company, who partnered on the piling component of the widening of a length of the M25 around London. They did this after they had already agreed on a firm price for the work. By partnering they were able to do the work in a much more efficient manner than originally envisaged, and they were able to save a very significant amount from the originally agreed cost.

When establishing partnerships, attention should be given to:

- maximizing the understanding of what is to be delivered by the partnership – the needs of the customer and the capability of the supplier must match perfectly if satisfaction and loyalty are to be the result;
- understanding what represents value for money – getting the commercial relationship right includes the clear understanding of scope and quality expectations and challenges;
- understanding the respective roles and ensuring an appropriate allocation of responsibilities – to the party best able to manage them;
- working in a supportive, constructive and a team-based relationship;
- having solid programmes of work, comprising agreed plans, timetables, targets, key milestones and decision points;
- identifying areas where the achievement of agreed goals requires training, and implementing joint training programmes between the partners;
- structuring the resolution of complaints, concerns or disputes rapidly and at the lowest practical level;

- enabling the incorporation of knowledge transfer and making sure this adds value;
- developing a stronger and stronger working relationship geared to delivering better and better products or services to the end-customer – based on continuous improvement principles.

# The role of purchasing in partnerships

A company selling wooden products had a very simple purchasing policy: it bought the cheapest wood it could find anywhere in the world. Down in the workshops they were scrapping doors and window frames as if they were going out of fashion – warping, knots in the wood, 'flaking', cracking, splits, etc. When the purchasing manager was informed, he visited the workshops and explained to the supervisors how cheap the wood was and instructed them to 'do the best you can – the customer will never notice'. On challenging this policy, the author was told that it would not change until someone proved to the purchasing manager, in a quantitative way, that the policy was wrong. That year the company 'lost' £1 million worth of wood – in scrap and rework. You can go out of business waiting for such proof.

Another example relates to workmanship in the waterproofing of wet areas in high-rise residential buildings in the Sydney market. In recent years, it has become all too common that either the waterproofing subcontractor or the tiler leaves poorly finished or damaged waterproofing membrane in shower recesses. The result being that relatively shortly after the occupants move in and start using the shower water damage appears in the adjoining room. The cost of rectifying this problem is in excess of $5000 for every occurrence while the cost of avoiding it in the first instance would have been between $20 and $100. This situation is the result of using cheap, poorly trained and unmotivated workmen to save money; however, any surpluses on the project quickly evaporate in rectification once occupants have moved in. Worse still, the company can never recover the lost goodwill caused as a result of the nuisance of both the initial leak and the subsequent rectification.

Very few organizations are self-contained to the extent that their products and services are all generated at one location, from basic materials. Some materials or services are usually purchased from outside organizations, and the primary objective of purchasing is to obtain the correct equipment, materials, and services in the right quantity, of the right quality, from the right origin, at the right time and cost. Purchasing can also play a vital role as the organization's 'window-on-the-world', providing information on any new products, processes, materials and services that become available. It can also advise on probable prices, deliveries, and performance of products under consideration by the research, design and development functions. In other words it should support any partnership in the supply chain.

Purchasing occurs at the interface between organizations and since the construction procurement process is highly fragmented and many services are difficult to fully define, this is one of the most critical areas of management in the sector. Major traps relate to difficulties in the definition of scope and quality. It has been found that to be successful, purchasing has to be responsible for the entire product area: specification, selection, price, quality, delivery, acceptability and reliability. In construction, there is a history of buying on price to the detriment of one of the other areas and if any part of this sequence is wrong customer satisfaction will suffer. Another critical area is timely delivery, this can be problematic for the supply of both products and services. The authors are aware of numerous instances where suppliers have not met their commitments with severe consequences for the process on site.

The purchasing or procurement system should be documented and include:

1. Assigning responsibilities for and within the purchasing procurement function.
2. Defining the manner in which suppliers are selected, to ensure that they are continually capable of supplying the requirements.
3. Specifying the purchasing documentation – written orders, specifications, etc. – required in any modern procurement activity.

Historically many organizations, particularly in the manufacturing industries, have operated an inspection-oriented quality system for bought-in parts and materials. Such an approach has many disadvantages. It is expensive, imprecise, and impossible to apply evenly across all material and parts, which all lead to variability in the degree of appraisal. Many organizations, such as Ford, have found that survival and future growth in both volume and variety demand that changes be made to this approach.

Construction is no different, a reliance on inspection (detection) to achieve quality is unproductive in two ways: it reduces any sense of responsibility at the workface, making inspectors responsible and, thereby, add wasteful cost. The inspector cannot be ever present; therefore, a responsible, careful, skilful and innovative workforce is essential for successful operations.

The prohibitive cost of holding large stocks of components and raw materials also pushed forward the 'just-in-time' (JIT) concept. As this requires that suppliers make frequent, on time, deliveries of small quantities of material, parts, components, etc., often straight to the point of use, in order that stocks can be kept to a minimum, the approach requires an effective supplier network – one producing goods and services that can be trusted to conform to the real requirements with a high degree of confidence.

## Commitment and involvement

The process of improving supplier performance is complex and clearly relies very heavily on securing real commitment from the senior management of both organizations to a partnership. This may be aided by presentations made to groups of directors of the suppliers brought together to share the realization of the importance of their organizations' performance in the quality chains. The synergy derived from members of the partnership meeting together, being educated, and discussing mutual problems, will be tremendous. If this can be achieved, within the constraints of business and technical confidentiality, it is always a better approach than the arm's-length approach to purchasing still used by many companies.

The authors recall the benefits that accrued from a partnership between a structural engineer and his client, a builder in the Singapore high-rise residential sector. The builder approached the structural engineer who had worked for him for some years with a view to revising their agreement for work. At first the engineer was fearful that the builder was looking for an excuse to cut his fees – a request that many readers would not find unusual. In fact, the builder's approach was along quite different lines: he had been monitoring the reinforcing steel and concrete quantities in his buildings. His offer was to pay a bonus, a share of the savings in materials in any new buildings. The structural engineer quickly found that his bonuses exceeded his original fees. It is interesting to speculate on what had changed. The structural engineer can work to three different value propositions, to:

■ make sure that the building stands up;
■ minimize the materials in the building; and
■ minimize the overall construction cost of the building.

In fact, when fees are being squeezed and scope is not clearly defined any organization will look for ways to minimize costs to stay profitable. In this case, the client managed to align his interests with the second proposition. The third is the real value proposition; however, it is harder to judge and perhaps harder to manage.

## Policy

One of the first things to communicate to any external supplier is the purchasing organization's policy on quality of incoming goods and services. This can include such statements as:

- It is the policy of this company to ensure that the quality of all purchased materials and services meets its requirements.
- Suppliers who incorporate a quality management system into their operations will be selected. This system should be designed, implemented and operated accordingly to the International Standards Organization (ISO) 9000 series (see Chapter 12).
- Suppliers who incorporate statistical process control (SPC) and continuous improvement methods into their operations (see Chapter 13) will be selected.
- Routine inspection, checking, measurement and testing of incoming goods and services will *not* be carried out by this company on receipt.
- Suppliers will be audited and their operating procedures, systems, and SPC methods will be reviewed periodically to ensure a never-ending improvement approach.
- It is the policy of this company to pursue uniformity of supply, and to encourage suppliers to strive for continual reduction in variability. (This may well lead to the narrowing of specification ranges.)
- It is the policy of this company to work with suppliers who are committed to safe work practices and to achieving right-first-time outcomes.

## Quality management system assessment certification

Many customers examine their suppliers' quality management systems themselves, operating a second party assessment scheme (see Chapters 8 and 12). Inevitably this leads to high costs and duplication of activity, for both the customer and supplier. If a qualified, independent third party is used instead to carry out the assessment, attention may be focused by the customer on any special needs and in developing closer partnerships with suppliers. Visits and dialogue across the customer/supplier interface are a necessity for the true requirements to be met, and for future growth of the whole business chain. Visits should be concentrated, however, on improving understanding and capability, rather than on close scrutiny of operating procedures, which is best left to experts, including those within the supplier organizations charged with carrying out internal system audits and reviews.

## Supplier approval and single sourcing

Some organizations have as an objective to obtain at least two 'approved' suppliers for each material or service purchased on a regular basis. It may be argued, however, that single sourcing – the development of an extremely close relationship with just one supplier for each item or service – encourages greater commitment and a true partnership to be created. This clearly needs careful management, but it is a sound policy, based on the premise that it is better to work together with a supplier to remove problems, improve capability, and generate a mutual understanding of the *real* requirements, than to hop from one supplier to another and thereby experience a different set of problems each time. In construction this may be impossible to achieve for large companies working in several markets or in many projects simultaneously; however, the idea of limiting the number of suppliers and developing close working relationships with a few suppliers is a fundamental prerequisite for the development of innovation and continuous quality improvement.

To become an 'approved supplier' or partner, it is usually necessary to pass through a number of stages:

1. *Technical approval* – largely to determine if the product/service meets the technical requirements. This stage should be directed at agreeing a specification that is consistent with the supplier's process capability.
2. *Conditional approval* – at this stage it is known that the product/service meets the requirements, following customer in-process trials, and there is a good commercial reason for purchase.
3. *Full approval* – when all the requirements are being met, including those concerning the operation of the appropriate management systems, and the commercial arrangements have been agreed.

It is, of course, normal for organizations to carry out audits of their suppliers and to review periodically their systems and process capabilities. This is useful in developing the partnership and ensuring that the customer needs continue to be met.

# Just-in-time management

There are so many organizations throughout the world now that are practising just-in-time (JIT) management principles that the probability of encountering it is very high. JIT, like many modern management concepts, is credited to the Japanese, who developed and began to use it in the late 1950s. It took approximately 20 years for JIT methods to reach Western hard goods industries and a further 10 years before businesses realized the generality of the concepts. Construction, because of the very considerable amount of material that has to be brought to a site and the limited space to store it, has always implemented aspects of JIT – it just had to.

Basically JIT is a programme directed towards ensuring that the right quantities are purchased or produced at the right time, and that there is no waste. Anyone who perceives it purely as a material-control system, however, is bound to fail with JIT. JIT fits well under the TQM umbrella, for many of the ideas and techniques are very similar and, moreover, JIT will not work without TQM in operation. Writing down a definition of JIT for all types of organization is extremely difficult, because the range of products, services and organization structures lead to different impressions of the nature and scope of JIT. It is essentially:

- A series of operating concepts which allows systematic identification of operational problems in the smooth flow of materials and services, as they are required.
- A series of technology-based tools for correcting problems following their identification.

An important outcome of JIT is a disciplined programme for improving productivity and reducing waste. This programme leads to cost-effective production or operation and delivery of only the required goods or services, in the correct quantity, at the right time and place. This is achieved with the minimum amount of resources – facilities, equipment, materials, and people. The successful operation of JIT is dependent upon a balance between the suppliers' flexibility and the users' stability and planning reliability, and of course requires total management and employee commitment and teamwork.

## Aims of JIT

The fundamental aims of JIT are to produce or operate to meet the requirements of the customer exactly, without waste, immediately on demand. In some manufacturing companies JIT has been introduced as 'continuous flow production', which describes very well the objective of achieving conversion of purchased material or service receipt to delivery, i.e. from supplier to customer. If this extends into the supplier and customer chains, all operating with JIT, a perfectly continuous flow of material, information or service will be achieved. JIT may be used in non-manufacturing and in administration areas; for example, by using external standard as reference points. JIT can be used to manage design decision-making and the preparation of the design documents, as well as in production to support the construction assembly process on site.

The JIT concepts identify operational problems by tracking the following:

1. *Material movements* – when material stops, is delayed, diverts or turns backwards – these always correlate with an aberration in the 'process'.
2. *Material accumulations* – these are there as a buffer for problems, excessive variability, etc.
3. *Process flexibility* – an absolute necessity for flexible operation and design.
4. *Value-added efforts* – identify where no value is added, much of what is done does not add value and the customer will not pay for it.

## The operation of JIT

The tools to carry out the monitoring required are familiar quality and operations management methods, such as:

- Flowcharting to better understand processes.
- Process study and analysis to identify potential for improvement.
- Preventive maintenance to avoid unplanned disruptions.
- Equipment layout to optimize material flow on construction sites.
- Standardized design to reduce process risk.
- Statistical process control, applicable in the manufacture of construction materials – this has limited application in construction itself.
- Value analysis and value engineering to ensure that the focus is on achieving client needs in the most efficient manner.

But some techniques are more directly associated with the operation of JIT systems:

1. Batch or lot size reduction to produce smoother flow of materials and services.
2. Flexible workforce to maintain smooth flow and to cope with unanticipated requirements.

3. Kanban or cards with material visibility, though its use is limited to manufacturing.
4. Mistake-proofing, ensuring errors cannot happen.
5. Pull-scheduling, one completed task pulling the other behind it.
6. Set-up time reduction: in construction processes this translates into minimum crane time for assembly operations, rapid assembly and stripping of formwork, easy alteration to forms at changes in core configurations.
7. Standardized containers: this has relevance on construction sites for the transport and handling of materials.

In addition, joint development programmes with suppliers and customers will be required to establish long-term relationships and develop single sourcing arrangements that provide frequent deliveries in small quantities. These can only be achieved through close communications and meaningful certified quality.

There is clear evidence that JIT has been an important component of business success in the Far East and that it is used by Japanese companies operating in the West. Many European and American companies that have adopted JIT have made spectacular improvements in performance. These include:

- Increased flexibility (particularly of the workforce).
- Reduction in stock and work-in-progress, and the space it occupies.
- Simplification of products and processes.

Translated into the construction process, these ideas offer similar benefits to those achieved in manufacture. In construction, some of the areas to target are:

- Work completion – focus on tasks which are essential for the following trade rather than work that is good for one party at the expense of another.
- Keeping subcontractors on site well co-ordinated, at an even production rate.
- Decisions, design production and construction – to be synchronized, not allowing design to get ahead of essential decisions nor behind the information needs of the construction.

Such programmes are *always* characterized by a real commitment to continuous improvement. Organizations have been rewarded, however, by the low cost, low risk aspects of implementation provided a sensible attitude prevails. The golden rule is never remove resources – such as stock – before the organization is ready and able to correct the problems that will be exposed by doing so. Reduction of the water level to reveal the rocks, so that they may be demolished, is fine, provided we can quickly get our hands back on the stock while the problem is being corrected.

## Just-in-time in partnerships and the supply chain

The development of long-term partnerships with a few suppliers, rather than short-term ones with many, leads to the concept of *co-producers* in networks of trust providing dependable quality and delivery of goods and services. Each organization in the supply chain is often encouraged to extend JIT methods to its suppliers. The requirements of JIT mean that suppliers are usually located near the purchaser's premises, delivering small quantities regularly to match the usage rate. Administration is kept to a minimum and standard quantities in standard containers are usual. The requirement for suppliers to be located near the buying organization, which places those at some distance at a competitive disadvantage, causes lead times to be shorter and deliveries to be more reliable.

It can be argued that JIT purchasing and delivery are suitable mainly for assembly line operations, and less so for certain process and service industries, but the reduction in the inventory, and the smooth flow of materials as required and transport costs that it brings, should encourage innovations to lead to its widespread adoption. The main point is that there must be recognition of the need to develop closer relationships and to begin the dialogue – the sharing of information and problems – that lead to the product or service of the right quality, being delivered in the right quantity, at the right time.

# Resources

All organizations assemble resources, other than human, to support the effective operation of the processes that hopefully will deliver the strategy. These come in many forms but certainly include financial resources, buildings, equipment, materials, technology, information and knowledge. How these are managed will have a serious effect on the effectiveness and efficiency of any establishment, whether it be in manufacturing, service provision, or the public sector.

## Financial resources

Investment is key for the future development and growth of business. The ability to attract investment often determines the strategic direction of commercial enterprises. Similarly the acquisition of funding will affect the ability of public sector organizations in health, education or law establishments to function effectively. The development and implementation of appropriate financial strategies and processes will, therefore, be driven by the financial goals and performance of the business. Focus on, for example, improving earnings before interest and tax (EBIT – a measure of profitability) and economic value added (EVA – a measure of the degree to which the returns generated exceed the costs of financing the assets used) can in a private company be the drivers for linking the strategy to action. The construction of plans for the allocation of financial resources in support of the policies and strategies should lead to the appropriate and significant activities being carried out within the business to deliver the strategy.

Consolidation of these plans, coupled with an iterative review and approval, provides a mechanism of providing the best possible chance for success. Use of a 'balanced scorecard' approach (see Chapters 2, 7 and 8) can help in ensuring that the long-term impact of financial decisions on processes, innovation and customer satisfaction is understood and taken into account. The extent to which financial resources are being used to support strategy needs to be subject to continuous appraisal – this will include evaluating investment in the tangible and non-tangible assets, such as knowledge.

Review the use of a 'balanced scorecard' approach by case study company Gammon (p. 399), a general contractor operating in Hong Kong and in the case study about Landcom (p. 421), a semi-government land development agency in New South Wales, Australia.

Compared to other business sectors, in construction it is more difficult to assess the benefits of investing in change and improvement strategies. This is primarily due to the project basis of production; each project being different with a different production team makes it difficult to compare one to the other. Hence investment in change strategies is often resisted. One way of overcoming this resistance is to assess the cost of current inefficiencies. This may produce stronger arguments for investments in strategies and technologies that will lead to improvement.

In small and medium sized enterprises it is even more important that the financial strategy forms a key part of the strategic planning system, and key financial goals are identified, deployed, and regularly scrutinized.

## Other resources

Many different types of resources are deployed by different types of organizations. Most organizations are established in some sort of building, use equipment and consume materials. In these areas directors and managers must pay attention to:

- utilization of these resources;
- security of the assets;
- maintenance of building and equipment;
- managing material inventories and consumption (see earlier section on JIT);
- waste reduction and recycling;
- environmental aspects, including conservation of non-renewable resources and adverse impact of products and processes.

Technology is a splendid and vital resource in the modern age. Exciting alternative and emerging technologies need to be identified, evaluated and appropriately deployed in the drive towards achieving organizational goals. This will include managing the replacement of 'old technologies' and the innovations that will lead to the adoption of new ones. There are clear links here, of course, with process redesign and re-engineering (see Chapter 11). It is not possible to create the 'paper-less' courtroom, for example, without consideration of the processes involved. A murder trial typically involves a million pieces of paper, which are traditionally wheeled into courtrooms on trolleys. To replace this with computer systems and files on disks requires more than just a flick of a switch. The whole end-to-end process of the criminal justice system may come under scrutiny in order to deliver the paper-free trial, and this will involve many agencies in the process – police, prosecution service, courts, probation services, and the legal profession. Their involvement in the end-to-end process design will be vital if technology solutions are to add value and deliver the improvement in justice and reductions in costs that the systems in most countries clearly need.

Construction projects rely on vast quantities of information, tens of thousands of documents prepared by numerous independent and, yet, interdependent parties. The authors estimate that the cost of information in the entire construction process can be as much as 33 percent of the total moneys expended to procure a complex building. To convert the entire system to computer systems takes time, training, development and investment. It also requires smaller operators in the sector to become more computer literate and to increase their investment in technology. The involvement of everyone, from one end of the process to the other, is essential if IT-based systems are to achieve their potential in the construction sector.

Most organizations' strategies these days have some if not considerable focus on technology and information systems, as these play significant roles in how they supply products and services to and communicate with customers. They need to identify technology requirements through business planning processes and work with technology partners and IT system providers to exploit technology to best advantage, improve processes and meet business objectives. Whether this requires a dedicated IT team to develop the strategy will depend on the size and nature of the business but it will always be necessary to assess information resource requirements, provide the right balance, and ensure value for money is provided. This is often a tall order, it seems, in the provision of IT services! Close effective partnerships that deliver in this area are often essential.

In the piloting and evaluation of new technology the impact on customers and the business itself should be determined. The roll-out of any new systems involves people across the organization and communication cycles need to be used to identify any IT issues and feedback to partners (see Chapter 16). IT support should be designed in collaboration with users to confirm business processes, functionality and the expected utilization and availability. Responsibilities and accountabilities are important here, of course, and in smaller organizations this usually falls on line management.

Like any other resource, knowledge and information need managing and this requires careful consideration in its own right. Chapter 16 on communication, innovation and learning covers this in some detail.

In the design of quality management systems, resource management is an important consideration and is covered by the detail to be found in the ISO 9000:2000 family of standards (see Chapter 12).

## Chapter highlights

### Partnering
- Partnering is particularly important in construction where so many organizations collaborate to produce a completed project and their ability to innovate and produce high quality depends of the effectiveness of their collaboration.
- Organizations increasingly recognize the need to establish mutually beneficial relationships in partnerships. The philosophies behind TQM and 'Excellence' lay down principles and guidelines to support them.
- How partnerships are planned and managed must be in line with overall policies and strategies and support the operation of the processes.
- Establishing effective partnerships requires attention to identification of key strategic partners, design/development of relationships, structured value-adding supply chains, cultural fit and mutual development, shared knowledge and learning, improved processes, measured performance and feedback.

### Role of purchasing in partnerships
- The prime objective of purchasing is to obtain the correct equipment, materials, and services in the right quantity, of the right quality, from the right origin, at the right time and cost. Purchasing also acts as a 'window-on-the-world'.
- The separation of purchasing from selling has been eliminated in many retail organizations, to give responsibility for a whole 'product line'. Market information must be included in *any* buying decision.
- The purchasing system should be documented and assign responsibilities, define the means of selecting suppliers, and specify the documentation to be used.
- It is critical that purchasing policy is driven by the achievement of value rather than simply by minimizing price; the challenge is to identify the value to be created and to ensure that the partners are aligned towards its achievement.

- Improving supplier performance requires from the suppliers' senior management commitment, education, a policy, an assessed quality system, and supplier approval.
- Single sourcing – the close relationship with one supplier for each item or service – depends on technical, conditional, and full stages of approval.

## Just-in-time management

- JIT fits well under the TQM umbrella and is essentially a series of operating concepts that allow the systematic identification of problems, *and* tools for correcting them.
- JIT aims to produce or operate, in accordance with customer requirements, without waste, immediately on demand. Some of the direct techniques associated with JIT are batch or lot size reduction, flexible workforce, Kanban cards, mistake proofing, set-up time reduction, and standardized containers.
- The development of long-term relationships with a few suppliers or 'co-producers' is an important feature of JIT. These exist in a network of trust to provide quality goods and services.

## Resources

- All organizations assemble resources to support operation of the processes and deliver the strategy. These include finance, buildings, equipment, materials, technology, information and knowledge.
- Investment and/or funding is key for future development of all organizations and often determines strategic direction. Financial goals and performance will, therefore, drive strategies and processes. Use of a 'balanced scorecard' approach with continuous appraisal helps in understanding the long-term impact of financial decisions.
- In the management of buildings, equipment and materials, attention must be given to utilization, security, maintenance, inventory, consumption, waste and environmental aspects.
- Technology plays a key role in most organizations and management of existing alternative and emerging technologies need to be identified, evaluated and deployed to achieve organizational goals.
- There are clear links between the introduction of new or the replacement of old technologies and process redesign/engineering (see Chapter 11). The roll-out of any new systems also involves people across the organization and good communications are vital.

# 6

# Design for quality

# Design, innovation and improvement

Products, services and processes may be designed both to add value to customers and to become more profitable. But leadership and management style is also designed through the creation of symbols and processes which are reflected in internal communication methods and materials. Almost all areas of organizations have design issues inherent within them.

Design can be used to gain and hold on to competitive edge, save time and effort, deliver innovation, stimulate and motivate staff, simplify complex tasks, delight clients and stakeholders, dishearten competitors, achieve impact in a crowded market, and justify a premium price. Design can be used to take the drudgery out of the mundane and turn it into something inspiring, or simply make money. Design can be considered as a management function, a cultural phenomenon, an art form, a problem-solving process, a discrete activity, an end-product, a service or, often, a combination of several of these.

In the Collins *Cobuild English Language Dictionary*, design is defined as: 'the way in which something has been planned and made, including what it looks like and how well it works'. Using this definition, there is very little of an organization's activities that is not covered by 'planning' or 'making'. Clearly the consideration of what it looks like and how well it works in the eyes of the customer determines the success of products or services in the marketplace.

All organizations need to update their products, processes and services periodically. In markets such as electronics, audio and visual goods, and office automation, new variants of products are offered frequently – almost like fashion goods. While in other markets the pace of innovation may not be as fast and furious there is no doubt that the rate of change for product, service, technology and process design has accelerated on a broad front.

Innovation entails both the invention and design of radically new products and services, embodying novel ideas, discoveries and advanced technologies, *and* the continuous development and improvement of existing products, services, and processes to enhance their performance and quality. It may also be directed at reducing costs of production or operations throughout the life cycle of the product or service system.

Within the construction industry, rapid innovation is changing every aspect of the business, including the products and the services offered. These include an increase in the use of IT-based technologies in design, communication, management and manufacturing. In addition there are numerous examples of new technologies such as new equipment for materials handling and assembly, new products and materials, new financing arrangements and new procurement processes which involve the sharing of risk during the construction phase and the operation of newly built assets.

In many organizations innovation is predominantly either technology-led (e.g. in the design phase and the manufacture of engineered to order products, 3D technology and virtual reality are driving change in processes), or marketing-led (e.g. in the design of the actual end-product – dictated by user preferences). What is always striking about leading product or service innovators is that their developments are market-led, which is different from marketing-led. The latter means that the marketing function takes the lead in product and service developments. But most leading innovators identify and set out to meet the existing and potential demands profitably and, therefore, are market-led constantly striving to meet the requirements even more effectively through appropriate experimentation.

Everything we experience in or from an organization is the result of a design decision, or lack of one. This applies not just to the tangible things like products and services, but the intangibles too: the systems and processes which affect the generation of products

**85**

and delivery of services. Design is about combining function and form to achieve fitness for purpose: be it an improvement to a supersonic aircraft, a new structural type, the development of a new building material or product, a new management process, a staff incentive scheme or this book.

Once fitness for purpose has been achieved, of course, the goalposts change. Events force a reassessment of needs and expectations and customers want something different. In such a changing world, design is an ongoing activity, dynamic not static, a verb not a noun – *design is a process.*

# The design process

Commitment in the most senior management helps to build quality throughout the *design process* and to ensure good relationships and communication between various groups and functional areas both within the organization and across the supply chain. Designing customer satisfaction and loyalty into products and services contributes greatly to competitive success. Clearly, it does not guarantee it, because the conformance aspect of quality must be present and the operational processes must be capable of producing to the design. As in the marketing/operations interfaces, it is never acceptable to design a product, service, system or process that the customer wants but the organization is incapable of achieving.

The design process often concerns technological innovation in response to, or in anticipation of, changing market requirements and trends in technology. Those companies with impressive records of product- or service-led growth have demonstrated a state-of-the-art approach to innovation based on three principles:

- *Strategic balance* between product and process development to ensure that product and service innovation maintains market position while process innovation ensures that production risks in safety, quality and productivity are effectively controlled and reduced.
- *Top management approach* to design to set the tone and ensure that commitment is the common objective by visibly supporting the design effort. Direct control should be concentrated on critical decision points, since overmeddling by very senior people in day-to-day project management can delay and demotivate staff.
- *Teamwork*, to ensure that once projects are under way, specialist inputs, e.g. from marketing and technical experts, are fused and problems are tackled simultaneously. The teamwork should be urgent yet informal, for too much formality will stifle initiative, flair and the fun of design.

The extent of the design process should not be underestimated, but it often is. Many people associate design with *styling* of products, and this is certainly an important aspect. But for certain products and many service operations the *secondary design* considerations are vital. Anyone who has bought an 'assemble-it-yourself' kitchen unit will know the importance of the design of the assembly instructions, for example. Aspects of design that affect quality in this way are packaging, customer-service arrangements, maintenance routines, warranty details and their fulfilment, spare-part availability, etc.

An industry that has learned much about the secondary design features of its products is personal computers. Many of the problems of customer dissatisfaction experienced in this market have not been product design features but problems with user manuals, availability and loading of software, and applications. For technically complex products or service systems, the design and marketing of after-sales arrangements are an essential component of the design activity. The design of production equipment and its layout to allow ease of access for repair and essential maintenance, or simple use as intended, widens the management of design quality into suppliers and contractors and requires their total commitment.

In construction, design has a much larger role than is generally recognized; for example, on site, in the process of assembling a building, the selection and positioning of equipment is a design task. Similarly the selection of technology – the decision between using cast in place or prefabricated components – is also essentially a design task. Proper design detailing of buildings and structures plays a major role in the elimination of errors, defectives and waste. Correct initial design also obviates the need for costly and wasteful modifications to be carried out once construction has commenced. It is at the design stage that such important matters as variability of details, reproducibility, technical risk of failure due to workmanship, ease of use in operation, maintainability, etc. should receive detailed consideration. As many constructed projects are one-of-a-kind, the design phase is even more difficult to manage than in process and service industries, as on such projects the design team may not have worked together previously.

## Designing

If design quality is taking care of all aspects of the customer's requirements, including cost, production, safe and easy use, and maintainability of products and services, then *designing* must take place in all aspects of:

■ Identifying the need (including need for change).
■ Developing that which satisfies the need.
■ Checking the conformance to the need.
■ Ensuring that the need is satisfied.

Designing covers every aspect, from the identification of a problem to be solved, usually a market need, through the development of design concepts and prototypes to the generation of detailed specifications or instructions required to produce the artefact or provide the service. It is the process of presenting needs in some physical form, initially as a solution, and then as a specific configuration or arrangement of materials resources, equipment, and people. Design permeates strategically and operationally many areas of an organization and, while design professionals may control detailed product styling, decisions on design involve many people from other functions. Total quality management supports such a cross-functional interpretation of design.

In the construction environment, this broad conceptualization of the design function is essential as design impacts on every stage of the production process: safety during construction, constructability, the cost and ease of prefabrication of engineered products, and the reliable achievement of product quality on site.

Design like any other activity, must be carefully managed. A flowchart of the various stages and activities involved in the design and development process appears in Figure 6.1.

By structuring the design process in this way, it is possible to:

■ Control the various stages.
■ Check that they have been completed.
■ Decide which management functions need to be brought in and at what stage.
■ Estimate the level of resources needed.

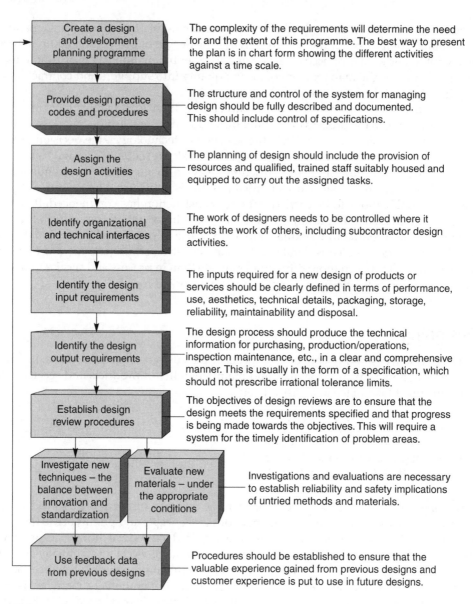

The complexity of the requirements will determine the need for and the extent of this programme. The best way to present the plan is in chart form showing the different activities against a time scale.

The structure and control of the system for managing design should be fully described and documented. This should include control of specifications.

The planning of design should include the provision of resources and qualified, trained staff suitably housed and equipped to carry out the assigned tasks.

The work of designers needs to be controlled where it affects the work of others, including subcontractor design activities.

The inputs required for a new design of products or services should be clearly defined in terms of performance, use, aesthetics, technical details, packaging, storage, reliability, maintainability and disposal.

The design process should produce the technical information for purchasing, production/operations, inspection maintenance, etc., in a clear and comprehensive manner. This is usually in the form of a specification, which should not prescribe irrational tolerance limits.

The objectives of design reviews are to ensure that the design meets the requirements specified and that progress is being made towards the objectives. This will require a system for the timely identification of problem areas.

Investigations and evaluations are necessary to establish reliability and safety implications of untried methods and materials.

Procedures should be established to ensure that the valuable experience gained from previous designs and customer experience is put to use in future designs.

**Figure 6.1**
The design and development process

The control of the design process must be carefully handled to avoid stifling the creativity of the designer(s), which is crucial in making design solutions a reality. It is clear that the design process requires a range of specialized skills, and the way in which these skills are managed, the way they interact, and the amount of effort devoted to the different stages of the design and development process is fundamental to the quality, producibility, and price of the service or final product. A team approach to the management of design is critical to the success of a project. The input of manufacturers, engineering fabricators and site assemblers is as crucial as the input from the end-user market.

It is never possible to exert the same tight control on the design effort as on other operational efforts, yet the cost and the time used are often substantial, and both must appear somewhere within the organization's budget.

Certain features make control of the design process difficult:

1. Construction is more design intensive than manufacturing because, generally, unless you are in a business such as the mass production of housing or a product system, each project is designed anew.
2. No design will ever be 'complete' in the sense that, with effort, some modification or improvement cannot be made.
3. Few designs are entirely novel. An examination of most 'new' products, services or processes will show that they employ existing techniques, components or systems to which have been added novel elements.
4. The longer the time spent on a design, the less the increase in the value of the design tends to be unless a technological breakthrough is achieved. This diminishing return from the design effort must be carefully managed. However, this has to be balanced with the need for adequate design resolution and sound documentation because production risk increases when the design is not properly resolved and effectively communicated.
5. The design process is information intensive and the timing of decision-making, both by the clients and the design team, is critical to the efficiency of the entire process. It is not practicable to manage the design process in the same manner as we do the production process – on the basis of tasks. For every task there may be up to ten information flows and the ratio of information flows to tasks is highly variable. Also there are a great number of concurrent and interdependent activities which need skill and experience in their effective resolution.
6. External and/or internal customers will impose limitations on design time and cost. It is as difficult to imagine a design project whose completion date is not implicitly fixed, either by a promise to a customer, the opening of a trade show or exhibition, a seasonal 'deadline', a production schedule or some other constraint, as it is to imagine an organization whose funds are unlimited, or a product whose price has no ceiling.

## Total design processes

Quality of design, then, concerns far more than the product or service design and its ability to meet the customer requirements. It is also about the activities of design and development. The appropriateness of the actual *design process* has a profound influence on the performance of any organization, and much can be learned by examining successful companies and how their strategies for research, design, and development are linked to the efforts of marketing and operations. In some quarters this is referred to as 'total design', and the term 'simultaneous engineering' has been used. This an integrated approach to a new product or service introduction, similar in many ways to quality function deployment (QFD – see next section) in using multifunction teams or task forces to ensure that research, design, development, manufacturing, purchasing, supply, and marketing all work in parallel from concept through to the final launch of the product or service into the marketplace, including servicing and maintenance.

# Quality function deployment (QFD) – the house of quality

The 'house of quality' is the framework of the approach to design management known as QFD. It originated in Japan in 1972 at Mitsubishi's Kobe shipyard, but it has been developed in numerous ways by Toyota and its suppliers, and many other organizations. The house of quality (HoQ) concept, initially referred to as quality tables, has been used successfully by manufacturers of integrated circuits, synthetic rubber, construction equipment, engines, home appliances, clothing, and electronics, mostly Japanese. Ford and General Motors use it, and other organizations, including AT&T, Bell Laboratories, Digital Equipment, Hewlett-Packard, Procter & Gamble, ITT, Rank Xerox, and Jaguar have applications. In Japan, its design applications include public services and retail outlets. The application of QFD in construction is limited to companies that have specialized in a specific market sector; for example, in Japan and Brazil it has been used to design apartment layouts, and it has application for areas of mass production of products and materials.

QFD is a 'system' for designing a product or service, based on customer requirements, with the participation of members of all functions of the supplier organization. It translates the customer's requirements into the appropriate technical requirements for each stage. The activities included in QFD are:

1. Market research
2. Basic research
3. Innovation
4. Concept design
5. Prototype testing
6. Final-product or service testing
7. After-sales service and troubleshooting

These are performed by people with different skills in a team whose composition depends on many factors, including the products or services being developed and the size of the operation. In many industries, such as cars, video equipment, electronics, and computers, 'engineering' designers are seen to be heavily into 'designing'. But in other industries and service operations designing is carried out by people who do not carry the word 'designer' in their job title. The failure to recognize the design inputs they make, and to provide appropriate training and support, will limit the success of the design activities and result in some offering that does not satisfy the customer. This is particularly true of internal customers.

## The QFD team in operation

The first step of a QFD exercise is to form a cross-functional QFD team. Its purpose is to take the needs of the market and translate them into such a form that they can be satisfied within the operating unit and delivered to the customers.

As with all organizational problems, the structure of the QFD team must be decided on the basis of the detailed requirements of each organization. One thing, however, is clear – close liaison must be maintained at all times between the design, marketing and operational functions represented in the team.

The QFD team must answer three questions – WHO, WHAT and HOW, i.e.:

WHO are the customers?
WHAT does the customer need?
HOW will the needs be satisfied?

WHO may be decided by asking 'Who will benefit from the successful introduction of this product, service, or process?' Once the customers have been identified, WHAT can be ascertained through interview/questionnaire/focus group processes, or from the knowledge and judgement of the QFD team members. HOW is more difficult to determine, and will consist of the attributes of the product, service, or process under development. This will constitute many of the action steps in a 'QFD strategic plan'.

WHO, WHAT and HOW are entered into a QFD matrix or grid of 'house of quality' (HoQ), which is a simple 'quality table'. The WHATs are recorded in rows and the HOWs are placed in the columns.

The HoQ provides structure to the design and development cycle, often likened to the construction of a house, because of the shape of matrices when they are fitted together. The key to building the house is the focus on the customer requirements, so that the design and development processes are driven more by what the customer needs than by innovations in technology. This ensures that more effort is used to obtain vital customer information. It may increase the initial planning time in a particular development project, but the overall time, including design and redesign, taken to bringing a product or service to the market will be reduced.

This requires that marketing people, design staff (including architects and engineers), and production/operations personnel work closely together from the time the new service, process, or product is conceived. It will need to replace in many organizations the 'throwing it over the wall' approach, where a solid wall exists between each pair of functions (Figure 6.2).

The HoQ provides an organization with the means for interdepartmental or interfunctional planning and communications, starting with the so-called customer attributes (CAs). These are phrases customers use to describe product, process, and service characteristics.

A complete QFD project will lead to the construction of a sequence of house of quality diagrams, which translate the customer requirements into specific operational process steps. For example, the 'feel' that customers like on the steering wheel of a motor car may translate into a specification for 45 standard degrees of synthetic polymer hardness, which in turn translates into specific manufacturing process steps, including the use of certain catalysts, temperatures, processes, and additives. Similarly in construction, the acoustic privacy that home-owners want is translated into a measurable decibel transfer rate and specific construction systems to achieve it.

The first steps in QFD lead to a consideration of the product as a whole and subsequent steps to consideration of the individual components. For example, in the design of an

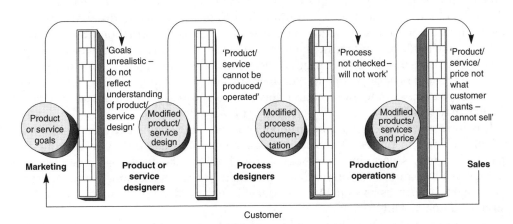

**Figure 6.2**
'Throw it over the wall'

apartment block, the overall building would be the starting point, but subsequent QFD exercises would tackle the kitchen, bathroom, bedrooms and lounge/dining room in detail. Each of the areas would have specific customer requirements but these would all need to be compatible with the overall service concept. An example of the application of this approach to apartment design in Malaysia has been described in some detail by Abdul-Rahman.[1] If you are interested in more detail about QFD, refer to the book *Total Quality Management* by John Oakland[2], Chapter 6, published by Butterworth-Heinemann.

# Design management with the analytical design planning technique

The Analytical Design Process (ADePT) developed by Simon Austin and his colleagues at Loughborough University in the UK is a powerful, complex problem-solving tool that is ideal for the management of the design process. Design is much more complex to manage than construction, yet relatively little research or development has been undertaken in the construction sector on issues of design management. Design is iterative and many decisions and tasks are undertaken concurrently. Furthermore, in spite of the fact that projects are increasing is size and complexity, clients are seeking to have projects delivered in less time and they are demanding high levels of reliability of outcome. Current practice is to consider the design process in terms of the timing of deliverables (drawings and specifications) which are scheduled in a similar manner to construction tasks, yet this does not recognize the fact that the actual design process is determined by the flow of information and decisions rather than of documents. Hence, while network analysis and CPM are generally used to plan and schedule design work on large to medium sized projects, these tools do nothing to recognize the ill-defined and iterative nature of the design process. The ADePT approach consists of three basic steps: in the first information flows are modelled, in the second a model of the building process based on work breakdown structures is created and finally a Dependency Structure Matrix (DSM) is used to facilitate the prioritizing of decisions.

## Modelling the design process

First of all Austin and his colleagues analysed and represented information flows using the 'Integration Definition Function Modelling' (IDEF-0) which is discussed in more detail in Chapter 10 in the section on process modelling. Information flows between activities are modelled and the actual transformation of work within activities is simply assumed. Within this modelling process each block of activity can be hierarchically subdivided to show finer detail on another diagram, ensuring any single diagram does not become too cumbersome. Information flows for each discipline area are distinguished as they may require different management approaches.

It is interesting to note that the number of information flows between tasks varies widely within any process. Rarely, there is a single information flow in and out, but more commonly several in and several out with no regular relationship between the numbers of incoming and outgoing information flows. Computer aided software engineering tools such as System Architect enable balanced IDEF-0 models to be constructed automatically.

Every information flow is prioritized according to a three-point scale from critical (3) to 'nice to have' (1) based on three parameters: its importance or 'strength of dependency', the sensitivity of subsequent activities to changes in the information and the ability to

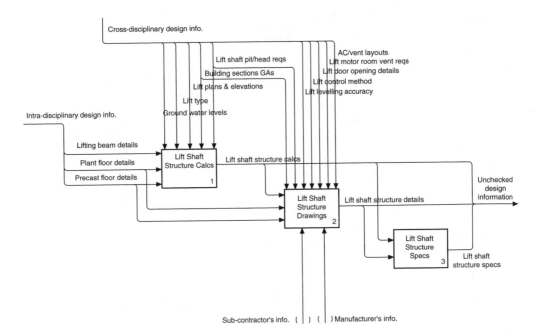

**Figure 6.3**
An example of a design process diagram for 'Lift Shaft Structure Design'

create reliable estimates for the information. Expert judgement is required to define these classifications.

## Building a process model for the project

Information flows are defined for every project; however, generic process models have been developed and it has been found that for buildings, for example, over 90 percent of the required process was contained in the generic model. The same will apply to other types of projects. The building model has a hierarchical or work breakdown structure (Figure 6.4), the first level of which subdivides the process into design undertaken by the professional disciplines and then breaks down into the building, subsystems and components. In other project management applications the process can be divided by function or system.

The project planning for a particular building will require some modification of the generic process model by the consultant team. Redundant sections are deleted, some added and others altered.

# Optimizing the process using the Dependency Structure Matrix

In the 1960s, Steward[3] developed a theory that a complex problem such as design could be solved more efficiently by representing the interrelationships between activities in the form of a Design Structure Matrix or Dependency Structure Matrix (DSM).

Figure 6.5 shows a matrix for a very simple design problem, activities are listed down the left-hand side of the matrix in the order they are planned to be undertaken and in the same order across the top. The rows in turn represent each activity and the columns

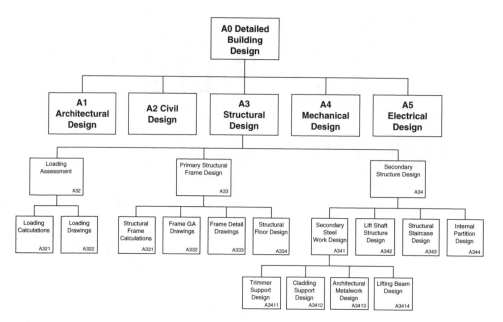

**Figure 6.4**
The design process hierarchy for structural design

**Figure 6.5**
An example of a simple DSM

the precedent ones. Information flows (dependencies) are shown by placing a 1, 2 or 3 where the row (activity) intersects the relevant column (precedent activity) to indicate the criticality of each information flow. This then logs all the information flows within the process and notes the criticality of each; furthermore marks below the diagonal indicate that the particular activity is dependent on a previous activity, whereas marks above the diagonal indicate dependency on information that has yet to be produced. Critical dependencies above the diagonal will hold up progress and hence relevant tasks have to be brought forward or as a last resort the information has to be assumed and subsequently checked.

For example, in Figure 6.5 it can be seen that Task 3 depends on Task 13 and this in turn depends on Task 19. Hence Tasks 13 and 19 need to be brought forward. The risk with making an assumption say about information from Task 19 is that if it is wrong then all the intervening work between 5 and 19 will have to be redone.

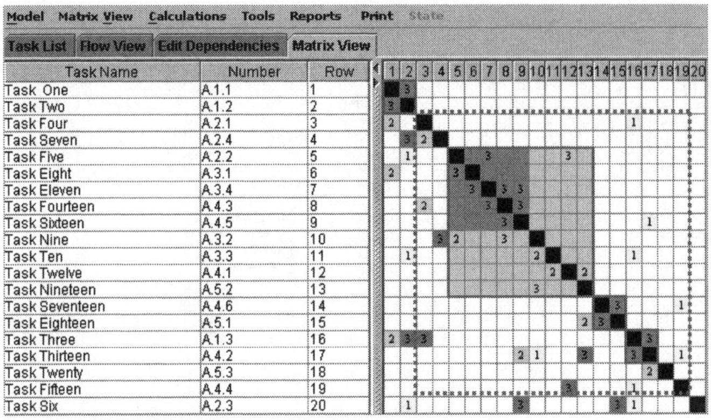

| Task List | Flow View | Edit Dependencies | Matrix View |

| Task Name | Number | Row | 1 2 3 4 5 6 7 8 9 10 11 12 13 14 15 16 17 18 19 20 |
| --- | --- | --- | --- |
| Task One | A.1.1 | 1 | |
| Task Two | A.1.2 | 2 | |
| Task Four | A.2.1 | 3 | |
| Task Seven | A.2.4 | 4 | |
| Task Five | A.2.2 | 5 | |
| Task Eight | A.3.1 | 6 | |
| Task Eleven | A.3.4 | 7 | |
| Task Fourteen | A.4.3 | 8 | |
| Task Sixteen | A.4.5 | 9 | |
| Task Nine | A.3.2 | 10 | |
| Task Ten | A.3.3 | 11 | |
| Task Twelve | A.4.1 | 12 | |
| Task Nineteen | A.5.2 | 13 | |
| Task Seventeen | A.4.6 | 14 | |
| Task Eighteen | A.5.1 | 15 | |
| Task Three | A.1.3 | 16 | |
| Task Thirteen | A.4.2 | 17 | |
| Task Twenty | A.5.3 | 18 | |
| Task Fifteen | A.4.4 | 19 | |
| Task Six | A.2.3 | 20 | |

**Figure 6.6**
Optimized example matrix

Such a situation was experienced by one of the authors early in his career when geotechnical advice regarding foundation materials for a 30-storey building was based on the immediately adjoining vacant site, an existing building occupied the site in question. After demolition it was found that the foundation conditions were worse than anticipated and the entire structural design had to be redone to design a more flexible structure, one that could absorb the predicted differential movements. In fact it would have been possible though expensive and inconvenient to drill from the basement of the existing building so that the design could have been based on known information rather than assumptions.

The need for estimates is eliminated by reordering the activities within the matrix so that all critical dependencies are below the diagonal or as close to it as possible, thus producing the optimum sequence (Figure 6.6). This maximizes the availability of information, and minimizes the amount of wasteful iteration and rework. It can be seen that the sequence is altered and that 15 activities contribute to three iterative blocks, the largest containing nine tasks. In this improved order the estimate of the information for Task 3 will only involve the reworking of two tasks. Optimizing a matrix identifies the interdependent activities that are within an iterative block and the block's location in the overall order. In real projects the iterative blocks can be very large, including hundreds of activities and the challenge of reducing them is considerable; however, the effort is well worth while as it creates a significantly more efficient design process.

The other strategy for reducing the amount of necessary iteration to a minimum is to declassify critical information by making decisions early, by building contingency into the design or by finding a way of creating sufficiently reliable estimates. Strategies might include increasing the expertise in the design team by bringing in an additional expert and getting all the experts in the team to work together to identify those decisions in the design process that must be made in order to improve the process flow.

To support this process of optimizing the decision-making process Austin and his team have developed a commercially available software called PlanWeaver. This helps to identify the declassifications that have the greatest effect on reducing the size of the block and hence scale of iteration.

Finally once the order of design activities is optimized on the basis of information flows and decisions, the design schedule can be represented in the form of a traditional bar chart. PlanWeaver can download the optimized sequence of activities into proprietary

Design for quality

95

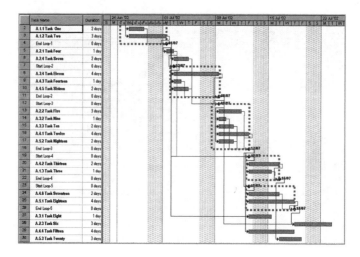

**Figure 6.7**
Bar chart for the example

planning software to produce a typical program. This process raises a number of issues. Conventional project management software represents sequential processes and does not allow elements of work containing iteration to be scheduled. Thus, feedback is not identified, resulting in co-ordination failures and rework both during design development and in production.

With PlanWeaver the optimized sequence is linked into existing planning software, and durations and resources are added. The output from the DSM is entered in a way that incorporates the iteration within the process. This is done by grouping tasks that form a block under a 'rolled-up' activity and removing interrelationships from within the loop so that they can be programmed to occur in parallel. The group's relationships with previous and subsequent tasks remain. The overall duration of the group of tasks must allow for the information exchanges necessary to achieve co-ordination.

While the end result, a bar chart, looks the same as that of conventional systems, the way you get there with ADePT is fundamentally different. Full account has been taken of the process's complex and interdependent nature.

The overall process delivers a number of clear benefits. The careful analysis of information flows and the classification of decisions require close consultation among the design team and this creates an integrated team approach. The final schedule gives clear guidance as to the organization of the design process, indicating when the whole team must meet to resolve critical interdependent decisions, when the logical milestones occur from a decision flow perspective; and thus provide a basis for cleanly integrating design and construction with transparent decision points and hold points. For more detail about the AdePT process refer to Reference 4.

# Specifications and standards

There is a strong relationship between standardization and specification. To ensure that a product or a service is *standardized* and may be repeated a large number of times in exactly the manner required, *specifications* must be written so that they are open to only one interpretation. The requirements, and therefore the quality, must be built into the design specification. There are national and international standards which, if used, help to ensure that specifications will meet certain accepted criteria of technical or managerial performance, safety, etc.

Standardization does not guarantee that the best design or specification is selected. It may be argued that the whole process of standardization slows down the rate and direction of technological development, and affects what is produced. If standards are used correctly, however, the process of drawing up specifications should provide opportunities to learn more about particular innovations and to change the standards accordingly.

These ideas are well illustrated by the construction sector's approach worldwide to the adoption of performance-based specifications wherever possible. Performance-based standards encourage innovation against measurable and transparent technical requirements. This allows the opportunity for manufacturers with new products and innovative solutions to have their ideas accredited and gain market entry. In areas like waterproofing, however, a building contractor might prefer to be very prescriptive in specifying the precise technical solution he wants. This is a particularly important area of construction where, based on everyday experience, we know that the risk of failure is high and its consequences of water leaking through the roof or out of a bathroom into adjoining rooms is simply unacceptable. In such areas, correct design and implementation is critical to managing an important area of risk for the general contractor.

It is possible to strike a balance between innovation and standardization; however, a sound approach to innovation clearly recognizes areas of design innovation that add value for the customer and areas of standardization that reduce risk in the production process. Clearly, it is desirable for designers to adhere where possible to past-proven materials and methods, in the interests of reliability, maintainability and variety control. Hindering designers from using recently developed materials, components, or techniques, however, can cause the design process to stagnate technologically. A balance must be achieved by analysis of materials, products and processes proposed in the design, against the background of their known reproducibility and reliability. If breakthrough innovations are proposed, then analysis or testing should be indicated objectively, justifying their adoption in preference to the established alternatives.

It is useful to define a specification. The International Standards Organization (ISO) defines it in ISO 8402 (1986) as 'The document that prescribes the requirements with which the product or service has to conform.' A document not giving a detailed statement or description of the requirements to which the product, service or process must comply cannot be regarded as a specification, and this is true of much sales literature.

The specification conveys the customer requirements to the supplier to allow the product or service to be designed, engineered, produced, or operated by means of conventional or stipulated equipment, techniques, and technology. The basic requirements of a specification are that it gives:

- Performance requirements of the product or service in measurable terms.
- Parameters – such as dimensions, concentration, turn-round time – which describe the product or service adequately (these should be quantified and include the units of measurement).
- Materials to be used by stipulating properties or referring to other specifications.
- Method of production or delivery of the service.
- Inspection/testing/checking requirements.
- References to other applicable specifications or documents.

To fulfil its purpose the specifications must be written in terminology that is readily understood, and in a manner that is unambiguous and so cannot be subject to differing interpretation. This is not an easy task, and one which requires all the expertise and knowledge available.

It is in relation to the clear communication of process specifications that the use of 3D and virtual reality (VR) technologies are showing great potential. At many stages of the design and construction process, complex information has to be communicated to the partners in the supply chain or to customers and their design teams. Often, end clients and other stakeholders are not able to conceptualize the design elements of a project; however, through the use of visualization tools their ability to interact with the design team is greatly enhanced. In other instances, the process design and detailing of parts of a structure can be very complex and VR simulation can assist both in optimizing the process through virtual prototyping and then in communicating the process to the people executing the work. For example, in the design of the Medical Office building for the Carmino Medical Group, 3D and VR technologies were used in the conceptual design, product and process design phases to enhance decision making and communication with all stakeholders.[5]

Good specifications are usually the product of much discussion, deliberation and sifting of information and data, and represent tangible output from a QFD team.

# Quality in the services sector

The emergence of the services sector has been suggested by economists to be part of the natural progression in which economic dominance changes first from agriculture to manufacturing and then to services. It is argued that if income elasticity of demand is higher for services than it is for goods, then as incomes rise, resources will shift toward services. The continuing growth of services verifies this, and is further explained by changes in culture, fitness, safety, demography and lifestyles.

In considering the design of services it is important to consider the differences between goods and services. Some authors argue that the marketing and design of goods and services should conform to the same fundamental rules, whereas others claim that there is a need for a different approach to service because of the recognizable differences between the goods and services themselves.

Constructed products are interesting in that where a customer is involved with the process from design through to construction, they experience the process as a service and the output as a product. Furthermore the quality of the service will influence their perception of the quality of the product. Great service will make them accepting of minor quality errors, whereas, poor service will expand even tiny quality problems into ones of monumental proportion, creating obstacles to payment and the complete loss of goodwill.

Furthermore, throughout the extremely fragmented construction supply chain, the achievement of successful outcomes relies on excellence in service quality among all the partners in the process. The focus on service to immediate customers (refer to Chapter 1) is essential, whether we are looking at the relationships within the design process or the relationships among subcontract suppliers on site. An interesting aspect of service within the site construction process is that while the contractual relationships of subcontractors are all with the general contractor, their service relationships are with preceding and subsequent trades. This is in marked contrast to manufacturing where subcontract suppliers deliver their fabricated components to the lead manufacturer's facility where they are integrated into the final product by the employees of the lead manufacturer. In construction, the relational inconsistency between contractual links and production process flow may have been an obstacle to managers in subcontracting and in general contracting organizations recognizing the importance of service quality among the subcontractors within the process.

In terms of design, it is possible to recognize three distinct elements in the service package – the physical elements or facilitating goods, the explicit service or sensual benefits, and implicit service or psychological benefits. In addition, the particular characteristics of service delivery systems may be itemized:

- Intangibility
- Simultaneity
- Heterogeneity

It is difficult, if not impossible, to design the intangible aspects of a service, since consumers often must use experience or the reputation of a service organization and its representatives to judge quality.

Simultaneity occurs because the consumer must be present before many services can take place. Hence, services are often formed in small and dispersed units, and it is difficult to take advantage of economies of scale. The rapid developments in computing and communications technologies are changing this in sectors such as banking, but contact continues to be necessary for many service sectors. Design considerations here include the environment and the systems used. Service facilities, procedures, and systems should be designed with the customer in mind, as well as the 'product' and the human resources. Managers need a picture of the total span of the operation, so factors which are crucial to success are not neglected. This clearly means that the functions of marketing, design, and operations cannot be separated in services, and this must be taken into account in the design of the operational controls, such as the diagnosing of individual customer expectations. A QFD approach here may be helpful to analyse the process in more detail when dealing with a standard product.

Heterogeneity of services occurs in consequence of explicit and implicit service elements relying on individual preferences and perceptions. Differences exist in the outputs of organizations generating the same service, within the same organization, and even the same employee on different occasions. Clearly, unnecessary variation needs to be controlled, but the variation attributed to estimating, and then matching, the consumers' requirements is essential to customer satisfaction and loyalty and must be designed into the systems. This inherent variability does, however, make it difficult to set precise quantifiable standards for all the elements of the service.

In the design of services it is useful to classify them in some way. The authors in their research are working with the SERVQUAL assessment tool.[6,7,8] This has been used in a study of the relationship between service quality and customer perceptions of product quality, and is currently being used to research service quality between subcontractors and between the general contractor and subcontractors within the construction process on building sites.

Parasuraman's five dimensions are:

- *Reliability* – ability to perform the promised service dependably and accurately.
- *Responsiveness* – willingness to help customers and provide prompt service.
- *Assurance* – knowledge and courtesy of employees and their ability to inspire trust and confidence.
- *Empathy* – caring, individualized attention the firm provides to its customers.
- *Tangibles* – physical facilities, equipment, and appearance of personnel.

As a part of their work Parasuraman and his co-researchers developed a generic survey instrument and this is widely recognized as an excellent tool for measuring *service quality*.

SERVQUAL scores *service quality* using 22 standardized statements to canvass customer views on the dimensions of *service quality*. Statements from the instrument are shown in Table 6.1.

Responses to these questions using a nine-point Likert scale are used to enable customer satisfaction to be assessed and benchmarked.

**Table 6.1**   SERVQUAL survey statements[6]

**Reliability**

1. Providing service as promised
2. Dependability in handling customers' service problems
3. Performing services right the first time
4. Providing services at the promised time
5. Maintaining error free records (e.g. financial)

**Responsiveness**

6. Keeping customers informed of when services will be performed
7. Prompt service to customers
8. Willingness to help customers
9. Readiness to respond to customers' requests

**Assurance**

10. Instilling confidence in customers
11. Make customers feel safe in their transactions

12. Being consistently courteous
13. Having the knowledge to answer questions

**Empathy**

14. Giving customers individual attention
15. Dealing with customers in a caring fashion
16. Having the customers' best interests at heart
17. Understanding the needs of their customers
18. Convenient business hours

**Tangibles**

19. Modern equipment
20. Visually appealing facilities
21. Having a neat, professional appearance
22. Visually appealing materials associated with the service

**Table 6.2**   A classification of selected services

| Service | Labour intensity | Contact | Interaction | Customization | Nature of act | Recipient of service |
|---------|------------------|---------|-------------|---------------|---------------|----------------------|
| Architect | High | Low | High | Adapt | Intangible | Things |
| Cleaning firm | High | Low | Low | Fixed | Tangible | People |
| Engineering design | High | Low | High | Adapt | Intangible | Things |
| Equipment hire | Low | Low | Low | Choice | Tangible | Things |
| Housing manufacture | Low | Low | High | Adapt | Tangible | Things |
| Subcontract labour | High | Low | Low | Adapt | Tangible | Things |
| Maintenance | Low | Low | Low | Choice | Tangible | Things |
| Management consultant | High | High | High | Adapt | Intangible | People |
| Nursery | High | Low | Low | Fixed | Tangible | People |
| Product manufacture | Low | Low | Low | Fixed | Tangible | Things |
| Repair firm | Low | Low | Low | Adapt | Tangible | Things |
| Subcontractor | Low | Low | Low | Adapt | Tangible | Things |

It is apparent that services are part of almost all organizations and not confined to the service sector. What is clear is that the service classifications and different attributes must be considered in any service design process. The authors are grateful to the contribution made by John Dotchin and Simon Austin to this section.

Several other service attributes have particular significance for the design of service operations:

1. *Labour intensity* – the ratio of labour costs incurred to the value of assets and equipment used (people versus equipment-based services).
2. *Contact* – the proportion of the total time required to provide the service for which the consumer is present in the system.
3. *Interaction* – the extent to which the consumer actively intervenes in the service process to change the content of the service; this includes customer participation to provide information from which needs can be assessed, and customer feedback from which satisfaction levels can be inferred.
4. *Customization* – which includes *choice* (providing one or more selections from a range of options, which can be single or *fixed*) and *adaptation* (the interactions process in which the requirement is decided, designed and delivered to match the need).
5. *Nature of service act* – either tangible, i.e. perceptible to touch and can be owned, or intangible, i.e. insubstantial.
6. *Recipient of service* – either people or things.

Table 6.2 gives a list of some typical construction sector services with their assigned attribute types.

# Failure mode, effect and criticality analysis (FMECA)

In the design of products, services and processes it is possible to determine possible modes of failure and their effects on the performance of the product or operation of the process or service system. Failure mode and effect analysis (FMEA) is the study of potential failures to determine their effects. If the results of an FMEA are ranked in order of seriousness, then the word CRITICALITY is added to give FMECA. The primary objective of a FMECA is to determine the features of product design, production or operation and distribution that are critical to the various modes of failure, in order to reduce failure. It uses all the available experience and expertise, from marketing, design, technology, purchasing, production/operation, distribution, service, etc., to identify the importance levels or criticality of potential problems and stimulate action to reduce these levels. FMECA should be a major consideration at the design stage of a product or service.

The elements of a complete FMECA are:

■ *Failure mode* – the anticipated conditions of operation are used as the background to study the most probable failure mode, location and mechanism of the product or system and its components.
■ *Failure effect* – the potential failures are studied to determine their probable effects on the performance of the whole product, process, or service, and the effects of the various components on each other.
■ *Failure criticality* – the potential failures on the various parts of the product or service system are examined to determine the severity of each failure effect in terms of lowering of performance, safety hazard, total loss of function, etc.

FMECA may be applied to any stage of design, development, production/operation or use, but since its main aim is to prevent failure, it is most suitably applied at the design

**Table 6.3** Probability and seriousness of failure and difficulty of detection

| Value | 1 | 2 | 3 | 4 | 5 | 6 | 7 | 8 | 9 | 10 |
|---|---|---|---|---|---|---|---|---|---|---|
| P | low chance of occurrence_____almost certain to occur |
| S | not serious, minor nuisance_____ total failure, safety hazard |
| D | easily detected_____unlikely to be detected |

stage to identify and eliminate causes. With more complex product or service systems, it may be appropriate to consider these as smaller units or subsystems, each one being the subject of a separate FMECA.

Special FMECA pro-formas are available and they set out the steps of the analysis as follows:

1. Identify the product or system components, or process function.
2. List all possible failure modes of each component.
3. Set down the effects that each mode of failure would have on the function of the product or system.
4. List all the possible causes of each failure mode.
5. Assess numerically the failure modes on a scale from 1 to 10. Experience and reliability data should be used, together with judgement, to determine the values, on a scale 1–10, for:
   *P*    the probability of each failure mode occurring (1 = low, 10 = high);
   *S*    the seriousness or criticality of the failure (1 = low, 10 = high);
   *D*    the difficulty of detecting the failure before the product or service is used
   by the consumer (1 = easy, 10 = very difficult). See Table 6.3.
6. Calculate the product of the ratings, $C = P \times S \times D$, known as the criticality index or risk priority number (RPN) for each failure mode. This indicates the relative priority of each mode in the failure prevention activities.
7. Indicate briefly the corrective action required and, if possible, which department or person is responsible and the expected completion date.

When the criticality index has been calculated, the failures may be ranked accordingly. It is usually advisable, therefore, to determine the value of C for each failure mode before completing the last columns. In this way the action required against each item can be judged in the light of the ranked severity and the resources available.

## Moments of truth

(MoT) is a concept that has much in common with FMEA. The idea was created by Jan Carlzon,[10] CEO of Scandinavian Airlines (SAS) and was made popular by Albrecht and Zemke.[11] An MoT is the moment in time when a customer first comes into contact with the people, systems, procedures, or products of an organization, which leads to the customer making a judgement about the quality of the organization's services or products.

In MoT analysis the points of potential dissatisfaction are identified proactively, beginning with the assembly of process flowchart type diagrams. Every small step taken by a customer in his/her dealings with the organization's people, products, or services is recorded. It may be difficult or impossible to identify all the MoTs, but the systematic approach should lead to a minimalization of the number and severity of unexpected failures, and this provides the link with FMEA.

---

In the Mirvac case study (p. 415) there is an excellent example of an MoT, the company has a policy of checking a dwelling immediately before the purchaser is handed the keys, the final inspection is recorded on a check sheet and fixed to the door. If the record is missing, the customer relations officer will not allow the new owners into the unit, in case the unit is not entirely ready; they do not want to risk creating a bad impression of their new home.

---

# The links between good design and managing the business

Research carried out by the European Centre for Business Excellence[9] has led to a series of specific aspects that should be addressed to integrate design into the business or organization. These are presented under various business criteria below.

## Leadership and management style

- 'Listening' is designed into the organization.
- Management communicates the importance of good design in good partnerships and vice versa.
- A management style is adopted that fosters innovation and creativity, and that motivates employees to work together effectively.

## Customers, strategy and planning

- The customer is designed into the organization as a focus to shape policy and strategy decisions.
- Designers and customers communicate directly.
- Customers are included in the design process.
- Customers are helped to articulate and participate in the understanding of their own requirements.
- Systems are in place to ensure that the changing needs of the customers inform changes to policy and strategy.
- Design and innovation performance measures are incorporated into policy and strategy reviews.
- The design process responds quickly to customers.

## People – their management and satisfaction

- People are encouraged to gain a holistic view of design within the organization.
- There is commitment to design teams and their motivation, particularly in cross-functional teamwork (e.g. quality function deployment teams).
- The training programme is designed, with respect to design, in terms of people skills training (e.g. interpersonal, management teamwork) and technical training (e.g. resources, software).
- Training helps integrate design activities into the business.

- Training impacts on design (e.g. honing creativity and keeping people up to date with design concepts and activity).
- Design activities are communicated (including new product or service concepts).
- Job satisfaction is harnessed to foster good design.
- The results of employee surveys are fed back into the design process.

## Resource management

- Knowledge is managed proactively, including investment in technology.
- Information is shared in the organization.
- Past experience and learning is captured from design projects and staff.
- Information resources are available for planning design projects.
- Suppliers contribute to innovation, creativity and design concepts.
- Concurrent engineering and design is integrated through the supply chains.

## Process management

- Design is placed at the centre of process planning to integrate different functions within the organization and form partnerships outside the organization.
- 'Process thinking' is used to resolve design problems and foster teamwork within the organization and with external partners.

## Impact on society and business performance

- Consideration is given to how the design of a product or service impacts on:
  - the environment;
  - the recyclability and disposal of materials;
  - packaging and wastage of resources;
  - the (local) economy (e.g. reduction of labour requirements);
  - the understanding of the impact of design on the business results, both financial and non-financial.

This same research showed that strong links exist between good design and proactive flexible deployment of business policies and strategies. These can be used to further improve design by encouraging the sharing of best practice within and across industries, by allowing designers and customers to communicate directly, by instigating new product/service introduction policies, project audits and design/innovation measurement policies and by communicating the strategy to employees. The findings of this work may be summarized by thinking in terms of the 'value chain', as shown in Figure 6.8.

---

The built environment has a greater impact on the environment than any other sector in the modern economy. Look at the Sekisui Heim (p. 469), CDL (p. 389) and Takenaka (p. 449) case studies in particular, these companies have articulated their social responsibilities very clearly and they see this as a central element of their TQM commitments.

---

Effective people management skills are essential for good design – these include the ability to listen and communicate, to motivate employees and encourage teamwork, as well as the ability to create an organizational climate which is conducive to creativity and continuous innovation.

Total Quality in the Construction Supply Chain

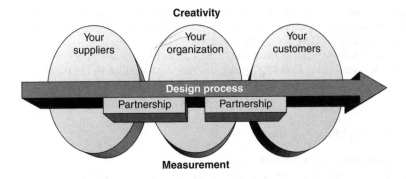

**Figure 6.8**
The value chain and design process

The only way to ensure that design actively contributes to business performance is to make sure it happens 'by design', rather than by accident. In short, it needs co-ordinating and managing right across the organization. The case studies at the end of the book describe companies that have excelled in the design of their management processes and products and through this broad commitment they have created viable and sustainable businesses in the construction sector.

# Acknowledgement

The authors are grateful for the contribution made by Simon Austin of Longborough University to the preparation of this chapter.

# References

1. Abdul-Rahman, H., Kwan, C.L. and Woods, P.C. (1999) 'Quality function deployment in construction design: application in low-cost housing design', *International Journal of Quality & Reliability Management*, Vol. 16, No. 6, pp. 591–605. MCB University Press, 0256-671X.
2. Oakland, J.S. (1993) *Total Quality Management*, 2nd edition, Butterworth-Heinemann.
3. Steward, D.V. (1965) 'Partitioning and Tearing Systems of Equations', *SIAM Journal on Numerical Analysis*, Vol. 2, No. 2, pp. 345–365.
4. Austin, S., Baldwin, A., Li, B. and Waskett, P. (2000) 'Analytical Design Planning Technique (ADePT): A Dependency Structure Matrix Tool to Schedule the Building Design Process', *Construction Management and Economics*, Vol. 18, pp. 173–182.
5. Khanzode, A., Fischer, M. and Reed, D. Case study of the implementation of the lean project delivery system using virtual building technologies on a larger healthcare project, *Proceedings 13th International Group for Lean Construction Conference*, (Kenley R., ed.) Sydney, pp. 153–160, 2005. http://www.iglc.net/conferences/2005/papers/session04/18_036_Khanzode_Fischer_Reed.pdf.
6. Parasuraman, A., Zeithaml, V.A. and Berry, L.L. (1988) 'SERVQUAL: A Multiple-item Scale for Consumer Perceptions of Service Quality', *Journal of Retailing*, Vol. 64, No. 1, pp. 12–40.
7. Parasuraman, A., Zeithaml, V.A. and Berry, L.L. (1991) 'Refinement and Reassessment of the SERVQUAL Scale', *Journal of Retailing*, Vol. 67, No. 4, pp. 420–450.
8. Parasuraman, A., Zeithaml, V.A. and Berry, L.L. (1994) 'Alternative Scales for Measuring Service Quality: A Comparative Assessment Based on Psychometric and Diagnostic Criteria', *Journal of Retailing*, Vol. 70, No. 3, pp. 201–230.

9. *Designing Business Excellence*, European Centre for Business Excellence (the Research and Education Division of Oakland Consulting plc, 33 Park Square, Leeds LS1 2PF)/British Quality Foundation/Design Council, 1998.
10. Carlzon, J. (1987) *Moments of Truth*, Ballinger, Cambridge, Mass.
11. Albrecht, K. and Zenke, R. (1985) *Service America! – Doing Business in the New Economy*, Dow Jones-Irwin, Homewood lll (USA).

## Chapter highlights

### Design, innovation and improvement

- Design is a multifaceted activity which covers many aspects of an organization.
- All businesses need to update their products, processes and services.
- Innovation entails both invention and design, *and* continuous improvement of existing products, services, and processes.
- Leading product/service innovations are market-led, not marketing-led.
- Everything in or from an organization results from design decisions.
- Design is an ongoing activity, dynamic not static, a verb not a noun – design is a process.

### The design process

- Commitment at the top is required to building in quality throughout the design process. Moreover, the operational processes must be capable of achieving the design.
- State-of-the-art approach to innovation is based on a strategic balance of old and new, top management approach to design, and teamwork.
- The 'styling' of products must also be matched by secondary design considerations, such as operating instructions and software support.
- Designing takes in all aspects of identifying the need, developing something to satisfy the need, checking conformance to the need and ensuring the need is satisfied.
- The design process must be carefully managed and can be flowcharted, like any other process, into: planning, practice codes, procedures, activities, assignments, identification of organizational and technical interfaces and design input requirements, review investigation and evaluation of new techniques and materials, and use of feedback data from previous designs.
- Total design or 'simultaneous engineering' is similar to quality function deployment and uses multifunction teams to provide an integrated approach to product or service introduction.

### Quality function deployment (QFD) – the house of quality

- The 'house of quality' is the framework of the approach to design management known as quality function deployment (QFD). It provides structure to the design and development cycle, which is driven by customer needs rather than innovation in technology.
- QFD is a system for designing a product or service, based on customer demands, and bringing in all members of the supplier organization.
- A QFD team's purpose is to take the needs of the market and translate them into such a form that they can be satisfied within the operating unit.
- The QFD team answers the following question. WHO are the customers? WHAT do the customers need? HOW will the needs be satisfied?

- The answers to the WHO, WHAT and HOW questions are entered into the QFD matrix or quality table, one of the seven new tools of planning and design.

## Design management with the analytical design planning technique
- Design processes are complex and many activities are concurrent, the relationship between tasks is better defined on the basis of information than simple task sequencing or on the basis of outputs such as plans and specifications.
- The ADePT design management tool provides a planning approach that recognizes the dependency between tasks on the basis of information flows.
- The output of the ADePT process is in the form of a typical bar chart of activities with interdependent activities blocked together, this format is ideal for planning the design process in detail.
- The relationship between design and construction can be designed in detail through information flows and hold points, enabling the fast tracking of projects with a far greater degree of control than is normally the case.

## Specifications and standards
- There is a strong relation between standardization and specifications. If standards are used correctly, the process of drawing up specifications should provide opportunities to learn more about innovations and change standards accordingly.
- The aim of specifications should be to reflect the true requirements of the product/service that are capable of being achieved.

## Quality design in the service sector
- In the design of services three distinct elements may be recognized in the service package: physical (facilitating goods), explicit service (sensual benefits), and implicit service (psychological benefits). Moreover, the characteristics of service delivery may be itemized as intangibility, simultaneity, and heterogeneity.
- The five dimensions of service quality (reliability, responsiveness, assurance, empathy and tangibles) are a very useful framework for assessing service quality weaknesses and for benchmarking service quality.
- A standard set of survey questions has been developed and validated and hence provides a mature and well-tried performance measurement and benchmarking tool for industry.
- The service attributes that are important in designing services include labour intensity, contact, interaction, customization, nature of service act, and the direct recipient of the act.
- Use of this framework allows services to be grouped under the six classifications.

## Failure mode, effect and criticality analysis (FMECA)
- FMEA is the study of potential product, service or process failures and their effects. When the results are ranked in order of criticality, the approach is called FMECA. Its aim is to reduce the probability of failure.
- The elements of a complete FMECA are to study failure mode, effect and criticality. It may be applied at any stage of design, development, production/operation or use.
- Moments of truth (MoT) is a similar concept to FMEA. It refers to the moments in time when customers first come into contact with an organization, leading to judgements about quality.

**The links between good design and managing the business**

- Research has led to a series of specific aspects to address in order to integrate design into an organization.
- The aspects may be summarized under the headings of: leadership and management style; customers, strategy and planning; people – their management and satisfaction; resource management; process management; impact on society and business performance.
- The research shows that strong links exist between good design and proactive flexible deployment of business policies and strategies – design needs co-ordinating and managing right across the organization.

# 7

# Performance measurement frameworks

# Performance measurement and the improvement cycle

Traditionally, performance measures and indicators have been derived from cost-accounting information, often based on outdated and arbitrary principles. These provide little motivation to support attempts to introduce TQM and, in some cases, actually inhibit continuous improvement because they are unable to map process performance. In the organization that is to succeed over the long term, performance must begin to be measured by the improvements seen by the customer.

In the cycle of never-ending improvement, measurement plays an important role in:

- Tracking progress against organizational goals.
- Identifying opportunities for improvement.
- Comparing performance against internal standards.
- Comparing performance against external standards.

Measures are used in *process control*, e.g. control charts (see Chapter 13), and in *performance improvement*, e.g. quality improvement teams (see Chapters 14 and 15), so they should give information about how well processes and people are doing and motivate them to perform better in the future.

The authors and their colleagues have seen many examples of so-called performance measurement systems that frustrated improvement efforts. Various problems include systems that:

1. Produce irrelevant or misleading information.
2. Track performance in single, isolated dimensions.
3. Generate financial measures too late, e.g. quarterly, for mid-course corrections or remedial action.
4. Do not take account of the customer perspective, both internal and external.
5. Distort management's understanding of how effective the organization has been in implementing its strategy.
6. Promote behaviour and undermine the achievement of the strategic objectives.

Typical harmful summary measures of local performance are purchase price, machine or plant efficiencies, direct labour costs, and ratios of direct to indirect labour. In the construction sector, typically, the primary measures used are cost against budget and time against scheduled time. These are incompatible with quality and productivity improvement measures because they cannot provide feedback that will motivate improvement. Measures such as process and throughput times, supply chain performance, inventory reductions, and increases in flexibility, which are first and foremost *non-financial*, can do that. Financial summaries provide valuable information, of course, but they should not be used for control. Effective decision-making requires direct measures for operational feedback and improvement.

One example of a 'measure' with these shortcomings is return on investment (ROI). ROI can be computed only after profits have been totalled for a given period. It was designed therefore as a single-period, long-term measure, but it is often used as a short-term one. Perhaps this is because most executive bonus 'packages' in the West are based on short-term measures. ROI tells us what happened, not what is happening or what will happen, and, for complex and detailed projects, ROI is inaccurate and irrelevant.

Many managers have a poor or incomplete understanding of their processes and products or services, and, looking for an alternative stimulus, become interested in financial indicators. The use of ROI, for example, for evaluating strategic requirements and performance can lead to a discriminatory allocation of resources. In many ways the financial indicators used in many organizations have remained static while the environment in which they operate has changed dramatically.

Traditionally, the measures used have not been linked to the processes where the value-adding activities take place. What has been missing is a performance measurement framework that provides feedback to people in all areas of business operations. Of course, TQM stresses the need to start with the process for fulfilling customer needs.

The critical elements of a good performance measurement framework (PMF) are:

- leadership and commitment;
- full employee involvement;
- good planning;
- sound implementation strategy;
- measurement and evaluation;
- control and improvement;
- achieving and maintaining standards of excellence.

The Deming cycle of continuous improvement – Plan, Do, Check, Act – clearly requires measurement to drive it, and yet it is a useful design aid for the measurement system itself:

PLAN:    establish performance objective and standards.
DO:      measure actual performance.
CHECK:   compare actual performance with the objectives and standards – determine
         the gap.
ACT:     take the necessary actions to close the gap and make the necessary
         improvements.

Before we use performance measurement in the improvement cycle, however, we should attempt to answer four basic questions:

1. Why measure?
2. What to measure?
3. Where to measure?
4. How to measure?

## Why measure?

It has been said often that it is not possible to manage what cannot be measured.

Whether this is strictly true or not there are clear arguments for measuring. In a quality-driven, never-ending improvement environment, the following are some of the main reasons *why measurement is needed* and why it plays a key role in quality and productivity improvement.

- To ensure customer requirements *have* been met.
- To be able to set sensible *objectives* and comply with them.
- To provide *standards* for establishing comparisons.
- To provide *visibility* and provide a 'scoreboard' for people to *monitor* their own performance levels.

- To highlight *quality problems* and determine which areas require *priority attention*.
- To give an indication of the *costs of poor quality*.
- To justify the *use of resources*.
- To provide *feedback* for driving the improvement effort.

It is also important to know the impact of TQM on improvements in business performance, on sustaining current performance, and perhaps on reducing any decline in performance. In the construction environment there is a need to develop performance measurement frameworks for projects as well as for enterprises. This is also important at the process level for processes both in design and construction that are to be targeted for improvement.

## What to measure?

A good starting point for deciding what to measure is to look at what are the key goals of senior management; what problems need to be solved; what opportunities are there to be taken advantage of; and what do customers perceive to be the key ingredients that influence their satisfaction. In the case studies there are numerous examples of performance measurement in different areas of enterprise and project management. These examples reflect the primary business goals of senior management in each case and they embrace the entire spectrum of issues that are addressed in a company's core values.

In the business of process improvement, process understanding, definition, measurement and management are tied inextricably together. In order to assess and evaluate performance accurately, appropriate measurement must be designed, developed and maintained by people who *own* the processes concerned. They may find it necessary to measure effectiveness, efficiency, quality, impact, and productivity. In these areas there are many types of measurement, including direct output or input figures, the cost of poor quality, economic data, comments and complaints from customers, information from customer or employee surveys, etc., generally continuous variable measures (such as time) or discrete attribute measures (such as absentees).

No one can provide a generic list of what should be measured but, once it has been decided in any one organization what measures are appropriate, they may be converted into indicators. These include ratios, scales, rankings, and financial and time-based indicators. Whichever measures and indicators are used by the process owners, they must reflect the true performance of the process in customer/supplier terms, and emphasize continuous improvement. Time-related measures and indicators have great value.

Current and recent research by the authors in the area of performance measurement has been both at the enterprise and the process level and these serve to illustrate both the approach to performance measurement and some specific measures that are effective. Frameworks for enterprise performance measurement in design companies and on construction projects and enterprises that were developed with industry participation are included as two case studies. Performance measures for site safety management and quality management have also been developed and are currently being trialed on construction sites in Sydney.[1,2]

## Where to measure?

If true measures of the effectiveness of TQM are to be obtained, there are three components that must be examined – the human, technical and business components.

The human component is clearly of major importance and the key tests are that wherever measures are used they must be:

1. *Transparent* – understood by all the people being measured.
2. *Non-controversial* – accepted by the individuals concerned.
3. *Internally consistent* – compatible with the rewards and recognition systems.
4. *Objective* – designed to offer minimal opportunity for manipulation.
5. *Motivational* – trigger a response to improve outcomes.

Technically, the measures must be the ones that truly represent the controllable aspects of the processes, rather than simple output measures that cannot be related to process management. They must also be correct, precise and accurate.

The business component requires that the measures are objective, timely, and result-oriented, and above all they must mean something to those working in and around the process, *including the customers*.

# How to measure?

Measurement, as any other management system, requires the stages of design, analysis, development, evaluation, implementation and review. The system must be designed to measure *progress*, otherwise it will not engage the improvement cycle. Progress is important in five main areas: effectiveness, efficiency, productivity, quality and safety, and impact.

## Effectiveness

Effectiveness may be defined as the percentage actual output over the expected output:

$$\text{Effectiveness} = \frac{\text{Actual output}}{\text{Expected output}} \times 100 \text{ percent}$$

Effectiveness then looks at the *output* side of the process and is about the implementation of the objectives – doing what you said you would do. Effectiveness measures should reflect whether the organization, group or process owner(s) are achieving the desired results, accomplishing the right things. Measures of this may include:

- Quality, e.g. a grade of product, or a level of service.
- Quantity, e.g. tonnes, lots, bedrooms cleaned, accounts opened.
- Timeliness, e.g. speed of response, product lead times, cycle time.
- Cost/price, e.g. unit costs.

## Efficiency

Efficiency is concerned with the percentage resource actually used over the resources that were planned to be used:

$$\text{Efficiency} = \frac{\text{Resources actually used}}{\text{Resources planned to be used}} \times 100 \text{ percent}$$

Clearly, this is a process *input* issue and measures the performance of the process system management; however, in construction because of the variable nature of construction work it also reflects on the skill of the initial estimation. It is, of course, possible to use resources 'efficiently' while being *ineffective*, so performance efficiency improvement must be related to certain output objectives.

All process inputs may be subjected to efficiency measurement, so we may use labour/staff efficiency, equipment efficiency (or utilization), materials efficiency, information efficiency, etc. Inventory data and throughput times are often used in efficiency and productivity ratios.

## *Productivity*

Productivity measures should be designed to relate the process outputs to its inputs:

$$\text{Productivity} = \frac{\text{Outputs}}{\text{Inputs}}$$

and this may be quoted as expected or actual productivity:

$$\text{Expected productivity} = \frac{\text{Expected output}}{\text{Resources expected to be consumed}}$$

$$\text{Actual productivity} = \frac{\text{Actual output}}{\text{Resources actually consumed}}$$

There is a vast literature on productivity and its measurement, but simple ratios such as tonnes per man-hour (expected and actual), sales output per telephone operator-day, and many others like this are in use. Productivity measures may be developed for each combination of inputs, e.g. sales/all employee costs.

## *Quality and safety*

This has been defined elsewhere of course (see Chapter 1). The *non-quality*-related measures include the simple counts of defect or error rates (perhaps in numbers per square metre or per thousand dollars spent), percentage outside specification or Cp/Cpk values, deliveries not on time or, more generally, as the costs of poor quality such as the measure of cost of rectification as a percentage of gross expenditure on site. When the positive costs of prevention of poor quality are included, these provide a balanced measure of the costs of quality (see next section).

In a research collaboration, one of the authors is working in the areas of safety and quality performance on construction sites. He is exploring three areas of process management: management actions compared to those planned, management reaction to problems, and outcome measures that indicate performance improvements in the specific focus area.[3,4] The authors are also involved in research on several projects to evaluate the service quality of suppliers within the on-site construction supply chain as well as the project culture. The former through the use of a SERVQUAL[5]-based questionnaire and the latter through the use of the Competing Values Framework developed by Cameron.[6] In all cases, these measures are being correlated against more traditional outcomes such as the incidence of errors and the cost of rectifying them.

The quality measures should also indicate positively whether we are doing a good job in terms of customer satisfaction, implementing the objectives, and whether the designs, systems, and solutions to problems are meeting the requirements. These really are voice-of-the-customer measures.

## Impact

Impact measures should lead to key performance indicators for the business or organization, including monitoring improvement over time. Value-added management (VAM) requires the identification and elimination of all non-value-adding wastes, including time. Value added is simply the volume of sales (or other measure of 'turnover') minus the total input costs, and provides a good direct measure of the impact of the improvement process on the performance of the business. A related ratio, percentage return on value added (ROVA) is another financial indicator that may be used.

$$\text{ROVA} = \frac{\text{Net profits before tax}}{\text{Value added}} \times 100 \text{ percent}$$

Other measures or indicators of impact on the business are *growth* in sales, assets, numbers of passengers/students, etc., and *asset-utilization* measures such as return on investment (ROI) or capital employed (ROCE), earnings per share, etc.

Some of the impact measures may be converted to people productivity ratios, e.g.:

$$\frac{\text{Value added}}{\text{Number of employees (or employee costs)}}$$

Activity-based costing (ABC) is an information system that maintains and processes data on an organization's activities and cost objectives. It is based on the activities performed being identified and the costs being traced to them. ABC uses various 'cost drivers' to trace the cost of activities to the cost of the products or services. The activity and cost-driver concepts are the heart of ABC. Cost drivers reflect the demands placed on activities by products, services or other cost targets. Activities are processes or procedures that cause work and thereby consume resources. This clearly measures impact, both on and by the organization.

# Costs of quality

Manufacturing a quality product, providing a quality service, or doing a quality job – one with a high degree of customer satisfaction – is not enough. The cost of achieving these goals must be carefully managed, so that the long-term effect on the business or organization is a desirable one. These costs are a true measure of the quality effort. A competitive product or service based on a balance between quality and cost factors is the principal goal of responsible management and may be aided by a competent analysis of the costs of quality (COQ).

The analysis of quality-related costs is a significant management tool that provides:

- A method of assessing the effectiveness of the management of quality.
- A means of determining problem areas, opportunities, savings, and action priorities.

The costs of quality are no different from any other costs. Like the costs of maintenance, design, sales, production/operations, and other activities, they can be budgeted, measured and analysed. Having said this, a major difficulty in construction is capturing the totality of the costs. The construction process is highly fragmented and many parties including design professionals, general contractor supervisors and workers and subcontractor supervisors and workers incur costs. Unless costs are to be recovered from another party either under the contract or as a variation, no one is interested in recording the costs. Yet a detailed knowledge of costs is potentially one of the main drivers for improvement and the authors are experimenting with industry partners to develop an accounting framework that cost effectively captures the significant costs.

Having specified the quality of design, the operating units have the task of matching it. The necessary activities will incur costs that may be separated into prevention costs, appraisal costs and failure costs, the so-called P-A-F model first presented by Feigenbaum. Failure costs can be further split into those resulting from internal and external failure.

# Prevention costs

These are associated with the design, implementation and maintenance of the quality management system. Prevention costs are planned and are incurred before actual operation. Prevention includes:

## Product or service requirements

The determination of requirements and the setting of corresponding specifications (which also takes account of process capability) for incoming materials, processes, intermediates, finished products and services.

## Quality planning

The creation of quality, reliability, and operational, production, supervision, process control, inspection and other special plans, e.g. pre-production trials, required to achieve the quality objective.

## Quality assurance

The creation and maintenance of the quality system.

## Inspection equipment

The design, development and/or purchase of equipment for use in inspection work.

## Training

The development, preparation and maintenance of training programmes for operators, supervisors, staff, and managers both to achieve and maintain capability.

## Miscellaneous

Clerical, travel, supply, shipping, communications and other general office management activities associated with quality.

Resources devoted to prevention give rise to the *'costs of doing it right the first time'*.

# Appraisal costs

These costs are associated with the supplier's and customer's evaluation of purchased materials, processes, intermediates, products and services to assure conformance with the specified requirements. Appraisal includes:

## Verification

Checking of incoming material, process set-up, first-offs, running processes, intermediates and final products, including produce or service performance appraisal against agreed specifications.

## Quality audits

To check that the quality system is functioning satisfactorily.

## Inspection equipment

The calibration and maintenance of equipment used in all inspection activities.

## Vendor rating

The assessment and approval of all suppliers, of both products and services.
Appraisal activities result in the 'costs of checking it is right'.

# Internal failure costs

These costs occur when the results of work fail to reach designed quality standards and are detected before transfer to the customer takes place. Internal failure includes the following:

## Scrap

Defective products or materials that cannot be repaired, used or sold.

## Identification and organization of rework*

The inspection, listing and organization of workers to return to the correct locations and redo the work.

## Rework or rectification*

The correction of defective material or errors to meet the requirements.

## Reinspection*

The re-examination of rectified work by independent design professionals and construction supervisors.

* Recent research by one of the authors measured the total cost of error identification, rectification and reinspection on high-rise residential housing developments in Sydney; it was found to be 5.5 percent of the total cost of work undertaken during the study period. The total work value was over $60 million spread over four projects and some 3600 instances of rework were assessed and costed. It was found that the direct cost of the rework was equivalent to the indirect cost of organizing and reinspecting it; while the general contractor's cost in organizing the work was equivalent to the subcontractor's costs in doing the work. It is noteworthy that the costs assessed under just these three items was equivalent to the profit margins of most major contractors in the sector.[7]

## Downgrading

A product that is usable but does not meet specifications may be downgraded and sold as 'second quality' at a low price. While it may be argued that this is not relevant in construction, recent experience in the Sydney property market has shown that defective buildings become stigmatized and dwellings in them lose value. The products of recognized 'poor quality' builders are discounted because of the builder's poor reputation even though they may not appear to be defective.

## Failure analysis

The activity required to establish the causes of internal product or service failure.

# External failure costs

These costs occur when products or services fail to reach design quality standards but are not detected until after transfer to the consumer. External failure includes:

## Repair and servicing

The repair of defective construction work has to be done on site and the cost can be prohibitive. One of the authors recently visited a building where the cost of repairing failed waterproofing in bathrooms was more than $20 000 per bathroom, while the cost of avoidance would have been less than $100 per occurrence.

## Warranty claims

Failed products that are replaced or services reperformed under some form of guarantee. Once again, the labour involved in replacement is usually far greater than the value of the defective part.

## Complaints

All work and costs associated with handling and servicing of customers' complaints.

## Returns

The handling and investigation of rejected or recalled products or materials including transport costs. While this has limited relevance in construction, the authors know of cases where the only way a builder was able to pacify an angry owner of a defective building was to purchase the building back at market value.

## Liability

The result of product or service liability litigation and other claims, which may include a change of contract.

## Loss of good will

The impact on reputation and image, which impinges directly on future prospects for sales.

External and internal failure produce the *'costs of getting it wrong'*.

Order re-entry, unnecessary travel and telephone calls, and conflict are just a few examples of the wastage or failure costs often excluded. Every organization should be aware of the costs of getting it wrong, and management needs to obtain some idea how much failure is costing each year.

Clearly, this classification of cost elements may be used to interrogate any internal transformation process. Using the internal customer requirements concept as the standard for failure, these cost assessments can be made wherever information, data, materials, service or artefacts are transferred from one person or one department to another. It is the 'internal' costs of lack of quality that lead to the claim that approximately one-third of *all* our efforts are wasted.

The relationship between the quality-related costs of prevention, appraisal, and failure and increasing quality awareness and improvement in the organization is shown in Figure 7.1. Where the quality awareness is low the total quality-related costs are high, the failure costs predominating. As awareness of the cost to the organization of failure gets off the ground, through initial investment in training, an increase in appraisal costs usually results. As the increased appraisal leads to investigations and further awareness, further investment in prevention is made to improve design features, processes and systems. As the preventive action takes effect, the failure *and* appraisal costs fall and the total costs reduce.

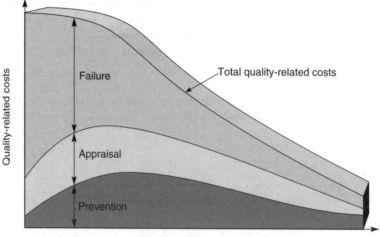

**Figure 7.1**
Increasing quality awareness

The first presentations of the P-A-F model suggested that there may be an optimum operating level at which the combined costs are at the minimum. The authors, however, have not yet found one organization in which the total costs have risen following investment in prevention.

# The process model for quality costing

The P-A-F model for quality costing has a number of drawbacks. In TQM, prevention of problems, defects, errors, waste, etc., is one of the prime functions, but it can be argued that everything a well-managed organization does is directed at preventing quality problems. This makes separation of *prevention costs* very difficult. There are clearly a range of prevention activities in any organization that are integral to ensuring quality but may never be included in the schedule of quality-related costs.

It may be impossible and unnecessary to categorize costs into the three categories of P-A-F. For example, a design review may be considered a prevention cost, an appraisal cost, or even a failure cost, depending on how and where it is used in the process. Another criticism of the P-A-F model is that it focuses attention on cost reduction and plays down, or in some cases even ignores, the positive contribution made to price and sales volume by improved quality.

The most serious criticism of the original P-A-F model presented by Feigenbaum and used in, for example, British Standard 6143 (1981) 'Guide to the determination and use of quality related costs', is that it implies an acceptable 'optimum' quality level above which there is a trade-off between investment in prevention and failure costs. Clearly, this is not in tune with the never-ending improvement philosophy of TQM. The key focus of TQM is on process improvement, and a cost categorization scheme that does not consider process costs, such as the P-A-F model, has limitations. (BS 6143-2 was republished in 1990 as 'Guide to the economies of quality: prevention, appraisal and failure model'.)

In a total quality-related costs system that focuses on processes rather than products or services, the operating costs of generating customer satisfaction will be of prime importance. The so-called 'process cost model', described in the revised BS 6143-1 (1992), sets out a method for applying quality costing to any process or service. It recognizes the importance of process ownership and measurement, and uses process modelling to simplify classification. The categories of the cost of quality (COQ) have been rationalized into the cost of conformance (COC) and the cost of non-conformance (CONC):

COQ = COC + CONC

The COC is the process cost of providing products or services to the required standards, by a given specified process in the most effective manner, i.e. the cost of the ideal process where every activity is carried out according to the requirements first time, every time. The CONC is the failure cost associated with the process not being operated to the requirements, or the cost due to variability in the process. Part 2 of BS 6143 (1991) still deals with the P-A-F model, but without the 'optimum'/minimum cost theory (see Figure 7.1).

Process cost models can be used for any process within an organization and developed for the process by flowcharting or use of the ICOR methodology (see Chapter 10). This will identify the key process steps and the parameters that are monitored in the process.

The process cost elements should then be identified and recorded under the categories of product/service (outputs), and people, systems, plant or equipment, materials, environment, information (inputs). The COC and CONC for each stage of the process will comprise a list of all the parameters monitored.

At this stage, the use of detailed modelling for quality costs is not used on construction sites and because of the fragmentation of the supply chain it is unlikely to be effective in that setting. However, the application of this technique in the production and design processes of specialist suppliers shows more promise. These errors could be assessed in this way and both the conformance and non-conformance costs are in-house and could be realistically captured.

## Steps in process cost modelling

Process cost modelling is a methodology that lends itself to stepwise analysis, while the following example is for the retrieval of medical records, it illustrates the process clearly and could be applied to any routine process in a volume production or service setting within the construction sector. The following are the key stages in building the model.

1. Choose a key process to be analysed, identify and name it, e.g. Retrieval of Medical Records (Acute Admissions).
2. Define the process and its boundaries.
3. Construct the process diagram:
   (a) identify the outputs and customers (for example, see Figure 7.2);
   (b) identify the inputs and suppliers (for example, see Figure 7.3);
   (c) identify the controls and resources (for example, see Figure 7.4).
4. Flowchart the process and identify the process owners (for example, see Figure 7.5). Note, the process owners will form the improvement team.
5. Allocate the activities as COC or CONC (see Table 7.1).

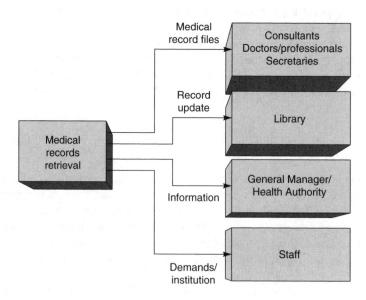

**Figure 7.2**
Building the model: identify outputs and customers

6. Calculate or estimate the quality costs (COQ) at each stage (COC + CONC). Estimates may be required where the accounting system is unable to generate the necessary information.
7. Construct a process cost report (see Table 7.2). The report summary and results are given in Table 7.3.

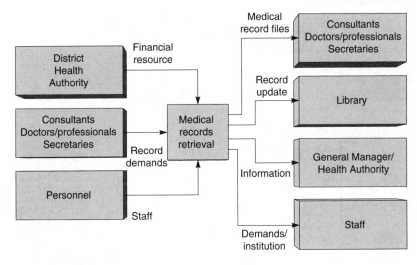

**Figure 7.3**
Building the model: identify inputs and suppliers

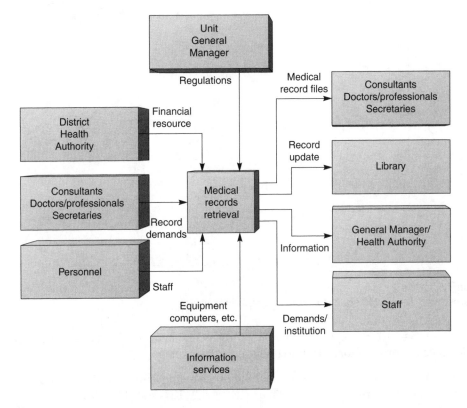

**Figure 7.4**
Building the model: identify controls and resources

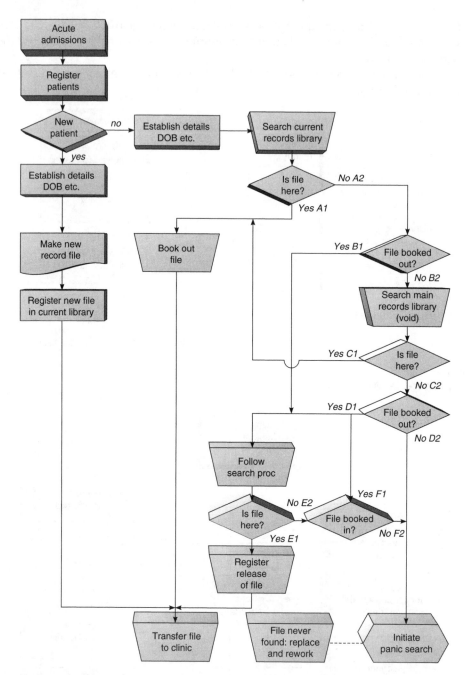

**Figure 7.5**
Present practice flowchart for acute admissions medical records retrieval

There are three further steps carried out by the process owners – the improvement team – which take the process forward into the improvement stage:

8. Prioritize the failure costs and select the process stages for improvement through reduction in costs of non-conformance (CONC). This should indicate any requirements for investment in prevention activities. An excessive cost of conformance (COC) may suggest the need for process redesign.

**Table 7.1** Building the model: allocate activities as COC or CONC

| Key activities | COC | CONC |
|---|---|---|
| Search for files | Labour cost incurred finding a record while adhering to standard procedure | Labour cost incurred finding a record while unable to adhere to standard procedure |
| Make up new files | New patient files | Patients whose original files cannot be located |
| Rework | | Cost of labour and materials for all rework files/records never found as a direct consequence of . . . |
| Duplication | | Cost incurred in duplicating existing files |

**Table 7.2** Building the model: process cost report

**Process cost report**
**Process: medical records retrieval (acute admissions)**
**Process owner: various**
**Time allocation: 4 days (96 hrs)**

| Process COC | Process CONC | Cost details Act | Synth | Definition | Source | |
|---|---|---|---|---|---|---|
| | Labour cost incurred finding records | # ref. Sample | | Cost of time required to find missing records | Medical records | £210 |
| | Cost incurred making up replacement files | | # | Labour and material costs multiplied by number of files replaced | Medical records | £108 |
| | Rework | | # | Labour and material cost of all rework | Medical records | £80 |
| | Duplication | | # | | Medical records | £24 |

9. Review the flowchart to identify the scope for reductions in the cost of conformance. Attempts to reduce COC require a thorough process understanding, and a second flowchart of what the new process should be may help (see Chapter 10).
10. Monitor conformance and non-conformance costs on a regular basis, using the model and review for further improvements. The process cost model approach should be seen as more than a simple tool to measure the financial implications of the gap between the actual and potential performance of a process. The emphasis given to the process, improving the understanding, and seeing in detail where the costs occur, should be an integral part of quality improvement.

# A performance measurement framework

A performance measurement framework (PMF) is proposed, based on the strategic planning and process management models outlined in Chapters 4 and 10. The framework

**Table 7.3** Process cost model: report summary

Labour cost
  14 hrs × £12.00/hr = £168
  £168 + overhead and contribution factor 25%
  = £210

Replacement costs
  No. of files unfound 9
  Cost to replace each file £12.00
  Overall cost £108

Rework costs
  2 × Pathology reports to be word processed £80

Duplication costs
  No. of files duplicated 2
  Cost per file £12.00
  Overall cost £24

TOTAL COST £422

RESULTS
Acute admissions operated 24 hrs/day 365 days/year
This project established a cost of non-conformance of approx. £422
This equates to £422 × 365/4 = £38 507.50
Or two personnel fully employed for 12 months

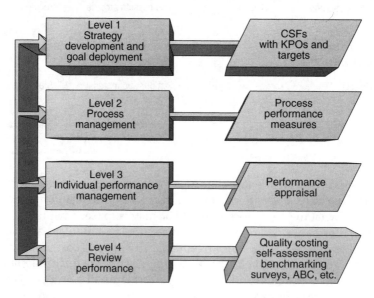

**Figure 7.6**
Performance measurement framework

has four elements related to: strategy development/goal deployment, process management, individual performance management, and review performance (Figure 7.6). This reflects an amalgamation of the approaches used by a range of organizations in performance measurement.

As we have seen in earlier chapters, the key to strategic planning and goal deployment is the identification of a set of critical success factors (CSFs) and associated key performance indicators (KPIs). These factors should be derived from the organization's mission, and represent a balanced mix of stakeholder issues. Action plans over both the short and medium term should be developed, and responsibility clearly assigned for performance. The strategic goals of the organization should then be clearly communicated to all individuals, and translated into measures of performance at the process/functional level.

The key to successful performance measurement at the process level is the identification and translation of customer requirements and strategic objectives into an integrated set of process performance measures. The documentation and management of processes has been found to be vital in this translation process. Even when a functional organization is retained, it is necessary to treat the measurement of performance between departments as the measurement of customer/supplier performance.

Performance measurement at the individual level usually relies on performance appraisal, i.e. formal planned performance reviews, and performance management, i.e. day-to-day management of individuals. A major drawback with some performance appraisal systems, of course, is the lack of their integration with other aspects of performance measurement.

Performance review techniques are used by many world-class organizations to identify improvement opportunities, and to motivate performance improvement. These companies typically use a wide range of such techniques and are innovative in performance measurement in their drive for continuous improvement.

The links between performance measurement at the four levels of the framework are based on the need for measurement to be part of a systematic process of continuous improvement, rather than for 'control'. The framework provides for the development and use of measurement, rather than prescriptive lists of measures that should be used. It is, therefore, applicable in all types of organization.

The elements of the performance measurement are distinct from the budgetary control process, and also from the informal control systems used within organizations. Having said that performance measurement should not be treated as a separate isolated system. Instead measurement is documented as and when it is used at the organizational, process and individual levels. In this way it can facilitate the alignment of the goals of all individuals, teams, departments and processes with the strategic aims of the organization and incorporate the voice of the stakeholders in all planning and management activities.

A number of factors have been found to be critical to the success of performance measurement systems. These factors include the level of top management support for non-financial performance measures, the identification of the vital few measures, the involvement of all individuals in the development of performance measurement, the clear communication of strategic objectives, the inclusion of customers and suppliers in the measurement process, and the identification of the key drivers of performance. These factors will need to be taken into account by managers wishing to develop a new performance measurement system, or refine an existing one.

In most world-class organizations there are no separate performance measurement systems. Instead, performance measurement forms part of wider organizational management processes. Although elements of measurement can be identified at many different points within organizations, measurement itself usually forms the 'check' stage of the continuous improvement PDCA cycle. This is important since measurement data that is collected but not acted upon in some way is clearly a waste of resources.

The four elements of the framework in Figure 7.6 are:

Level 1 *Strategy development and goal deployment* leading to mission/vision, *critical success factors* and *key performance outcomes* (KPOs).

Level 2 *Process management and process performance measurement* through *key performance indicators* (KPIs) (including input, in-process and output measures, management of internal and external customer/supplier relationships and the use of management control systems).

Level 3 *Individual performance management and performance appraisal.*

Level 4 *Review performance* (including internal and external benchmarking, self-assessment against quality award criteria and quality costing).

## Level 1 – Strategy development and goal deployment

The first level of the performance measurement framework is the development of organizational strategy, and the consequent deployment of goals throughout the organization. Steps in the strategy development and goal deployment measurement process are (see also Chapter 4):

1. Develop a mission statement based on recognizing the needs of all organizational stakeholders, customers, employees, shareholders and society. Based on the mission statement, identify those factors critical to the success of the organization achieving its stated mission. The CSFs should represent all the stakeholder groups, customers, employees, shareholders and society.
2. Define performance measures for each CSF – i.e. key performance outcomes (KPOs). There may be one or several KPOs for each CSF. Definition of KPO should include:
   (a) title of KPO;
   (b) data used in calculation of KPO;
   (c) method of calculation of KPO;
   (d) sources of data used in calculation;
   (e) proposed measurement frequency;
   (f) responsibility for the measurement process.
3. Set targets for each KPO. If KPOs are new, targets should be based on customer requirements, competitor performance or known organizational criteria. If no such data exists, a target should be set based on best guess criteria. If the latter is used, the target should be updated as soon as enough data is collected to be able to do so.
4. Assign responsibility at the organizational level for achievement of desired performance against KPO targets. Responsibility should rest with directors and very senior managers.
5. Develop plans to achieve the target performance. This includes both action plans for one year, and longer-term strategic plans.
6. Deploy mission, CSFs, KPOs, targets, responsibilities and plans to the core business processes. This includes the communication of goals, objectives, plans, and the assignment of responsibility to appropriate individuals.
7. Measure performance against organizational KPOs, and compare to target performance.
8. Communicate performance and proposed actions throughout the organization.
9. At the end of the planning cycle compare organizational capability to target against all KPOs, and begin again at step 2 above.
10. Reward and recognize superior organizational performance.

Strategy development and goal deployment is clearly the responsibility of senior management within the organization, although there should be as much input to the process as possible by employees to achieve 'buy-in' to the process.

The system outlined above is similar to the policy deployment approach known as Hoshin Kanri, developed in Japan and adapted in the West.

## Level 2 – Process management and measurement

The second level of the performance measurement framework is process management and measurement, the steps of which are:

1. If not already completed, identify and map processes. This information should include identification of:
   (a) process customers and suppliers (internal and external);
   (b) customer requirements (internal and external);
   (c) core and non-core activities;
   (d) measurement points and feedback loops.
2. Translate organizational goals, action plans and customer requirements into process performance measures (input, in-process and output) – key performance indicators (KPIs). This includes definition of measures, data collection procedures, and measurement frequency.
3. Define appropriate performance targets, based on known process capability, competitor performance and customer requirements.
4. Assign responsibility and develop plans for achieving process performance targets.
5. Deploy measures, targets, plans and responsibility to all subprocesses.
6. Operate processes.
7. Measure process performance and compare to target performance.
8. Use performance information to:
   (a) implement continuous improvement activities;
   (b) identify areas for improvement;
   (c) update action plans;
   (d) update performance targets;
   (e) redesign processes, where appropriate;
   (f) manage the performance of teams and individuals (performance management and appraisal) and external suppliers;
   (g) provide leading indicators and explain performance against organizational KPIs.
9. At the end of each planning cycle compare process capability to customer requirements against all measures, and begin again at step 2.
10. Reward and recognize superior process performance, including subprocesses, and teams.

The same approach should be deployed to subprocesses and to the activity and task levels.

The above process should be managed by the process owners, with inputs wherever possible from the owners of subprocesses. The process outlined should be used whether an organization is organized and managed on a process or functional departmental basis. If functionally organized, the key task is to identify the customer/supplier

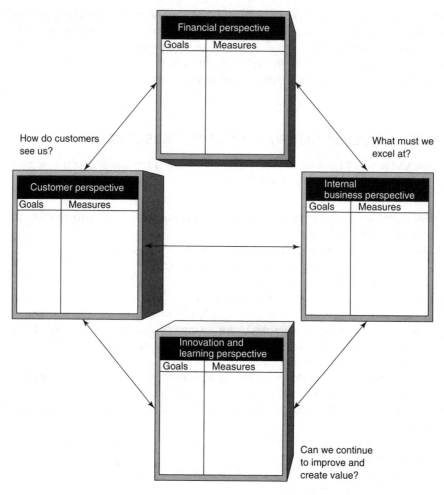

**Figure 7.7**
The balanced scorecard linking performance

relationships between functions, and for functions to see themselves as part of a customer/supplier chain.

## Key performance outcomes (KPOs)

The derivation of KPOs and KPIs may follow the 'balanced scorecard' model, proposed by Kaplan, which divides measures into financial, customer, internal business and innovation and learning perspectives (Figure 7.7).

A balanced scorecard derived from the Business Excellence Model described in Chapters 2 and 8 would include key performance results, customer results (measured via the use of customer satisfaction surveys and other measures, including quality and delivery), people results (employee development and satisfaction), and society results (including community perceptions and environmental performance). In the areas of customers, people, and society there needs to be a clear distinction between perception measures and other performance measures.

Refer to the case studies on Landcom (p. 421), the New South Wales semi-government land development agency and Gammon (p. 399). The Hong Kong builder Gammon uses the balanced scorecard to assess performance and reward excellence. Landcom triple bottom line to assess their own projects, they also use it as the basis of defining performance expectations and subsequently proposals from contractors wishing to build housing estates on their land. The Landcom case study provides a detailed example of the development of performance measurement within the triple bottom line framework.

Financial performance for external reporting purposes may be seen as a result of performance across the other KPOs, the non-financial KPOs and KPIs assumed to be the leading indicators of performance. The only aspect of financial performance that is cascaded throughout the organization is the budgetary process, which acts as a constraint rather than a performance improvement measure.

In summary then, organizational KPOs and KPIs should be derived from the balancing of internal capabilities against the requirements of identified stakeholder groups. This has implications for both the choice of KPOs/KPIs and the setting of appropriate targets. There is a need to develop appropriate action plans and clearly define responsibility for meeting targets if they are to be taken seriously.

Performance measures used at the process level differ widely between different organizations. Some organizations measure process performance using a balanced scorecard approach, while others monitor performance across different dimensions according to the process. Whichever method is used, measurements should be identified as input (supplier), in-process, and output (or results-customers).

It is usually at the process level that the greatest differences can be observed between the measurement used in manufacturing and services organizations. However, all organizations should measure quality, delivery, customer service/satisfaction, and cost.

Depending on the process, measurement frequency varies from daily, for example in the measurement of delivery performance, to annual, for example in the measurement of employee satisfaction, which has implications for the PDCA cycle time of the particular process(es). Measurement frequency at the process level may, of course, be affected by the use of information technology. Cross-functional process performance measurement is a vital component in the removal of 'functional silos', and the consequent potential for suboptimization and failure to take account of customer requirements. The success of performance measurement at the process level is dependent on the degree of management of processes and on the clarity of the deployment of strategic organizational objectives.

## *Measuring and managing the whats and the hows*

Busy senior management teams find it useful to distil as many things as possible down to one piece of paper or one spreadsheet. The use of KPOs, with targets, as measures for CSFs, and the use of KPIs for processes may be combined into one matrix which is used by the senior management team to 'run the business'.

Figure 7.8 (also shown in Chapter 4, Figure 4.9) is an example of such a matrix which is used to show all the useful information and data needed:

- the CSFs and their owners – the *whats;*
- the KPIs and their targets;

**Figure 7.8**
CSF/core process reporting matrix

|  | Conduct research | Manage int. systems | Manage financials | Manage our accounts | Develop new business | Develop products | Manage people | Core processes | CSFs: We must have | Measures | Year targets | Target CSF owner |
|---|---|---|---|---|---|---|---|---|---|---|---|---|
|  | | × | × | × | × | | × | | Satisfactory financial and non-financial performance | Sales volume. Profit. Costs versus plan. Shareholder return. Associate/employee utilization figures | Turnover £2m. Profit £200k. Return for shareholders. Days/month per person | |
|  | × | × | | × | × | × | × | | A growing base of satisfied customers | Sales/customer. Complaints/recommendations. Customer satisfaction | >£200k = 1 client £100k–£200k = 5 clients £50k–£100k = 6 clients <£50k = 12 clients | |
|  | | | × | × | | | × | | A sufficient number of committed and competent people | No. of employed staff/associates. Gaps in competency matrix. Appraisal results. Perceptions of associates and staff | 15 employed staff 10 associates including 6 new by end of year | |
|  | × | | | × | | | × | | Research projects properly completed and published | Proportion completed on time, in budget with customers satisfied. Number of publications per project | 3 completed on time, in budget with satisfied customers | |
|  | | | ‡ | ‡ | | | ‡ | | ‡ = Priority for improvement | | | |
|  | | | | | | | | | Process owner | | | |
|  | | | | | | | | | Process performance | | | |
|  | | | | | | | | | Measures and targets | | | |

- the core business processes and their sponsors – the *hows*;
- the process performance measures – KPIs.

It also shows the impacts of the core processes on the CSFs. This is used in conjunction with a 'business management calendar', which shows when to report/monitor performance, to identify process areas for improvement. This slick process offers senior teams a way of:

- gaining clarity about what is important and how it is measured;
- remaining focused on what is important and what the performance is;
- knowing where to look if problems occur.

# Level 3 – Individual performance and appraisal management

The third level of the performance measurement framework is the management of individuals. Performance appraisal and management is usually the responsibility of the direct managers of individuals whose performance is to be appraised. At all stages in the process, the individuals concerned must be included to ensure 'buy-in'.

Steps in performance and management appraisal are:

1. If not already completed, identify and document job descriptions based on process requirements and personal characteristics. This information should include identification of:
   (a) activities to be undertaken in performing the job;

(b) requirements of the individual with respect to the identified activities, in terms of experience, skills and training;

(c) requirements for development of the individual, in terms of personal training and development.

2. Translate process goals and action plans, and personal training and development requirements into personal performance measures.

3. Define appropriate performance targets based on known capability and desired characteristics (or desired characteristics alone if there is no prior knowledge of capability).

4. Develop plans towards achievement of personal performance targets.

5. Document 1 to 4 using appropriate forms, which should include space for the results of performance appraisal.

6. Manage performance. This includes:
   (a) planning tasks on a daily/weekly basis;
   (b) managing performance of the tasks;
   (c) monitoring performance against task objectives using both quantitative (process) and qualitative information on a daily and/or weekly basis;
   (d) giving feedback to individuals of their performance in carrying out tasks;
   (e) giving recognition to individuals for superior performance.

7. Formally appraise performance against range of measures developed, and compare to target performance.

8. Use comparison with target to:
   (a) identify areas for improvement;
   (b) update action plans;
   (c) update performance targets;
   (d) redesign jobs, where appropriate. This impacts step 1 of the process.

9. After a suitable period, ideally more than once a year, compare capability to job requirements and begin again at step 2.

10. Reward and recognize superior performance.

The above activities should be undertaken by the individual whose performance is being managed, together with their immediate superior.

The major differences in approaches in the management of individuals lies in the reward of effort as well as achievement and the consequently different measures used, and in the use of information in continuous improvement required to reward and recognize performance, including teamwork. Unlike management by objectives (MBO), where the focus is on measurement of results – which are often beyond the control of the individual whose performance is appraised – good performance management systems attempt to measure a combination of process/task performance (effort and achievement) and personal development.

The frequency of formal performance appraisal is generally defined by the frequency of the appraisal process usually with a minimum frequency of six months. Between the formal performance appraisal reviews, most organizations rely on the use of other performance management techniques to manage individuals. Measures of team performance, or of participation in teams should be included in the appraisal systems where possible, to improve team performance. In many organizations, the performance appraisal system is probably the least successfully implemented element of the framework. Appraisal systems are often designed to motivate individuals to achieve process and personal development objectives, but not to perform in teams. One of the limitations of appraisal processes is the frequency of measurement, which could be increased, but few organizations would consider doing so.

Three of the case studies take quite different approaches to performance appraisal and it is worth reviewing them to gain a deeper insight into the differences. Gammon (p. 399) use triple bottom line assessments of projects and individual performance, and determine rewards on this basis, Takenaka (p. 449) are very cautious about linking individual performance appraisal and reward to project outcomes as it is very easy to send the wrong signals to individuals, and Graniterock (p. 375) uses the individual appraisal process as more of a career and educational planning and review process.

## Level 4 – Performance review

The fourth level of the performance measurement framework is the use of performance review techniques. Steps in review are as follows:

1.  Identify the need for review, which may come from:
    (a)  poor performance at the organizational or process levels against KPO/KPIs;
    (b)  identified superior performance of competitors;
    (c)  customer inputs;
    (d)  the desire to better direct improvement efforts;
    (e)  the desire to concentrate attention on the need for performance improvement.
2.  Identify method of performance review to be used. This involves determining whether the review should be carried out internally within the organization, or externally, and the method that should be carried out. Some techniques are mainly internal, e.g. self-assessment, quality costing; while others, e.g. benchmarking, involve obtaining information from sources external to the organization. The choice should depend on:
    (a)  how the need for review was identified (see 1);
    (b)  the aim of the review, e.g. if the aim is to improve performance relative to competitors, external benchmarking may be a better option that internally measuring the cost of quality;
    (c)  the relative costs and expected benefits of each technique.
3.  Carry out the review.
4.  Feed results into the planning process at the organizational or process level.
5.  Determine whether to repeat the exercise. If it is decided to repeat the exercise, the following points should be considered:
    (a)  frequency of review;
    (b)  at what levels to carry out future reviews, e.g. organization-wide or process by process;
    (c)  decide whether the review technique should be incorporated into regular performance measurement processes, and if so how this will be managed.

Review methods often require the use of a level of resources greater than that normally associated with performance measurement, often due to the need to develop data collection procedures, train people in their use, and the cost of data collection itself. However, review techniques usually give a broader view of performance than most individual measures.

The use of review techniques is most successful when it is based on a clearly identified need, perhaps due to perceived poor performance against existing performance measures or against competitors, and the activity itself is clearly planned and the results used in performance improvement. This is often the difference between the success and failure of quality costing and benchmarking in particular. The use of most of the review techniques has been widely documented, but often without regard to their integration into the wider processes of measurement and management.

### Review techniques

Techniques identified for review include:

1. Quality costing, using either presentation-appraisal-failure, or process costing methods.
2. Self-assessment against Baldrige, EFQM Excellence Model, or internally developed criteria.
3. Benchmarking, internal or external.
4. Customer satisfaction surveys.
5. Activity-based costing (ABC).

# The implementation of performance measurement systems

It has already been established that a good measurement system will start with the customer and measure the right things. The value of any measure clearly needs to be compared with the cost of producing it. There will be appropriate measures for different parts of the organization, but everywhere they must relate process performance to the needs of the process customer. All critical parts of the process must be measured, and it is often better to start with simple measures and improve them.

There must be a recognition of the need to distinguish between different measures for different purposes. For example, an operator may measure time, various process parameters, and amounts, while at the management level measuring costs and delivery timeliness may be more appropriate.

Participation in the development of measures enhances their understanding and acceptance. Process owners can assist in defining the required performance measures, provided that senior managers have communicated their mission clearly, determined the critical success factors, and identified the critical processes.

If all employees participate, and own the measurement processes, there will be lower resistance to the system, and a positive commitment towards future changes will be engaged. This will derive from the 'volunteered accountability', which will in turn make the individual contribution more visible. Involvement in measurement also strengthens the links in the customer/supplier chains and gives quality improvement teams much clearer objectives. This should lead to greater short-term and long-term productivity gains.

There are a number of possible reasons why measurement systems fail:

1. They do not define performance operationally.
2. They do not relate performance to the process.
3. The boundaries of the process are not defined.
4. The measures are misunderstood or misused or measure the wrong things.
5. There is no distinction between control and improvement.
6. There is a fear of exposing poor and good performance.
7. It is seen as an extra burden in terms of time and reporting.
8. There is a perception of reduced autonomy.

9. Too many measurements are focused internally and too few are focused externally.
10. There is a fear of the introduction of tighter management controls.

These and other problems are frequently due to poor planning at the implementation stage or a failure to assess current systems of measurement. Before the introduction of a total quality-based performance measurement system, an audit of the existing systems should be carried out. Its purpose is to establish the effectiveness of existing measures, their compatibility with the quality drive, their relationship with the processes concerned, and their closeness to the objectives of meeting customer requirements. The audit should also highlight areas where performance has not been measured previously, and indicate the degree of understanding and participation of the employees in the existing systems and the actions that result.

Generic questions that may be asked during the audit include:

- Is there a performance measurement system in use?
- Has it been effectively communicated throughout the organization?
- Is it systematic?
- Is it efficient?
- Is it well understood?
- Is it applied?
- Is it linked to the mission and objectives of the organization?
- Is there a regular review and update?
- Is action taken to improve performance following the measurement?
- Are the people who own the processes engaged in measuring their own performance?
- Have employees been properly trained to conduct the measurement?

Following such an audit, there are 12 basic steps for the introduction of TQM-based performance measurement. Half of these are planning steps and the other half implementation.

# Planning

1. Identify the purpose of conducting measurement, i.e. is it for:
   (a) reporting, e.g. ROI reported to shareholders;
   (b) controlling, e.g. using process data on control charts;
   (c) improving, e.g. monitoring the results of a quality improvement team project.
2. Choose the right balance between individual measures (activity- or task-related) and group measures (process- and subprocess-related) and make sure they reflect process performance.
3. Plan to measure all the key elements of performance, not just one, e.g. time, cost, and product quality variables may all be important.
4. Ensure that the measures will reflect the voice of the internal/external customers.
5. Carefully select measures that will be used to establish standards of performance.
6. Allow time for the learning process during the introduction of a new measurement system.

# Implementation

7. Ensure full participation during the introductory period and allow the system to mould through participation.
8. Carry out cost/benefit analysis on the data generation, and ensure measures that have high 'leverage' are selected.

9. Make the effort to spread the measurement system as widely as possible, since effective decision-making will be based on measures from *all* areas of the business operation.
10. Use *surrogate* measures for subjective areas where quantification is difficult, e.g. improvements in morale may be 'measured' by reductions in absenteeism or staff turnover rates.
11. Design the measurement systems to be as flexible as possible, to allow for changes in strategic direction and continual review.
12. Ensure that the measures reflect the quality drive by showing small incremental achievements that match the never-ending improvement approach.

In summary, the measurement system must be designed, planned and implemented to reflect customer requirements, give visibility to the processes and the progress made, communicate the total quality effort and engage the never-ending improvement cycle. So it must itself be periodically reviewed.

# References

1. Marosszeky, M., Karim, K., Davis, S. and Naik, N. (2004) Lessons Learnt in Developing Effective Performance Measures for Construction Safety Management, *Proceedings 12th Conference of the International Group for Lean Construction* (Ed: Carlos T. Formoso), Denmark.
2. Marosszeky, M. and Karim, K. (2002) Enterprise process monitoring using key performance indicators. In: *Building in Value* (R. Best and G. de Valence, eds). Butterworth-Heinemann. ISBN 0750651490 June.
3. Karim, K., Davis, S., Naik, N. and Marosszeky, M. (January 2004) Designing an Effective Framework for Performance Measurement, *CIOB Special Publication – Quality Management*.
4. Marosszeky, M., Karim, K., Perera, P. and Davis, S. (2005) 'Improving Work Flow Reliability Through Quality Control Mechanisms', *Proceedings 13th Conference of the International Group for Lean Construction* (Ed: Russel Kenley), Sydney.
5. Parasuraman, A., Zeithaml, V.A. and Berry, L.L. (1994) 'Alternative scales for measuring service quality: a comparative assessment based on psychometric and diagnostic criteria', *Journal of Retailing*, Vol. 70, No. 3, pp. 201–230.
6. Cameron, K. and Quinn R.E. (1999) *Diagnosing and Changing Organizational Culture: Based on the competing values framework*, Addison Wesley, New York.
7. Thomas, R., Marosszeky, M., Karim, K., Davis, S. and McGeorge, D. (2003) *Enhancing Project Completion*, Australian Centre for Construction Innovation, ISBN 0733420613, pp. 28.

## *Chapter highlights*

### Performance measurement and the improvement cycle

- Traditional performance measures based on cost-accounting information provide little to support TQM, because they do not map process performance and improvements seen by the customer.
- Measurement is important in tracking progress, identifying opportunities, and comparing performance internally and externally. Measures, typically non-financial, are used in process control and performance improvement.
- Some financial indicators, such as ROI, are often inaccurate, irrelevant and too late to be used as measures for performance improvement.
- The Deming cycle of Plan, Do, Check, Act is a useful design aid for measurement systems, but first four basic questions about measurement should be asked, i.e. why, what, where, and how.

- In answering the question 'how to measure?' progress is important in five main areas: effectiveness, efficiency, productivity, quality and impact.
- Activity-based costing (ABC) is based on the activities performed being identified and costs traced to them. ABC uses cost drivers, which reflect the demands placed on activities.

## Costs of quality

- A competitive product or service based on a balance between quality and cost factors is the principal goal of responsible management.
- The analysis of quality-related costs may provide a method of assessing the effectiveness of the management of quality and of determining problem areas, opportunities, savings, and action priorities.
- Total quality costs may be categorized into prevention, appraisal, internal failure, and external failure costs, the P-A-F model.
- Prevention costs are associated with doing it right the first time, appraisal costs with checking it is right, and failure costs with getting it wrong.
- When quality awareness in an organization is low, the total quality-related costs are high, the failure costs predominating. After an initial rise in costs, mainly through the investment in training and appraisal, increasing investment in prevention causes failure, appraisal and total costs to fall.

## The process model for quality costing

- The P-A-F model or quality costing has a number of drawbacks, mainly due to estimating the prevention costs, and its association with an 'optimized' or minimum total cost.
- An alternative – the process costs model – rationalizes cost of quality (COQ) into the costs of conformance (COC) and the cost of non-conformance (CONC). COQ = COC + CONC at each process stage.
- Process cost modelling calls for choice of a process and its definition; construction of a process diagram; identification of outputs and customers, inputs and suppliers, controls and resources; flowcharting the process and identifying owners; allocating activities as COC or CONC; and calculating the costs. A process cost report with summaries and results is produced.
- The failure costs of CONC should be prioritized for improvements.

## A performance measurement framework

- A suitable performance measurement framework (PMF) has four elements related to strategy development, goal deployment, process management, individual performance management and review.
- The key to successful performance measurement at the strategic level is the identification of a set of critical success factors (CSFs) and associated key performance indicators (KPIs).
- The key to success at the process level is the identification and translation of customer requirements and strategic objectives into a process framework, with process performance measures.
- The key to success at the individual level is performance appraisal and planned formal reviews, through integrated performance management.
- The key to success in the review stage is the use of appropriate innovative techniques to identify improvement opportunities followed by good implementation.

- A number of factors are critical to the success of performance measurement systems including top management support for non-financial performance measures, the identification of the vital few measures, the involvement of all individuals in the development of performance measurement, the clear communication of strategic objectives, the inclusion of customers and suppliers in the measurement process, and the identification of the key drivers of performance.

### The implementation of performance measurement systems
- The value of any measure must be compared with the cost of producing it. All critical parts of the process must be measured, but it is often better to start with the simple measures and improve them.
- Process owners should take part in defining the performance measures, which must reflect customer requirements.
- Prior to introducing TQM measurement, an audit of existing systems should be carried out to establish their effectiveness, compatibility, relationship and closeness to the customer.
- Following the audit, there are 12 basic steps for implementation, six of which are planning steps. The measurement system, then, must be designed, planned and implemented to reflect customer requirements, give visibility to the processes and progress made, communicate the total quality effort and drive continuous improvement. It must also be periodically reviewed.

# 8

# Self-assessment audits and reviews

# Assessments of quality

In this chapter we look at the ideas of self-assessment and assessment by independent parties who may be expert and experienced in the task. To be effective any assessment must be undertaken against a well-structured and well-documented framework of criteria, otherwise it cannot be compared to assessments conducted by others nor can it be repeated at different times with any consistency.

Organizations everywhere are under constant pressure to improve their business performance, and they measure themselves against world-class standards and focus their efforts on the customer. To help in this process, many are turning to total quality models such as the European Foundation for Quality Management's (EFQM) Excellence Model promoted in the UK by the British Quality Foundation (see also Chapter 2).

Businesses in the construction sector are no different, though relatively few organizations have evaluated their performance against these standards, the case studies in this book are about the small number of construction sector enterprises that have an outstanding record of achievement when measured against such criteria. They have succeeded against enterprises from other sectors to win some of the most prestigious awards.

# Frameworks for self-assessment

'Total quality' (TQ) is the goal of many organizations but it has been difficult until relatively recently to find a universally accepted definition of what this actually means. For some people TQ means statistical process control (SPC) or quality management systems, for others teamwork and involvement of the workforce. More recently, in some organizations, it has been replaced by the terms business excellence or six sigma. In fact, a number of people from the construction sector have asked the authors what the difference is between these many conceptualizations of process improvement.

Clearly there are many different views on what constitutes the 'excellence' organization and, even with an understanding of a framework, there exists the difficulty of calibrating the performance or progress of any organization towards it.

The so-called excellence models now available recognize that customer satisfaction, business objectives, safety, and environmental considerations are mutually dependent and are applicable in any organization. Clearly the application of these ideas involves investment primarily in people and time – time to implement new concepts, time to train, time for people to recognize the benefits and move forward into new or different organizational cultures. But how will organizations know when they are getting close to excellence or whether they are even on the right road, how will they measure their progress and performance?

There have been many recent developments and there will continue to be many more, in the search for a standard or framework, against which organizations may be assessed or measure themselves, and carry out the so-called 'gap analysis'. To many the ability to judge progress against an accepted set of criteria would be most valuable and informative.

Most TQM approaches strongly emphasize measurement, some insist on the use of cost of quality. The value of a structured discipline using a points system has been well established in quality and safety assurance systems (for example, ISO 9000 and vendor auditing). The extension of this approach to a total quality auditing process has been long established in the Japanese 'Deming Prize' which is perhaps the most demanding and intrusive auditing process and there are other excellence models and standards used throughout the world. While in the US the Baldrige 'Criteria for Performance Excellence' is best known, in Europe, the EFQM is better known and this is a more recent and slightly different conceptualization, both in terms of identifying 'enablers' and 'results' of quality focused management and through its recognition of the community as a key stakeholder with an interest in the outcomes of industry's efforts. Many companies have realized the necessity to assess themselves against the Baldrige and Deming criteria, if not to enter for the awards or prizes, then certainly as an excellent basis for self-audit and review, to highlight areas for priority attention and provide internal and external benchmarking. (See Chapter 2 for details of the Deming Prize and Baldrige Award criteria.)

## The European Excellence Model for self-assessment

In Europe it has also been recognized that the technique of self-assessment is very useful for any organization wishing to monitor and improve its performance. In 1992 the European Foundation for Quality Management (EFQM) launched a European Quality Award which is now widely used for systematic review and measurement of operations. The EFQM Excellence Model recognized that processes are the means by which a company or organization harnesses and releases the talents of its people to produce results performance.

One feature of the EFQM Excellence Model that differentiates it from others and makes it particularly suitable for organizations in the construction sector is its inclusion of *impact on society*. It is the only award to do so. Given the great and critical impact that the construction sector has on the community and the environment the EFQM Excellence Model has the most suitable structure against which construction sector-based organizations should assess themselves. It is no surprise then that several of the case study companies have placed a very major emphasis on increasing the safety of their workforce, and also on their environmental and whole-of-life performance.

Figure 8.1 displays graphically the principle of the full Excellence Model.* As described in Chapter 2, customer results, employee results, and favourable society results are archived through leadership driving policy and strategy, people partnerships, resources, and processes, which lead ultimately to excellence in key performance results – the enablers deliver the results which in turn drive innovation and learning. The EFQM have provided a weighting for each criteria, shown in Figure 8.1, which may be used in scoring self-assessments and making awards. The weightings are not rigid and may be modified to suit specific organizational needs, allowing an organization to bias its scoring towards the priority improvement goals that it has set.

The EFQM have thus built a model of criteria and a review framework against which an organization may face and measure itself, to examine any 'gaps'. Such a process is known as self-assessment and organizations such as the EFQM, and in the UK the BQF, publish guidelines for self-assessment, including specific ones directed at public sector organizations.

Many managers feel the need for a rational basis on which to measure progress in their organization, especially in those companies 'a few years into TQM' which would like the answers to questions such as: 'Where are we now?', 'Where do we need/want to be?', and 'What have we got to do to get there?' These questions need to be answered from internal employees' views, the customers' views, and the views of suppliers.

Self-assessment promotes business excellence by involving a regular and systematic review of processes and results. It highlights strengths and improvement opportunities, and drives continuous improvement.

---

* An excellent resource is *The Model in Practice – using the EFQM Excellence Model to deliver continuous improvement*, Volumes 1 & 2, published by the British Quality Foundation, London, 2002.

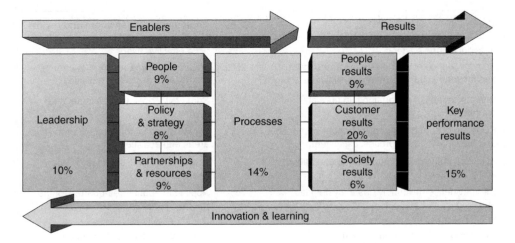

**Figure 8.1**
The EFQM Excellence Model

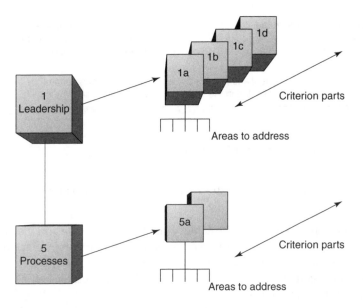

**Figure 8.2**
Structure of the criteria: enablers

# Enablers

In the Excellence Model, the enabler criteria of leadership, policy and strategy, people, partnerships and resources and processes focus on what is needed to be done to achieve results. The structure of the enabler criteria is shown in Figure 8.2. Enablers are assessed on the basis of the combination of two factors (see Figure 8.3, Chart 1, The enablers):

1. The degree of excellence of the *approach*.
2. The degree of *deployment* of the approach.

| Approach | Score | Deployment, assessment and review |
|---|---|---|
| Anecdotal or no evidence. | 0% | Little effective usage. |
| Some evidence of soundly based approaches and prevention based processes/systems. | 25% | Implemented in about one-quarter of the relevant areas and activities. |
| Some evidence of integration into normal operations. | | Some evidence of assessment and review. |
| Evidence of soundly based systematic approaches and prevention based processes/systems. | 50% | Implemented in about half the relevant areas and activities. |
| Evidence of integration into normal operations and planning well established. | | Evidence of assessment and review. |
| Clear evidence of soundly based systematic approaches and prevention based processes/systems. | 75% | Applied to about three-quarters of the relevant areas and activities. |
| Clear evidence of integration of approach into normal operations and planning. | | Clear evidence of refinement and improved business effectiveness through review cycles. |
| Comprehensive evidence of soundly based systematic approaches and prevention based processes/systems. | 100% | Implemented in all relevant areas and activities. |
| Approach has become totally integrated into normal working patterns. Could be used as a role model for other organizations. | | Comprehensive evidence of refinement and improved business effectiveness through review cycles. |

For *Approach, Deployment, Assessment* and *Review* the assessor may choose one of the five levels 0%, 25%, 50%, 75%, or 100% as presented in the chart, or interpolate between these values.

**Figure 8.3**
Scoring with the self-assessment process: Chart 1, The enablers

The elements of the detailed criteria in the enablers category are as follows:

1. **Leadership**
   How leaders develop and facilitate the achievement of the mission and vision, develop values required for long-term success and implement these via appropriate actions and behaviours, and are personally involved in ensuring that the organization's management system is developed and implemented. Self-assessment should demonstrate how leaders:
   (a) develop the mission, vision and values and are role models of a culture of excellence;
   (b) are personally involved in ensuring the organization's management system is developed, implemented and continuously improved;
   (c) are involved with customers, partners and representatives of society;
   (d) motivate, support and recognize the organization's people.

2. **Policy and strategy**
   How the organization implements its mission and vision via a clear stake-holder focused strategy, supported by relevant policies, plans, objectives, targets and processes. Self-assessment should demonstrate how policy and strategy are:
   (a) based on the present and future needs and expectations of stakeholders;
   (b) based on information from performance measurement, research, learning and creativity-related activities;

(c) developed, reviewed and updated;

(d) deployed through a framework of key processes;

(e) communicated and implemented.

3. **People**

How the organization manages, develops and releases the knowledge and full potential of its people at an individual, team-based and organization-wide level, and plans these activities in order to support its policy strategy and the effective operation of its processes. Self-assessment should demonstrate how:

(a) resources are planned, managed and improved;

(b) knowledge and competencies are identified, developed and sustained;

(c) people are involved and empowered;

(d) the organization has a dialogue;

(e) people are rewarded, recognized and cared for.

4. **Partnerships and resources**

How the organization plans and manages its external partnerships and internal resources in order to support its policy and strategy and the effective operation of its processes. External partnerships in the construction process are very important; to reflect this, additional emphasis should be placed on this aspect of management. Self-assessment should demonstrate how:

(a) external partnerships are managed;

(b) finances are managed;

(c) buildings, equipment and materials are managed;

(d) technology is managed;

(e) information and knowledge are managed.

5. **Processes**

How the organization designs, manages and improves its processes in order to support its policy strategy and fully satisfy, and generate increasing value for, its customers and stakeholders. Self-assessment should demonstrate how:

(a) processes are systematically designed and managed;

(b) processes are improved, as needed, using innovation in order to fully satisfy and generate increasing value for customers and other stakeholders;

(c) products and services are designed and developed based on customer needs and expectations;

(d) products and services are produced, delivered and serviced;

(e) customer relationships are managed and enhanced.

## Assessing the enablers criteria

The criteria are concerned with how an organization or business unit achieves its results. Self-assessment asks the following questions in relation to each criterion part:

- What is currently done in this area?
- How is it done? Is the approach systematic and prevention based?
- How is the approach reviewed and what improvements are undertaken following review?
- How widely used are these practices?

# Results

The EFQM Excellence Model's result criteria of customer results, people results, society results, and key performance results focus on what the organization has achieved and is

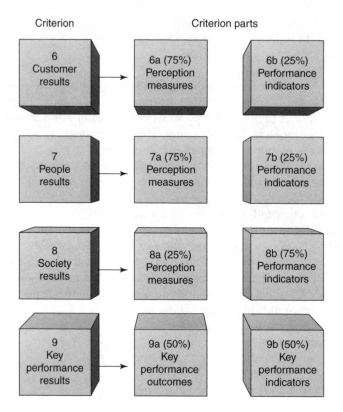

| Criterion | Criterion parts | |
|---|---|---|
| 6<br>Customer<br>results | 6a (75%)<br>Perception<br>measures | 6b (25%)<br>Performance<br>indicators |
| 7<br>People<br>results | 7a (75%)<br>Perception<br>measures | 7b (25%)<br>Performance<br>indicators |
| 8<br>Society<br>results | 8a (25%)<br>Perception<br>measures | 8b (75%)<br>Performance<br>indicators |
| 9<br>Key<br>performance<br>results | 9a (50%)<br>Key<br>performance<br>outcomes | 9b (50%)<br>Key<br>performance<br>indicators |

**Figure 8.4**
Structure of the criteria: results

achieving in relation to its:

- external customer;
- people;
- local, national, and international society, as appropriate;
- planned performance.

These can be expressed as discrete results, but ideally as trends over a period of years. The structure of the results criteria is shown in Figure 8.4.

'Performance excellence' is assessed relative to the organization's business environment and circumstances, based on information which sets out:

- the organization's actual performance;
- the organization's own targets;

and wherever possible:

- the performance of competitors or similar organizations;
- the performance of 'best in class' organizations.

Results are assessed on the basis of the combination of two factors (see Figure 8.5, Chart 2, The results):

- the degree of excellence of the results;
- the scope of the results.

| Results | Score | Scope |
|---|---|---|
| No results or anecdotal information. | 0% | Results address few relevant areas and activities. |
| Some results show positive trends and/or satisfactory performance. Some favourable comparisons with own targets/external organizations. Some results are caused by approach. | 25% | Results address some relevant areas and activities. |
| Many results show strongly positive trends and/or sustained good performance over the last 3 years. Favourable comparisons with own targets in many areas. Some favourable comparison with external organizations. Many results are caused by approach. | 50% | Results address many relevant areas and activities. |
| Most results show strong positive trends and/or sustained excellent performance over at least 3 years. Favourable comparisons with own targets in most areas. Favourable comparisons with external organizations in many areas. Most results are caused by approach. | 75% | Results address most relevant areas and activities. |
| Strongly positive trends and/or sustained excellent performance in all areas over at least 5 years. Excellent comparisons with own targets and external organizations in most areas. All results are clearly caused by approach. Positive indication that leading position will be maintained. | 100% | Results address all relevant areas and facets of the organization. |

For both *Results* and *Scope* the assessor may choose one of the five levels 0%, 25%, 50%, 75%, or 100% as presented in the chart, or interpolate between these values.

**Figure 8.5**
Scoring within the self-assessment process: Chart 2, The results

The elements of the detailed criteria in the results category are as follows:

6. **Customer results**
   What the organization is achieving in relation to its external customers.
   Self-assessment should demonstrate the organization's success in satisfying the needs and expectations of its external customers. Areas to consider:
   (a) results achieved for the measurement of customer perception of the organization's products, services and customer relationships;
   (b) internal performance indicators relating to the organization's customers.
7. **People results**
   What the organization is achieving in relation to its people. Self-assessment should demonstrate the organization's success in satisfying the needs and expectations of its people. Areas to consider:
   (a) results of people's perception of the organization;
   (b) internal performance indicators relating to people.

8. **Society results**

What the organization is achieving in relation to local, national and international society as appropriate. Self-assessment should demonstrate the organization's success in satisfying the needs and expectations of the community at large. Areas to consider:

(a) society's perception of the organization;
(b) internal performance indicators relating to the organization and society;
(c) impact of the organization's operations on the environment;
(d) whole-of-life efficiency of products.

9. **Key performance results**

What the organization is achieving in relation to its planned performance. Areas to consider:

(a) key performance outcomes, including financial and non-financial;
(b) key indicators of the organization's performance which might predict likely key performance outcomes.

## Assessing the results criteria

These criteria are concerned with what an organization has achieved and is achieving. Self-assessment addresses the following issues:

- the measures used to indicate performance;
- the extent to which the measures cover the range of the organization's activities;

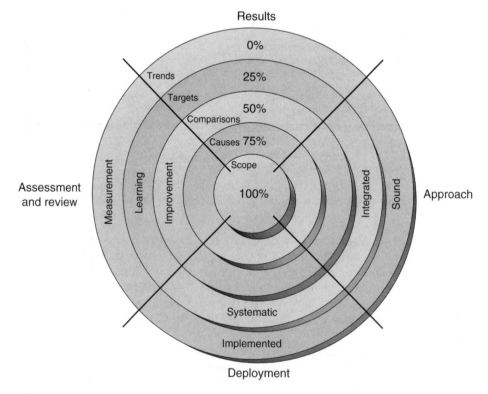

**Figure 8.6**
The RADAR screen

150

- the relative importance of the measures presented;
- the organization's actual performance;
- the organization's performance against targets;

and wherever possible:

- comparisons of performance with similar organizations;
- comparisons of performance with 'best in class' organizations.

Self-assessment against the Excellence Model may be performed generally using the so-called RADAR system:

- **Results**
- **Approach**
- **Deployment**
- **Assessment**
- **Review**

The RADAR 'screen' with the net level of detail is shown in Figure 8.6.

# Methodologies for self-assessment

The EFQM provide a flow diagram of the general steps involved in undertaking self-assessment. A simplified version of this is shown in Figure 8.7.

There are a number of approaches to carrying out self-assessment including:

- discussion group/workshop methods;
- surveys, questionnaires and interviews (peer involvement);
- pro formas;
- organizational self-analysis matrices (e.g. see Figure 8.8);
- an award simulation;

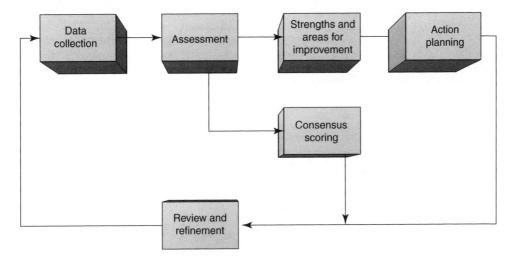

**Figure 8.7**
The key steps in self-assessment

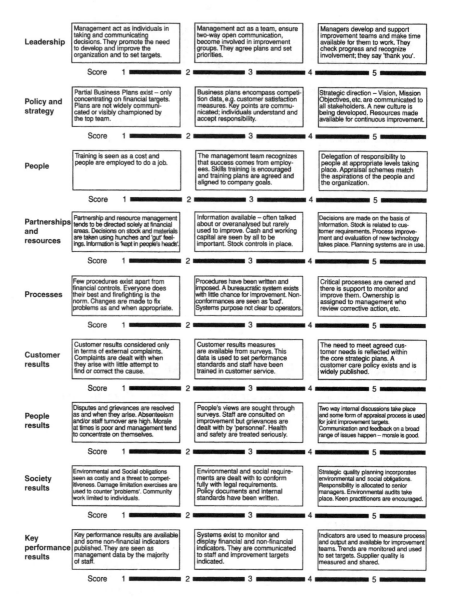

| | | | |
|---|---|---|---|
| **Leadership** | Management act as individuals in taking and communicating decisions. They promote the need to develop and improve the organization and to set targets. | Management act as a team, ensure two-way open communication, become involved in improvement groups. They agree plans and set priorities. | Managers develop and support improvement teams and make time available for them to work. They check progress and recognize involvement; they say 'thank you'. |

Score   1 ▬▬▬ 2 ▬▬▬ 3 ▬▬▬ 4 ▬▬▬ 5 ▬▬▬

| | | | |
|---|---|---|---|
| **Policy and strategy** | Partial Business Plans exist – only concentrating on financial targets. Plans are not widely communicated or visibly championed by the top team. | Business plans encompass competition data, e.g. customer satisfaction measures. Key points are communicated; individuals understand and accept responsibility. | Strategic direction – Vision, Mission Objectives, etc. are communicated to all stakeholders. A new culture is being developed. Resources made available for continuous improvement. |

Score   1 ▬▬▬ 2 ▬▬▬ 3 ▬▬▬ 4 ▬▬▬ 5 ▬▬▬

| | | | |
|---|---|---|---|
| **People** | Training is seen as a cost and people are employed to do a job. | The management team recognizes that success comes from employees. Skills training is encouraged and training plans are agreed and aligned to company goals. | Delegation of responsibility to people at appropriate levels taking place. Appraisal schemes match the aspirations of the people and the organization. |

Score   1 ▬▬▬ 2 ▬▬▬ 3 ▬▬▬ 4 ▬▬▬ 5 ▬▬▬

| | | | |
|---|---|---|---|
| **Partnerships and resources** | Partnership and resource management tends to be directed solely at financial areas. Decisions on stock and materials are taken using hunches and 'gut' feelings. Information is 'kept in people's heads'. | Information available – often talked about or overanalysed but rarely used to improve. Cash and working capital are seen by all to be important. Stock controls in place. | Decisions are made on the basis of information. Stock is related to customer requirements. Process improvement and evaluation of new technology takes place. Planning systems are in use. |

Score   1 ▬▬▬ 2 ▬▬▬ 3 ▬▬▬ 4 ▬▬▬ 5 ▬▬▬

| | | | |
|---|---|---|---|
| **Processes** | Few procedures exist apart from financial controls. Everyone does their best and firefighting is the norm. Changes are made to fix problems as and when appropriate. | Procedures have been written and imposed. A bureaucratic system exists with little chance for improvement. Non-conformances are seen as 'bad'. Systems purpose not clear to operators. | Critical processes are owned and there is support to monitor and improve them. Ownership is assigned to management who review corrective action, etc. |

Score   1 ▬▬▬ 2 ▬▬▬ 3 ▬▬▬ 4 ▬▬▬ 5 ▬▬▬

| | | | |
|---|---|---|---|
| **Customer results** | Customer results considered only in terms of external complaints. Complaints are dealt with when they arise with little attempt to find or correct the cause. | Customer results measures are available from surveys. This data is used to set performance standards and staff have been trained in customer service. | The need to meet agreed customer needs is reflected within the core strategic plans. A customer care policy exists and is widely published. |

Score   1 ▬▬▬ 2 ▬▬▬ 3 ▬▬▬ 4 ▬▬▬ 5 ▬▬▬

| | | | |
|---|---|---|---|
| **People results** | Disputes and grievances are resolved as and when they arise. Absenteeism and/or staff turnover are high. Morale at times is poor and management tend to concentrate on themselves. | People's views are sought through surveys. Staff are consulted on improvement but grievances are dealt with by 'personnel'. Health and safety are treated seriously. | Two way internal discussions take place and some form of appraisal process is used for joint improvement targets. Communication and feedback on a broad range of issues happen – morale is good. |

Score   1 ▬▬▬ 2 ▬▬▬ 3 ▬▬▬ 4 ▬▬▬ 5 ▬▬▬

| | | | |
|---|---|---|---|
| **Society results** | Environmental and Social obligations seen as costly and a threat to competitiveness. Damage limitation exercises are used to counter 'problems'. Community work limited to individuals. | Environmental and social requirements are dealt with to conform fully with legal requirements. Policy documents and internal standards have been written. | Strategic quality planning incorporates environmental and social obligations. Responsibility is allocated to senior managers. Environmental audits take place. Keen practitioners are encouraged. |

Score   1 ▬▬▬ 2 ▬▬▬ 3 ▬▬▬ 4 ▬▬▬ 5 ▬▬▬

| | | | |
|---|---|---|---|
| **Key performance results** | Key performance results are available and some non-financial indicators published. They are seen as management data by the majority of staff. | Systems exist to monitor and display financial and non-financial indicators. They are communicated to staff and improvement targets indicated. | Indicators are used to measure process and output and available for improvement teams. Trends are monitored and used to set targets. Supplier quality is measured and shared. |

Score   1 ▬▬▬ 2 ▬▬▬ 3 ▬▬▬ 4 ▬▬▬ 5 ▬▬▬

**Figure 8.8**
Organizational self-analysis matrix (Source: UK North West Quality Award Model)

- activity or process audits;
- hybrid approaches.

Whichever method is used, the emphasis should be on understanding the organization's strengths and areas for improvement, rather than the score. The scoring charts provide a consistent basis for establishing a quantitative measure of performance against the model and gaining consensus promotes discussion and development of the issues facing the organization. It should also gain the involvement, interest and commitment of the senior management, but the scores should not become an end in themselves. Tito Conti, often called 'the father of self-assessment', following the contribution he made to its establishment and development through the EFQM, when he was head

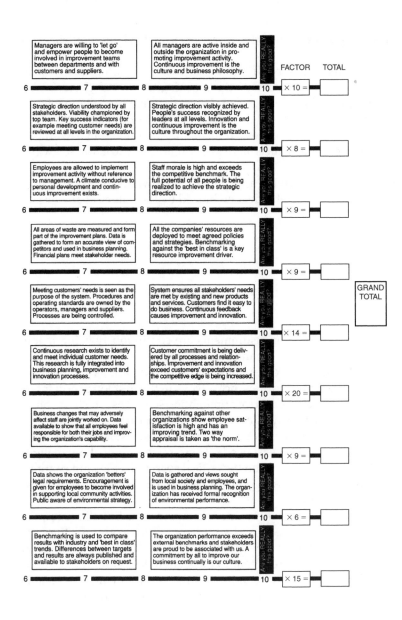

of Fiat, has expressed concern that organizations can become obsessed with self-assessment scores rather than focusing on the improvement opportunities identified.

## Using assessment

There is great overlap between the criteria used by the various awards and it may be necessary for an organization to rationalize them. The main components, however, must be the organization's processes, management systems, people management and results, customer results and key performance results. Self-assessment provides an organization with vital information in monitoring its progress towards its goals and business 'excellence'. The external assessments used in the processes of making awards must be based on these self-assessments that are performed as prerequisites for improvement.

Management jargon is increasingly confused by a vast literature, spiced with acronyms, the generation of which often bends the meaning of words. There is also often in the leadership of large organizations an ego-driven or publicity seeking wish to invent new buzz-words. It may be necessary to assess the status of the language to be used before launching a self-assessment process. If recipients are not familiar with certain language, many propositions will be meaningless. A preliminary teach-in or awareness process may even be necessary.

Whatever are the main 'motors' for driving an organization towards its vision or mission, they must be linked to the five stakeholders embraced by the values of any organization, namely:

- Customers
- Employees
- Suppliers
- Stakeholders
- Community

In any normal business or organization, measurements are continuously being made, often in retrospect, by the leaders of the organization to reflect the value put on the organization by its five stakeholders. Too often, these continuous readings are made by internal biased agents with short-term priorities, not always in the best long-term interests of the organization or its customers, i.e. narrow firefighting scenarios which can blind the organization's strategic eye. Third party agents, however, can carry out or facilitate periodic assessments from the perspective of one or more of the key stakeholders, with particular emphasis on forward priorities and needs. These reviews will allow realignment of the principal driving motors to focus on the critical success factors and continuous improvement, to maintain a balanced and powerful general thrust which moves the whole organization towards it mission.

The relative importance of the five stakeholders may vary in time but all are important. The first three, customers, employees, and suppliers, which comprise the core value chain, are the determinant elements. The application of total quality principles in these areas will provide satisfaction as a resultant to the shareholders and the community. Thus, added value will benefit the community and the environment. The ideal is a long way off in most organizations, however, and active attention to the needs of the shareholders and/or community remain a priority for one major reason – they are the 'customers' of most organizational activities and are vital stakeholders.

# Securing prevention by audit and review of the management systems

Error or defect prevention is the process of removing or controlling error/defect causes in the management systems. There are two major elements of this:

- Checking the systems.
- Error/defect investigation and follow-up.

These have the same objectives – to find, record and report *possible* causes of error, and to recommend future preventive or corrective action.

# Checking the systems

There are six methods in general use:

1. *Quality audits and reviews*, which subject each area of an organization's activity to a systematic critical examination. Every component of the total system is included, i.e. policy, attitudes, training, processes, decision features, operating procedures, documentation. Audits and reviews, as in the field of accountancy, aim to disclose the strengths and the main areas of vulnerability or risk – the areas for improvement.
2. *Quality survey*, a detailed, in-depth examination of a narrower field of activity, i.e. major key areas revealed by system audits, individual sites/plants, procedures or specific problems common to an organization as a whole.
3. *Quality inspection*, which takes the form of a routine scheduled inspection of a unit or department. The inspection should check standards, employee involvement and working practices, and that work is carried out in accordance with the agreed processes and procedures.
4. *Quality tour*, which is an unscheduled examination of a work area to ensure that, for example, the standards of operation are acceptable, obvious causes of defects or errors are removed, and in general quality standards are maintained.
5. *Quality sampling*, which measures by random sampling, similar to activity sampling, the error/defect potential. Trained observers perform short tours of specific locations by prescribed routes and record the number of potential errors or defects seen. The results may be used to portray trends in the general quality situation.
6. *Quality scrutinies*, which are the application of a formal, critical examination of the process and technological intentions for new or existing facilities, or to assess the potential for maloperation or malfunction of equipment and the consequential effects of quality. There are similarities between quality scrutinies and FMECA studies (see Chapter 6).

The design of a prevention programme, combining all these elements, is represented in Figure 8.9.

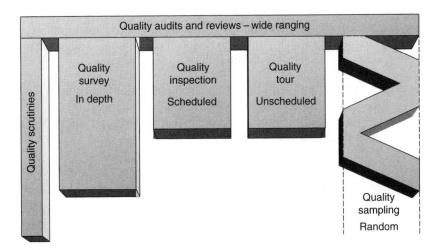

**Figure 8.9**
A prevention programme combining various elements of 'checking' the system

**Table 8.1** Following up errors

| System type | Aim | | General effects |
|---|---|---|---|
| Investigation | To prevent a similar error or defect | | *Positive:*<br>identification<br>notification<br>correction |
| Inquisition | To identify responsibility | | *Negative:*<br>blame<br>claims<br>defence |

## Error or defect investigations and follow-up

The investigation of errors and defects can provide valuable error prevention information. The general method is based on:

- *Collecting* data and information relating to the error or defect.
- *Checking* the validity of the evidence.
- *Selecting* the evidence without making assumptions or jumping to conclusions.

The result of the analysis are then used to:

- *Decide* the most likely cause(s) of the errors or defect.
- *Notify* immediately the person(s) able to take corrective action.
- *Record* the findings and outcomes.
- *Record* them to everyone concerned, to prevent recurrence.

The investigation should not become an inquisition to apportion blame, but focus on the positive preventive aspects. The types of follow-up to errors and their effects are shown in Table 8.1.

It is hoped that errors or defects are not normally investigated so frequently that the required skills are developed by experience, nor are these skills easily learned in a classroom. One suggested way to overcome this problem is the development of a programmed sequence of questions to form the skeleton of an error or defect investigation questionnaire. This can be set out with the following structure:

- *People* – duties, information, supervision, instruction, training, attitudes, etc.
- *Systems* – procedures, instructions, monitoring, control methods, etc.
- *Plant/equipment* – description, condition, controls, maintenance, suitability, etc.
- *Environment* – climatic, space, humidity, noise, etc.

# Internal and external management system audits and reviews

A good management system will not function without adequate audits and reviews. The system reviews, which need to be carried out periodically and systematically, are conducted to ensure that the system achieves the required effect, while audits are carried out to make sure that actual methods are adhering to the documented procedures. The reviews should use the findings of the audits, for failure to operate according to the plan often signifies difficulties in doing so. A re-examination of the processes and procedures actually being used may lead to system improvements unobtainable by other means.

A schedule for carrying out the *audits* should be drawn up, different activities perhaps requiring different frequencies. All procedures and systems should be audited at least once during a specified cycle, but not necessarily all at the same audit. For example, every three months a selected random sample of the processes could be audited, with the selection designed so that each process is audited at least once per year. There must be, however, a facility to adjust this on the basis of the audit results.

A quality management system *review* should be instituted, perhaps every 12 months, with the aims of:

- ensuring that the system is achieving the desired results;
- revealing defects or irregularities in the system;
- indicating any necessary improvements and/or corrective actions to eliminate waste or loss;
- checking on all levels of management;
- uncovering potential danger areas;
- verifying that improvements or corrective action procedures are effective.

Clearly, the procedures for carrying out the audits and reviews and the results from them should be documented, and themselves be subject to review. Useful guidance on quality management system audits is given in the international standard ISO 10011.

The assessment of a quality management system against a particular standard or set of requirements by internal audit and review is known as a *first party* assessment or approval scheme. If an *external* customer makes the assessment of a supplier against either its own or a national or international standard, a *second party* scheme is in operation. The assessment by an independent organization, not connected with any contract between customer and supplier, but acceptable to them both, is known as an *independent third party* assessment scheme. The latter often results in some form of certification or registration by the assessment body.

One advantage of the third party schemes is that they obviate the need for customers to make their own detailed checks, potentially saving both suppliers and customers time and money, and avoiding issues of commercial confidentiality. Just one knowledgeable organization has to be satisfied, rather than a multitude with varying levels of competence. This method can be used to certify suppliers for contracts without further checking, but good customer/supplier relations often include second party extensions to the third party requirements and audits.

Each certification body usually has its own recognized mark, which may be used by registered organizations of assessed capability in their literature, letter headings, and marketing activities. There are also publications containing lists of organizations whose quality management systems and/or products and services have been assessed. To be of value, the certification body must itself be recognized and, usually, assessed and registered with a national or international accreditation scheme.

Many organizations have found that the effort of designing and implementing a quality management system, good enough to stand up to external independent third party assessment, has been extremely rewarding in:

- involving staff and improving morale;
- better process control and improvement;

- reduced wastage and costs;
- reduced customer service costs.

This is also true of those organizations that have obtained third party registrations and supply companies which still insist on their own second party assessment. The reason for this is that most of the standards on quality management systems, whether national, international, or company-specific, are now very similar indeed. A system that meets the requirements of the ISO 9001 standard, for example, should meet the requirements of most other standards, with only the slight modifications and small emphases here and there required for specific customers. It is the authors' experience, and that of their colleagues, that an assessment carried out by one of the good independent certified assessment bodies is a rigorous and delving process.

Internal system audits and reviews should be positive and conducted as part of the preventive strategy and not as a matter of expediency resulting from problems. They should not be carried out only prior to external audits, nor should they be left to the external auditor – whether second or third party. An external auditor, discovering discrepancies between actual and documented systems, will be inclined to ask why the internal review methods did not discover and correct them.

Any management team needs to be fully committed to operating an effective quality management system for all the people within the organization, not just the staff in the 'quality department'. The system must be planned to be effective and achieve its objectives in an uncomplicated way. Having established and documented the processes it is necessary to ensure that they are working and that everyone is operating in accordance with them. The system once established is not static, it should be flexible to enable the constant seeking of improvements or streamlining.

## Quality auditing standard

The growing use of standards internationally emphasizes the importance of auditing as a management tool for this purpose. There are available several guides to management systems auditing (e.g. ISO 19011, ISO 9001) and the guidance provided in these can be applied equally to any one of the three specific and yet different auditing activities:

1. *First party or internal audits*, carried out by an organization on its own systems, either by staff who are independent of the systems being audited, or by an outside agency.
2. *Second party audits*, carried out by one organization (a purchaser or its outside agent) on another with which it either has contracts to purchase goods or services or intends to do so.
3. *Third party audits*, carried out by independent agencies, to provide assurance to existing and prospective customers for the product or service.

Audit objectives and responsibilities, including the roles of auditors and their independence, and those of the 'client' or auditee should be understood. The generic steps involved are as follows:

- *initiation*, including the audit scope and frequency;
- *preparation*, including review of documentation, the programme, and working documents;

- *execution*, including the opening meeting, examination and evaluation, collecting evidence, observations, and closing the meeting with the auditee;
- *report*, including its preparation, content and distribution;
- *completion*, including report submission and retention.

Attention should be given at the end of the audit to corrective action and follow-up and the improvement process should be continued by the auditee after the publication of the audit report. This may include a call by the client for a verification audit of the implementation of any corrective actions specified.

Any instrument which is developed for assessment, audit or review may be used at several stages in an organization's history:

- before starting an improvement programme to identify 'strengths' and 'areas for improvement', and focus attention (at this stage a parallel cost of quality exercise may be a powerful way to overcome scepticism and get 'buy-in');
- as part of a programme launch, especially using a 'survey' instrument;
- every one or two years after the launch to steer and benchmark.

The systematic measurement and review of operations is one of the most important management activities of any organization. Self-assessment, audit and review should lead to clearly discerned strengths and areas for improvement by focusing on the relationship between the people, processes, and performance. Within any successful organization these will be regular activities.

## Chapter highlights

### Frameworks for self-assessment
- Many organizations are turning to total quality models to measure and improve performance. Thee frameworks, including the Japanese Deming Prize, the US Baldrige Award and in Europe the EFQM Excellence Model.
- The nine components of the Excellence Model are: leadership, policy and strategy partnerships, people, resources, and processes (ENABLERS), customer results, people results, society results, and key performance results (RESULTS).
- The various award criteria provide rational bases against which to measure progress towards TQM in organizations. Self-assessment against, for example, the EFQM Excellence Model should be a regular activity, as it identifies opportunities for improvement in performance through processes and people.

### Methodologies for self-assessment
- Self-assessment against the Excellence Model may be performed using RADAR: results, approach, deployment, assessment and review.
- There are a number of approaches for self-assessment, including groups/workshops, surveys, pro-formas, matrices, award simulations, activity/process audits or hybrid approaches.

### Securing prevention by audit and review of the management systems
- There are two major elements of error or defect prevention: checking the system, and error/defect investigations and follow-up. Six methods of checking the quality systems are in general use: audits and reviews, surveys, inspections, tours, sampling, and scrutinies.

- Investigations proceed by collecting, checking and selecting data, and analysing it by deciding causes, notifying people, recording and reporting findings and outcomes.

**Internal and external quality management system audits and reviews**
- A good management system will not function without adequate audits and reviews. Audits make sure the actual methods are adhering to documented procedures. Reviews ensure the system achieves the desired effect.
- System assessment by internal audit and review is known as first party, by external customer as second party, and by an independent organization as third party certification. For the latter to be of real value the certification body must itself be recognized.

# Benchmarking

# The why and what of benchmarking

Product, service and process improvements can take place only in relation to established standards, with the improvements then being incorporated into new standards. *Benchmarking*, one of the most transferable aspects of Rank Xerox's approach to total quality management, and thought to have originated in Japan, measures an organization's operations, products and services against those of its competitors in a ruthless fashion. It is a means by which targets, priorities and operations that will lead to competitive advantage can be established.

As with performance measurement, the construction sector has had some difficulty in implementing benchmarking to full effect. Projects vary widely in terms of scope, process and technique, and because each project has its own unique set of circumstances in terms of the negotiated timeframe, price and contract, comparisons between projects are difficult to make. Furthermore the sector has historically focused on projects and when seen through this lens, construction is different to other sectors of the economy. Each project is seen as separate and unique and this is an obstacle to the transfer of management innovation from other sectors.

It is only relatively recently that leading academics and managers are developing a process view of construction. This perspective sees projects within a continuum of process change with key performance attributes being compared between projects. There are many drivers for benchmarking including the external ones:

- customers continually demand better quality, lower prices, shorter lead times, etc.;
- competitors are constantly trying to get ahead and steal markets;
- legislation – changes in our laws place ever greater demands for improvement.

Internal drivers include:

- targets that require improvements on our 'best ever' performance;
- technology – a fundamental change in processes is often required to benefit fully from introducing new technologies;
- self-assessment results, which provide opportunities to learn from adapting best practices.

The word 'benchmark' is a reference or measurement standard used for comparison, and benchmarking is the continuous process of identifying, understanding and adapting best practice and processes that will lead to superior performance.

Benchmarking is *not*:

- a panacea to cure the organization's problems, but simply a practical tool to drive up process performance;
- primarily a cost reduction exercise, although many benchmarking studies will result in improved financial performance;
- industrial tourism – study tours have their place, but proper benchmarking goes beyond 'tourism' to really understanding the enablers to outstanding results;
- spying – use of a benchmarking code of conduct ensures the work is done with the agreement and openness of all parties;
- catching up with the best – the aim is to reach out and extend the current best practice (by the time we have caught up, the benchmark will have moved anyway).

There may be many reasons for carrying out benchmarking. Some of them are set against various objectives in Table 9.1. The links between benchmarking and TQM are clear – establishing objectives based on industry best practice should directly contribute to better meeting of the internal and external customer requirements.

The benefits of benchmarking can be numerous but include:

- creating a better understanding of the current position;
- heightening sensitivity to changing customer needs;
- encouraging innovation;
- developing realistic stretch goals;
- establishing realistic action plans.

Data from the American Productivity and Quality Center's (APQC) International Benchmarking Clearinghouse suggests that an average benchmarking study takes six months to complete, occupies more than a quarter of the team members' time, and costs around $50 000. The same source identified that the average return was *five times* the cost of the study, in terms of reduced costs, increased sales, greater customer retention and enhanced market share.

There are four basic categories of benchmarking:

- *Internal* – the search for best practice of internal operations by comparison, e.g. multi-site comparison of process management performance in terms of planning reliability, safety and quality management outcomes.
- *Functional* – seeking functional best practice outside an industry, e.g. mining company benchmarking preventive maintenance of pneumatic/hydraulic equipment with Disney, construction company benchmarking salary payment processes with an insurance company or a mechanical contractor benchmarking service call-outs with a white goods maintenance company.
- *Generic* – comparison of outstanding processes irrespective of industry or function, e.g. restaurant chain benchmarking kitchen design with US nuclear submarine fleet to improve restaurant to kitchen space ratios, or Motorola benchmarking financial reporting with the banking sector.

**Table 9.1** Reasons for benchmarking

| Objectives | Without benchmarking | With benchmarking |
|---|---|---|
| Becoming competitive | - Internally focused<br>- Evolutionary change | - Understanding of competitiveness<br>- Ideas from proven practices |
| Industry best practices | - Few solutions<br>- Frantic catch-up activity | - Many options<br>- Superior performance |
| Defining customer requirements | - Based on history or gut feeling<br>- Perception | - Market reality<br>- Objective evaluation |
| Establishing effective goals and objectives | - Lacking external focus<br>- Reactive | - Credible, unarguable<br>- Proactive |
| Developing true measures of productivity | - Pursuing pet projects<br>- Strength and weaknesses not understood<br>- Route of least resistance | - Solving real problems<br>- Understanding outputs<br>- Based on industry best practices |

- *Competitive* – specific competitor to competitor comparisons for a product, service, or function of interest, e.g. retail outlets comparing price performance and efficiency of internet ordering systems, design firms benchmarking financial ratios through an independent benchmarking club or an Australian truss manufacturer benchmarking its processes with a similar business in the USA.

# The purpose and practice of benchmarking

The evolution of benchmarking in an organization is likely to progress through four focuses. Initially attention may be concentrated on competitive products or services, including, for example, design, development and operational features. This should develop into a focus on industry best practices and may include, for example, aspects of distribution or service. The real breakthroughs are when organizations focus on all aspects of the total business performance, across all functions and aspects, and addresses current *and projected* performance gaps. This should lead to the focus on processes and true continuous improvement.

At its simplest, competitive benchmarking, the most common form, requires every department to examine itself against the counterpart in the best competing companies. This includes a scrutiny of all aspects of their activities. Benchmarks which may be important for *customer satisfaction*, for example, might include:

- Product or service consistency.
- Correct and on-time delivery.
- Speed of response or new product development.
- Correct billing.

For internal *impact* the benchmarks may include:

- Waste, rejects or errors.
- Inventory levels/work-in-progress.
- Costs of operation.
- Staff turnover.

The task is to work out what has to be done to improve on the competition's performance in each of the chosen areas.

Benchmarking is very important in the administration areas, since it continuously measures services and practices against the equivalent operation in the toughest direct competitors or organizations renowned as leaders in the areas, even if they are in the same organization. An example of quantitative benchmarks in absenteeism is given in Table 9.2.

Technologies and conditions vary between different industries and markets, but the basic concepts of measurement and benchmarking are of general validity. The objective should be to produce products and services that conform to the requirements of the customer in a never-ending improvement environment. The way to accomplish this is to use the continuous improvement cycle in all the operating departments – nobody should be exempt. Benchmarking is not a separate science or unique theory of management, but rather another strategic approach to getting the best out of people and processes, to deliver improved performance.

**Table 9.2** Quantitative benchmarking in absenteeism

| Organization's absence level (%) | Productivity opportunity |
| --- | --- |
| Under 3 | This level matches an aggressive benchmark that has been achieved in 'excellent' organizations. |
| 3–4 | This level may be viewed within the organization as a good performance – representing a moderate productivity opportunity improvement. |
| 5–8 | This level is tolerated by many organizations but represents a major improvement opportunity. |
| 9–10 | This level indicates that a serious absenteeism problem exists. |
| Over 10 | This level of absenteeism is extremely high and requires immediate senior management attention. |

> Almost each of the case studies shows how a leading edge company has used benchmarking to drive process improvement. For example, look at how CDL (p. 389) uses benchmarking to compare between the performance of contractors on every project they procure, also how Gammon (p. 399) uses benchmarking in safety to drive performance and Graniterock (p. 375) has used benchmarking to drive improvement in product and service quality.

The purpose of benchmarking then is predominantly to:

- *change* the perspectives of executives and managers;
- *compare* business practices with those of world-class organizations;
- *challenge* current practices and processes;
- *create* improved goals and practices for the organization.

As a managed process for change, benchmarking uses a disciplined structured approach to identify *what* needs to change, *how* it can be changed, and the *benefits* of the change. It also creates the desire for change in the first place. Any process or practice that can be defined can be benchmarked but the focus should be on those which impact on customer satisfaction and/or business results – financial or non-financial.

For organizations which have not carried out benchmarking before, it may be useful initially to carry out a simple self-assessment of their readiness in terms of:

- how well processes are understood;
- how much customers are listened to;
- how committed the senior team is.

Table 9.3 provides a simple pro-forma for this purpose. The score derived gives a crude guide to the readiness of the organization for benchmarking:

32–48   Ready for benchmarking
16–31   Some further preparation required before the benefits of benchmarking can be fully derived

**Table 9.3**  Is the organization ready for benchmarking

**After studying the statements below tick one box for each to reflect the level to which the statement is true for the organization**

|  | Most | Some | Few | None |
|---|---|---|---|---|
| Processes have been documented with measures to understand performance | ☐ | ☐ | ☐ | ☐ |
| Employees understand the processes that are related to their own work | ☐ | ☐ | ☐ | ☐ |
| Direct customer interactions, feedback or studies about customers influence decisions about products and services | ☐ | ☐ | ☐ | ☐ |
| Problems are solved by teams | ☐ | ☐ | ☐ | ☐ |
| Employees demonstrate by words and deeds that they understand the organization's mission, vision and values | ☐ | ☐ | ☐ | ☐ |
| Senior executives sponsor and actively support quality improvement projects | ☐ | ☐ | ☐ | ☐ |
| The organization demonstrates by words and by deeds that continuous improvement is part of the culture | ☐ | ☐ | ☐ | ☐ |
| Commitment to change is articulated in the organization's strategic plan | ☐ | ☐ | ☐ | ☐ |
| Add the columns: | ☐ | ☐ | ☐ | ☐ |
|  | $\times 6 =$ | $\times 4 =$ | $\times 2 =$ | $\times 0 =$ |
| Multiply by the factor | ☐ | ☐ | ☐ | ☐ |
| Obtain the grand total? | | | | |

0–15   Some help is required to establish the foundations and a suitable platform for benchmarking

The benchmarking process has five main stages which are all focused on trying to measure comparisons and identify areas for action and change (Figure 9.1). The detail is as follows:

PLAN the study:

- Select processes for benchmarking.
- Bring together the appropriate team to be involved and establish roles and responsibilities.
- Identify benchmarks and measures for data collection.

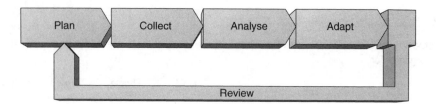

**Figure 9.1**
The five main stages of benchmarking

- Identify best competitors or operators of the process(es), perhaps using customer feedback or industry observers.
- Document the current process(es).

COLLECT data and information:

- Decide information and data collection methodology, including desk research.
- Record current performance levels.
- Identify benchmarking partners.
- Conduct a preliminary investigation.
- Prepare for any site visits and interact with target organizations.
- Use data collection methodology.
- Carry out site visits.

ANALYSE the data and information:

- Normalize the performance data, as appropriate.
- Construct a matrix to compare current performance with benchmarking competitors'/partners' performance.
- Identify outstanding practices.
- Isolate and understand the process enablers, as well as the performance measures.

ADAPT the approaches:

- Catalogue the information and create a 'competency profile' of the organization.
- Develop new performance level objectives/targets/standards.
- Vision alternative process(es) incorporating best practice enablers.
- Identify and minimize barriers to change.
- Develop action plans to adapt and implement best practices, make process changes, and achieve goals.
- Implement specific actions and integrate them into the organization.

REVIEW performance and the study:

- Monitor the results/improvements.
- Assess outcomes and learnings from the study.
- Review benchmarks.
- Share experiences and best practice learnings from implementation.
- Review relationships with target/partner organizations.
- Identify further opportunities for improving and sustaining performance.

In a typical benchmarking study involving several organizations, the study will commence with the *Plan* phase. Participants will be invited to a 'kick-off' meeting where they will share their aspirations and objectives for the study and establish roles and responsibilities. Participants will analyse their own organization to understand the strengths and areas for improvement. They will then agree appropriate measures for the study.

It is important to include participants from all levels in the organization to get the appropriate level of 'buy-in', so, for example, if a construction company were to institute benchmarking between projects, it is important to involve the key stakeholders at the head office, project engineering and management as well as the project supervisory level, otherwise implementation may become stalled down the track.

In the *Collect* phase, participants will collect data on their current performance, based on the agreed measures. The benchmarking partners will be identified, using a suitable screening process, and the key learning points will be shared. The site visits will then be planned and conducted, with appropriate training. Five to seven site visits might take place in each study.

Data collected from the site visits will be *Analysed* in the next phase to identify best practices and the enablers that deliver outstanding performance. The reports from this phase will capture the learning and key outcomes from the site visits and present them as the main process enablers, linked to major performance outcomes.

In the *Adapt* phase, the participants will attend a feedback session where the conclusions from the study will be shared, and they will be assisted in adapting them to their own organization. Reports to partners should be issued after this session. (A 'subject expert' is often useful in benchmarking studies, to ensure good learning and adaptation at this stage.)

The final phase of the study will be a post-completion *Review*.

This will give all the participants and partners valuable feedback and establish, above all else, what actions are required to sustain improved performance. Best practice databases may be created to enable further sharing and improvement among participants and other members of the organization.

# The role of benchmarking in change

One aspect of benchmarking is to enable organizations to gauge how well they are performing against others who undertake similar tasks and activities. But a more important aspect of best practice benchmarking is gaining an understanding of *how* other organizations achieve superior performance. A good benchmarking study, for example in customer satisfaction and retention, will provide its participants with data and ideas on how excellent organizations undertake their activities and demonstrate best practices that may be adopted, adapted and used.

This new knowledge will result in the benchmarking team being able to judge the gap between leading and less good performance, as well as planning considered actions to bring about changes to bridge that gap. These changes may be things that can be undertaken quickly, with little adaptation and at a minimum of cost and disruption. Such changes, often brought about by the affected operational team, are often called 'quick wins'. This type of change is incremental and carries low levels of risk but usually lower levels of benefit.

Quick wins will often give temporary or partial relief from the problems associated with poor performance and tend to address symptoms not the underlying 'diseases'. They can have a disproportionately favourable physiological impact upon the organizations. Used well, quick wins should provide a platform from which longer lasting changes may be made, having created a feeling of movement and success. All too often, however, once quick wins are implemented there is a tendency to move on to other areas, without either fully measuring the impact of the change or getting to the root cause of a performance issue.

Quick wins are clearly an important weapon in effecting change but must be followed up properly to deliver sustainable business improvement through the adoption of best practice. The changes needed to do this will usually be of a more fundamental nature and require investment in effort and money to implement. Such changes will need to be carefully planned and systematically implemented as a discrete change project or programme of projects. They carry substantial risk, if not systematically managed and controlled, but they have the potential for significant improvement in performance. These types of change projects are sometimes referred to as 'step change' or 'breakthrough' projects/programmes.

Whatever type of change is involved, a key ingredient of success is taking the people along. A first class communication strategy is required throughout and beyond any change activity, as well as the linked activity of stakeholder management. The benchmarking efforts need to fit into the change model deployed – such a framework is proposed in Figure 9.2. Many change models exist in diagrammatic form and are often, in

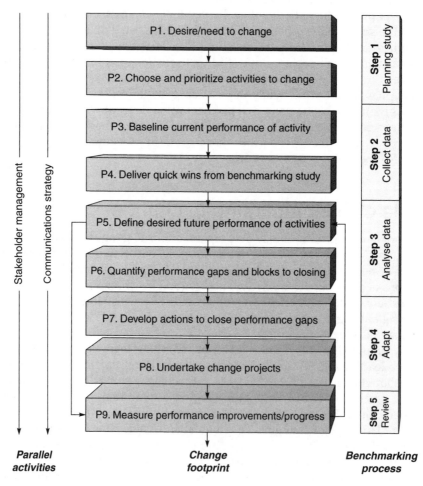

**Figure 9.2**
The benchmarking-change footprint

both intent and structure, quite similar. Such a model may be considered as a 'footprint' that will lead to the chosen destination, in this case the desired performance improvements through adoption of best practice. The footprint in Figure 9.2 demonstrates where benchmarking activities link into the general flow of change activity leading to better results.

The success and benefits derived from any benchmarking and change-related activity is directly related to the excellence of the preparation. It is necessary to consider both the 'hard' and 'soft' aspects represented in Figure 9.2 and to systematically plan to meet and overcome any difficulties and challenges identified.

# Communicating, managing stakeholders and lowering barriers

The importance of first class communication during change can never be overemphasized. A vital element of excellent communication is targeting the right audience with the right message in the right way at the right time. A scattergun approach to communication rarely has the intended impact.

In any benchmarking study it is a wise and well-founded investment in time and effort to define and understand the key stakeholders. The rise of the term 'stakeholder' in business language is relatively recent – used to describe any group or individual that has some, however small, vested interest or influence in the proposed change. Stakeholders are frequently referred to by generic groupings and may be either internal or external to an organization or business. The importance of forming, managing and maintaining good working relations with these groups is widely acknowledged and accepted.

The reality is that this activity is frequently not performed well in benchmarking. A disgruntled or ignored stakeholder with high direct organizational power or influence can easily derail the intent and hard work of others. Stakeholders with less direct power or influence can, at best, provide an unwelcome and costly distraction from the main objectives of a benchmarking study. The art of stakeholder management is to proactively head off any major confrontations. This means really understanding the stakeholders' needs and their potential to do both good and ill.

The burden of effective stakeholder management rests with the benchmarking team charged with stimulating change. They may need the ongoing patronage and support of people outside their direct control. In any good benchmarking study early thought will be given to who the stakeholders are and this will be valuable input to developing a robust stakeholder management strategy.

The elements of successful stakeholder management should include:

1. Defining and mapping the stakeholder groupings.
2. Analysing and prioritizing these groupings.
3. Researching the key players in the most important groupings.
4. Developing a management strategy.
5. Deploying the strategy by tactical actions.
6. Reviewing effectiveness of the strategy and improving the future approach.

Objective measurement is also key to targeting change activity wisely. Benchmarking project budgets are often limited and it is good practice to target such discretionary spend at changes and improvements that will deliver the best return for their investment.

Systematic measurement will provide a reliable baseline for making such decisions. By relating current performance against desired performance it should be possible to define both the gaps and appreciate the scale of improvements required to achieve the desired change.

Benchmarking studies add an extra dimension by understanding the levels of performance that best practices and leading organizations achieve. This allows realistic and sometimes uncomfortable comparisons with what an organization is currently able to achieve and what is possible. This is especially useful when setting stretch but realistic targets for future performance.

Baselining performance will allow teams to monitor and understand how successful they have been in delivering beneficial change. Used with care, as part of an overall communications strategy, successes on the road to achieving superior performance through change is a powerful motivator and useful influencing tool. Many organizations have clearly defined sets of performance measures, some self-imposed and some statute based. These should be used, if in existence. If the interest is in customer satisfaction and retention, for example, a generic but good starting point might be:

- internal measures (the lead/predictor measures) – production cycle times, unit costs, defect rate found (quality) and complaints resolved;
- external measures (the lag/reality measures) – customer satisfaction (perception), customer retention and complaints received.

The benchmarking activity may provide teams with ideas on how they might change the way goods or services are produced and delivered. They will need to prioritize this opportunity, however, to deliver best value for time and money invested and to ensure the organization does not become paralysed by initiative overload – while making improvements the day job has to continue!

The benchmarking data collected will give a clear steer to the areas that require the most urgent attention but decisions will still have to be made. Measurement and benchmarking are tools not substitutes for management and leadership – the data on its own cannot make the decisions.

# Choosing benchmarking-driven change activities wisely

As we have seen, benchmarking studies should fuel the desire to undertake change activities, but the excitement generated can allow the desire for change to take on a life of its own and irrational and impractical decisions can follow. These will inevitably result in full or partial failure to deliver the desired changes and waste of the valuable financial and people resource spent on the benchmarking itself.

Organizations should resist the temptation to start yet another series of improvement initiatives, without any consideration of their impact upon existing initiatives and the 'business as usual' activities. It is important to target the change wisely and a number of key questions need to be answered including:

- Do we fully understand the scale of the change?
- Do we have the financial resources to support the change?
- Do we have the people resources to undertake the change?
- Do we have the right skills available to undertake the change?

- Do we fully understand the operational impact during the change?
- Can beneficial changes be made without major disruption to the business?
- Will the delivered change support achievement of our business goals?
- What will the new changes do to existing change initiatives?
- Is the organization culturally ready for change?

Table 9.4 shows a simple decision-making tool to help consider the opportunities that are presented. The process may be viewed as a series of filters – it is assumed that the organization has defined business goals.

In a recently reported study[1] benchmarking was used to examine the contribution that business management systems (BMS) make to the achievement of organizational objectives. The Defence Evaluation and Research Agency (DERA), with the help of Oakland Consulting, conducted the study, which was based on the approach set out in this chapter. This included the following 12 key steps:

1. A benchmarking team was formed and educated.
2. Background research was conducted.
3. A decision was made on precisely what to benchmark (including metrics).
4. A questionnaire was produced and sent out to prospective companies.
5. The returns were analysed.
6. Partners were selected.
7. Partners were site visited.
8. Data collected during the site visit was analysed.
9. Good practice was distilled from the data.
10. A final report was produced.
11. Recommendations were made.
12. The project was reviewed.

**Table 9.4** Simple decision tool for choosing change activities

| No. | Filter test | Yes | No |
|-----|-------------|-----|-----|
| 1. | Does the benchmarking-driven proposed change support the achievement of one or more of the defined business goals? | Allow the opportunity to move forward for consideration | Decline the opportunity or defer taking forward and schedule a review |
| 2. | Does the change require financial and people resources above those agreed for the current budget round? | Prepare a business case within a project definition for consideration by senior management | Pass the opportunity to local operational management to undertake the changes as 'quick win' initiative |
| 3. | Will current improvement activity be adversely impacted by the envisaged new changes? | Consider the relative merits and benefits of new and existing change initiates and amalgamate or amend or cancel existing initiates | Allow change project to proceed and add to the controlled list of overall change projects |
| 4. | Is the required additional financial and people resource needed to undertake new change projects available? | Senior management agree and sign off project definition and project begins | Senior management prioritize change activity agreeing necessary slippage or deferment or cancellation of some change projects |

The benchmarking project was conducted on the basis of sharing best practice to the benefit of both DERA and its external partners. The details of the study and its findings will not be repeated here, but what is relevant to this chapter are the actions resulting from the study. The benchmarking project and its recommendations were key in the development of revisions to the DERA BMS. It was not the sole input but, as a result of the work, an improvement project was initiated with the aim of making the BMS more process based than it had previously been. A top level process model was derived, in parallel with the development of the future strategy for DERA by the senior management. This included, of course, the recent part-privatization of DERA to QinetiQ. From this model, key processes were further developed and better use of web-based technology stimulated.

Similarly, benchmarking studies in the BBC have provided insight on, for example, the potential for new technology to radically change existing programme-making processes. Indeed, benchmarking is an integral part of each process re-engineering project that Oakland Consulting undertakes. The external perspective provided by the benchmarking studies helped BBC employees to see how things could be different (thinking outside the box), and provided valuable input to the steps required to implement new processes (see also Chapters 10 and 11).

Benchmarking can also be used within a group of competing companies to foster best practice. In an Australian study, benchmarking criteria were developed for enterprise performance among design consultants by Marton Marosszeky and his colleagues at UNSW. This study developed measures to compare performance in a range of performance areas ranging from use of technology, income, and a number of financial efficiency measures. In such cases it is important to find a trusted impartial external party to collect and report performance data to the entire group.[2]

The drivers of change are everywhere but properly conducted systematic benchmarking studies can define clearer objectives and help their effective deployment through well-executed change management. Best practice benchmarking and change management clearly are bedfellows. If well understood and integrated they can deliver lasting improvements in performance, which satisfy all stakeholder needs. Benchmarking is an efficient way to promote effective change by learning from the successful experiences of others and putting that learning to good effect.

# Acknowledgement

The authors are grateful for the contribution made in the preparation of this chapter by their colleagues Dr Steve Tanner and Robin Walker.

# References

1. Morling, P. and Tanner, S.J. (2000) 'Benchmarking a public service business management system', *Total Quality Management*, Vol. 11, Nos 4, 5 and 6, pp. 417–426.
2. Marosszeky, M. and Karim, K. (2002) Enterprise process monitoring using key performance indicators. In: *Building in Value* (R. Best and G. de Valence, eds). Butterworth-Heinemann.

## Chapter highlights

### The why and what of benchmarking

■ Benchmarking measures an organization's products, services and processes to establish targets, priorities and improvements, leading in turn to competitive advantage and/or cost reductions.

- Benefits of benchmarking can be numerous and include creating a better understanding of the current position, heightening sensitivity to changing customer needs, encouraging innovation, developing stretch goals, and establishing realistic action plans.
- Data from APQC suggests an average benchmarking study takes six months to complete, occupies a quarter of the team members' time and costs around $50 000. The average return was five times the costs.
- The four basic types of benchmarking are internal, functional, generic and competitive, although the evolution of benchmarking in an organization is likely to progress through focus on continuous improvement.

## The purpose and practice of benchmarking
- The evolution of benchmarking is likely to progress through four focuses: competitive products/services, industry best practices, all aspects of the business in terms of performance gaps, and focus on processes and true continuous improvement.
- The purpose of benchmarking is predominantly to change perspective, compare business practices, challenge current practices and processes, and to create improved goals and practices, with the focus on customer satisfaction and business results.
- A simple scoring pro-forma may help an organization to assess whether it is ready for benchmarking, if it has not engaged in it before. Help may be required to establish the right platforms if low scores are obtained.
- The benchmarking process has five main stages: plan, collect, analyse, adapt and review. These are focused on trying to measure comparisons and identify areas for action and change.

## The role of benchmarking in change
- An important aspect of benchmarking is gaining an understanding of how other organizations achieve superior performance. Some of this knowledge will result in 'quick wins', with low risk but relatively low levels of benefit.
- Step changes are of a more fundamental nature, usually require further investment in time and money, will need to be carefully planned and systematically implemented, and typically carry a higher risk.
- A change model or 'footprint' should lead to the chosen destination – improved performance through the adoption of best practice – and show the role of benchmarking.

## Communicating, managing stakeholders and lowering barriers
- Communication is vital during change, and a vital element is targeting the right audience, with the right message, in the right way at the right time.
- Defining and understanding the key stakeholders is a wise investment of time. This should be followed by building and managing good relationships. This falls on the benchmarking team.
- Elements of successful stakeholder group management include: defining and mapping, analysing and prioritizing, researching key players/groups, developing and deploying a strategy, and reviewing effectiveness.
- Objective measurement is key to targeting change wisely and provides a reliable baseline for decisions. Baselining performance allows teams to monitor and understand success in delivering beneficial change.

**Choosing benchmarking-driven change activities wisely**

- Organizations should start benchmarking-driven improvement activities only with consideration of their impact on existing initiatives. Questions to be asked include those related to the scale of the change, the financial and people resources (including skills) required, the impact and disruption aspects, the degree of support to the business goals, and the cultural implications.
- Benchmarking may be used to drive revisions in business management systems, facilitate the application of new technologies, and generally help people to see how processes might be different.
- Properly conducted systematic benchmarking studies can aid the definition of clearer objectives and help their deployment through well-executed change management.

# Process management

# The process management vision

Organizations create value by delivering their products and/or services to customers. Everything they do in that whole chain of events is a process. So to perform well in the eyes of the customers and the stakeholders all organizations need very good process management – underperformance is primarily caused by poor processes.

This is recognized, of course, in the EFQM Excellence Model, in which the processes criterion is the central 'anchor' box linking the other enablers and the results together. Performance can be improved often by improving or changing processes but the devil is in the detail and successful exponents of process management understand all the dimensions related to:

- *Process strategy* – particularly deployment.
- *Operationalizing processes* – including definition/design/systems.
- *Process performance* – measurement and improvement.
- *People and leadership roles* – values, beliefs, responsibilities, accountabilities, authorities and rewards.
- *Information and knowledge* – capturing and leveraging throughout the supply chains.

Where process management is established and working, executives no longer see their organizations as sets of discrete vertical functions with silo-type boundaries. Instead they visualize things from the customer perspective – as a series of interconnected work and information flows that cut horizontally across the business. As the supply chain for a product becomes more fragmented through outsourcing, the challenge for management is to design, visualize and implement efficient processes across organizational boundaries. This characterizes the challenge facing many organizations in the construction sector – especially those engaged in the delivery of design and construction services. Effectively the customers are pictured as 'taking a walk' through some or all of these 'end-to-end' processes and interfacing with the company or service organization wherever it lies within the supply chain, experiencing how it generates demand for products and services, how it fulfils orders, services products, etc. (Figure 10.1). All these processes need managing – planning, measuring and improving – sometimes discontinuously.

In monitoring process performance, measurement will inevitably identify necessary improvement actions. In many process managed companies they have shifted the focus of the measurement systems from functional to process goals and even based remuneration and career advancement on process performance.

## Operationalizing process management

Top management in many organizations now base their approach to business on the effective management of 'key or core business processes'. These are well-defined and developed sequences of steps with clear rationale, which add value by producing required outputs from a variety of inputs. Moreover these management teams have aligned the core processes with their strategy, combining related activities and cutting out ones that do not add value. This has led in some cases to a fundamental change in the way the place is managed.

The changes required have caused these organizations to emerge as true 'process enterprises'. There are many such organizations, including Texas Instruments, ST Microelectronics, Philips, IBM, Celestica, BBC, Highways Agency, QinetiQ and Apollo, that are familiar to the authors and their colleagues.

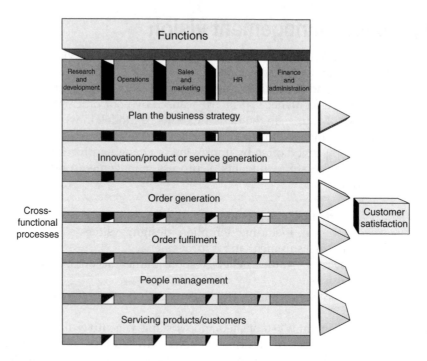

**Figure 10.1**
Cross-functional approach to managing core business processes

Companies comprising a number of different business units, such as outsourcing companies, face an early and important strategic decision when introducing process management – should all the business units follow the same process framework and standardization, or should they tailor processes to their own particular and diverse needs? Each organization must consider this question carefully and there can be no one correct approach.

Deployment of a common high-level process framework throughout the organization gives many benefits, including presenting 'one company' to the customers and suppliers, lower costs, and increased flexibility, particularly in terms of resource allocation.

In research on award-winning companies[1,2] the authors and their colleagues identified process management best practices as:

- Identifying the key business processes:
  - prioritizing on the basis of the value chain, customer needs and strategic significance, and using process models and definitions.
- Managing processes systematically:
  - giving process ownership to the most appropriate individual or group and resolving process interface issues through meetings or ownership models.
- Reviewing processes and setting improvement targets:
  - empowering process owners to set targets and collect data from internal and external customers.
- Using innovation and creativity to improve processes:
  - adopting self-managed teams, business process improvement and idea schemes.
- Changing processes and evaluating the benefits:
  - through process improvement or re-engineering teams, project management and involving customers, and suppliers.

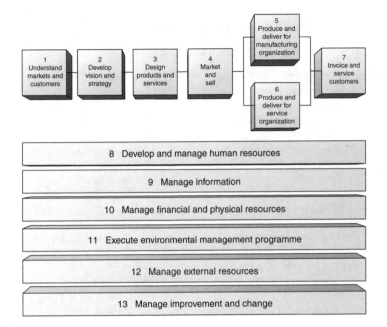

**Figure 10.2**
Process Classification Framework: overview

Too many businesses are still not process oriented, however; they focus instead on tasks, on jobs, the people who do them, and on structures.

# The Process Classification Framework

In establishing a core process framework, many organizations have found inspiration in the Process Classification Framework developed and copyrighted by the American Productivity and Quality Center (APQC) International Benchmarking Clearinghouse with the assistance of several major international corporations. The intent was to create a high-level generic enterprise model that encourages businesses and other organizations to see their activities from a cross-industry, process viewpoint rather than from a narrow functional viewpoint.

The Process Classification Framework supplies a generic view of business processes often found in multiple industries and sectors – manufacturing and service companies, health care, government, education, and others. As we saw earlier, many organizations now seek to understand their inner workings from a horizontal, process viewpoint, rather than from a vertical, functional viewpoint.

The Process Classification Framework seeks to represent major processes and sub-processes, not functions, through its structure (Figure 10.2) and vocabulary (Table 10.1). The framework does not list all processes within any specific organization. Likewise, not every process listed in the framework is present in every organization.

## About the framework

The Process Classification Framework was originally envisioned as a 'taxonomy' of business processes during the design of the American Productivity and Quality

**Table 10.1**   APQC Process Classification Framework vocabulary

## OPERATING PROCESSES

### 1. UNDERSTAND MARKETS AND CUSTOMERS

1.1 Determine customer needs and wants

   1.1.1 Conduct qualitative assessments

      1.1.1.1 Conduct customer interviews

      1.1.1.2 Conduct focus groups

   1.1.2 Conduct quantitative assessments

      1.1.2.1 Develop and implement surveys

   1.1.3 Predict customer purchasing behavior

1.2 Measure customer satisfaction

   1.2.1 Monitor satisfaction with products and services

   1.2.2 Monitor satisfaction with complaint resolution

   1.2.3 Monitor satisfaction with communication

1.3 Monitor changes in market or customer expectations

   1.3.1 Determine weaknesses of product/service offerings

   1.3.2 Identify new innovations that meet customer needs

   1.3.3 Determine customer reactions to competitive offerings

### 2. DEVELOP VISION AND STRATEGY

2.1 Monitor the external environment

   2.1.1 Analyze and understand competition

   2.1.2 Identify economic trends

   2.1.3 Identify political and regulatory issues

   2.1.4 Assess new technology innovations

   2.1.5 Understand demographics

   2.1.6 Identify social and cultural changes

   2.1.7 Understand ecological concerns

2.2 Define the business concept and organizational strategy

   2.2.1 Select relevant markets

   2.2.2 Develop long-term vision

   2.2.3 Formulate business unit strategy

   2.2.4 Develop overall mission statements

2.3 Design the organization structure and relationships between organizational units

2.4 Develop and set organizational goals

### 3. DESIGN PRODUCTS AND SERVICES

3.1 Develop new product/service concept and plans

   3.1.1 Translate customer wants and needs into product and/or service requirements

   3.1.2 Plan and deploy quality targets

   3.1.3 Plan and deploy cost targets

   3.1.4 Develop product life cycle and development timing targets

   3.1.5 Develop and integrate leading technology into product/service concept

3.2 Design, build, and evaluate prototype products and services

   3.2.1 Develop product/service specifications

   3.2.2 Conduct concurrent engineering

   3.2.3 Implement value engineering

   3.2.4 Document design specifications

   3.2.5 Develop prototypes

   3.2.6 Apply for patents

3.3 Refine existing products/services

   3.3.1 Develop product/service enhancements

   3.3.2 Eliminate quality/reliability problems

   3.3.3 Eliminate outdated products/services

3.4 Test effectiveness of new or revised products or services

3.5 Prepare for production

   3.5.1 Develop and test prototype production process

   3.5.2 Design and obtain necessary materials and equipment

   3.5.3 Install and verify process or methodology

3.6 Manage the product/service development process

### 4. MARKET AND SELL

4.1 Market products or services to relevant customer segments

   4.1.1 Develop pricing strategy

   4.1.2 Develop advertising strategy

   4.1.3 Develop marketing messages to communicate benefits

   4.1.4 Estimate advertising resource and capital requirements

   4.1.5 Identify specific target customers and their needs

   4.1.6 Develop sales forecast

   4.1.7 Sell products and services

   4.1.8 Negotiate terms

4.2 Process customer orders

   4.2.1 Accept orders from customers

   4.2.2 Enter orders into production and delivery process

### 5. PRODUCE AND DELIVER FOR MANUFACTURING-ORIENTED ORGANIZATIONS

5.1 Plan for and acquire necessary resources

   5.1.1 Select and certify suppliers

   5.1.2 Purchase capital goods

   5.1.3 Purchase materials and supplies

   5.1.4 Acquire appropriate technology

5.2 Convert resources or inputs into products

   5.2.1 Develop and adjust production delivery process (for existing process)

**Table 10.1** *(continued)*

## OPERATING PROCESSES

5.2.2 Schedule production
5.2.3 Move materials and resources
5.2.4 Make product
5.2.5 Package product
5.2.6 Warehouse or store product
5.2.7 Stage products for delivery
5.3 Deliver products
    5.3.1 Arrange product shipment
    5.3.2 Deliver products to customers
    5.3.3 Install product
    5.3.4 Confirm specific service requirements for individual customers
    5.3.5 Identify and schedule resources to meet service requirements
    5.3.6 Provide the service to specific customers
5.4 Manage production and delivery process
    5.4.1 Document and monitor order status
    5.4.2 Manage inventories
    5.4.3 Ensure product quality
    5.4.4 Schedule and perform maintenance
    5.4.5 Monitor environmental constraints

### 6. PRODUCE AND DELIVER FOR SERVICE-ORIENTED ORGANIZATIONS

6.1 Plan for and acquire necessary resources
    6.1.1 Select and certify suppliers
    6.1.2 Purchase materials and supplies
    6.1.3 Acquire appropriate technology
6.2 Develop human resource skills
    6.2.1 Define skill requirements
    6.2.2 Identify and implement training
    6.2.3 Monitor and manage skill development
6.3 Deliver service to the customer
    6.3.1 Confirm specific service requirements for individual customer
    6.3.2 Identify and schedule resources to meet service requirements
    6.3.3 Provide the service to specific customers
6.4 Ensure quality of service

### 7. INVOICE AND SERVICE CUSTOMERS

7.1 Bill the customer
    7.1.1 Develop, deliver, and maintain customer billing
    7.1.2 Invoice the customer
    7.1.3 Respond to billing inquiries
7.2 Provide after-sales service
    7.2.1 Provide post-sales service
    7.2.2 Handle warranties and claims
7.3 Respond to customer inquiries
    7.3.1 Respond to information requests
    7.3.2 Manage customer complaints

## MANAGEMENT AND SUPPORT PROCESSES

### 8. DEVELOP AND MANAGE HUMAN RESOURCES

8.1 Create and manage human resource strategies
    8.1.1 Identify organizational strategic demands
    8.1.2 Determine human resource costs
    8.1.3 Define human resource requirements
    8.1.4 Define human resource's organizational role
8.2 Cascade strategy to work level
    8.2.1 Analyze, design, or redesign work
    8.2.2 Define and align work outputs and metrics
    8.2.3 Define work competencies
8.3 Manage deployment of personnel
    8.3.1 Plan and forecast workforce requirements
    8.3.2 Develop succession and career plans
    8.3.3 Recruit, select and hire employees
    8.3.4 Create and deploy teams
    8.3.5 Relocate employees
    8.3.6 Restructure and rightsize workforce
    8.3.7 Manage employee retirement
    8.3.8 Provide outplacement support
8.4 Develop and train employees
    8.4.1 Align employee and organization development needs
    8.4.2 Develop and manage training programs
    8.4.3 Develop and manage employee orientation programs
    8.4.4 Develop functional/process competencies
    8.4.5 Develop management/leadership competencies
    8.4.6 Develop team competencies
8.5 Manage employee performance, reward and recognition
    8.5.1 Define performance measures
    8.5.2 Develop performance management approaches/feedback
    8.5.3 Manage team performance
    8.5.4 Evaluate work for market value and internal equity
    8.5.5 Develop and manage base and variable compensation
    8.5.6 Manage reward and recognition programs
8.6 Ensure employee well-being and satisfaction
    8.6.1 Manage employee satisfaction
    8.6.2 Develop work and family support systems
    8.6.3 Manage and administer employee benefits
    8.6.4 Manage workplace health and safety

Process management

Table 10.1 (*continued*)

## MANAGEMENT AND SUPPORT PROCESSES

8.6.5  Manage internal communications
8.6.6  Manage and support workforce diversity
8.7  Ensure employee involvement
8.8  Manage labor-management relationships
    8.8.1  Manage collective bargaining process
    8.8.2  Manage labor-management partnerships
8.9  Develop Human Resource Information Systems (HRIS)

### 9. MANAGE INFORMATION RESOURCES

9.1  Plan for information resource management
    9.1.1  Derive requirements from business strategies
    9.1.2  Define enterprise system architectures
    9.1.3  Plan and forecast information technologies/methodologies
    9.1.4  Establish enterprise data standards
    9.1.5  Establish quality standards and controls
9.2  Develop and deploy enterprise support systems
    9.2.1  Conduct specific needs assessments
    9.2.2  Select information technologies
    9.2.3  Define data life cycles
    9.2.4  Develop enterprise support systems
    9.2.5  Test, evaluate, and deploy enterprise support systems
9.3  Implement systems security and controls
    9.3.1  Establish systems security strategies and levels
    9.3.2  Test, evaluate, and deploy systems security and controls
9.4  Manage information storage and retrieval
    9.4.1  Establish information repositories (data bases)
    9.4.2  Acquire and collect information
    9.4.3  Store information
    9.4.4  Modify and update information
    9.4.5  Enable retrieval of information
    9.4.6  Delete information
9.5  Manage facilities and network operations
    9.5.1  Manage centralized facilities
    9.5.2  Manage distributed facilities
    9.5.3  Manage network operations
9.6  Manage information services
    9.6.1  Manage libraries and information centers
    9.6.2  Manage business records and documents
9.7  Facilitate information sharing and communication
    9.7.1  Manage external communications systems
    9.7.2  Manage internal communications systems
    9.7.3  Prepare and distribute publications
9.8  Evaluate and audit information quality

### 10. MANAGE FINANCIAL AND PHYSICAL RESOURCES

10.1  Manage financial resources
    10.1.1  Develop budgets
    10.1.2  Manage resource allocation
    10.1.3  Design capital structure
    10.1.4  Manage cash flow
    10.1.5  Manage financial risk
10.2  Process finance and accounting transactions
    10.2.1  Process accounts payable
    10.2.2  Process payroll
    10.2.3  Process accounts receivable, credit, and collections
    10.2.4  Close the books
    10.2.5  Process benefits and retiree information
    10.2.6  Manage travel and entertainment expenses
10.3  Report information
    10.3.1  Provide external financial information
    10.3.2  Provide internal financial information
10.4  Conduct internal audits
10.5  Manage the tax function
    10.5.1  Ensure tax compliance
    10.5.2  Plan tax strategy
    10.5.3  Employ effective technology
    10.5.4  Manage tax controversies
    10.5.5  Communicate tax issues to management
    10.5.6  Manage tax administration
10.6  Manage physical resources
    10.6.1  Manage capital planning
    10.6.2  Acquire and redeploy fixed assets
    10.6.3  Manage facilities
    10.6.4  Manage physical risk

### 11. EXECUTE ENVIRONMENTAL MANAGEMENT PROGRAM

11.1  Formulate environmental management strategy
11.2  Ensure compliance with regulations
11.3  Train and educate employees
11.4  Implement pollution prevention program
11.5  Manage remediation efforts
11.6  Implement emergency response programs
11.7  Manage government agency and public relations
11.8  Manage acquisition/divestiture environmental issues
11.9  Develop and manage environmental information system
11.10  Monitor environmental management program

Total Quality in the Construction Supply Chain

Table 10.1    (*continued*)

## MANAGEMENT AND SUPPORT PROCESSES

**12. MANAGE EXTERNAL RELATIONSHIPS**
12.1  Communicate with shareholders
12.2  Manage government relationships
12.3  Build lender relationships
12.4  Develop public relations program
12.5  Interface with board of directors
12.6  Develop community relations
12.7  Manage legal and ethical issues

**13. MANAGE IMPROVEMENT AND CHARGE**
13.1  Measure organizational performance
    13.1.1  Create measurement systems
    13.1.2  Measure product and service quality
    13.1.3  Measure cast of quality
    13.1.4  Measure costs
    13.1.5  Measure cycle time
    13.1.6  Measure productivity
13.2  Conduct quality assessments
    13.2.1  Conduct quality assessments based on external criteria

    13.2.2  Conduct quality assessments based on internal criteria
13.3  Benchmark performance
    13.3.1  Develop benchmarking capabilities
    13.3.2  Conduct process benchmarking
    13.3.3  Conduct competitive benchmarking
13.4  Improve processes and systems
    13.4.1  Create commitment for improvement
    13.4.2  Implement continuous process improvement
    13.4.3  Re-engineer business processes and systems
    13.4.4  Manage transition to change
13.5  Implement TQM
    13.5.1  Create commitment for TQM
    13.5.2  Design and implement TQM systems
    13.5.3  Manage TQM life cycle

Process management

Center's International Benchmarking Clearinghouse. That design process involved more than 80 organizations with a strong interest in advancing the use of benchmarking in the USA and around the world. The Process Classification Framework can be a useful tool in understanding and mapping business processes. In particular, a number of organizations have used the framework to classify both internal and external information for the purpose of cross-functional and cross-divisional communication.

The Process Classification Framework is an evolving document and the Center will continue to enhance and improve it on a regular basis. To that end, the Center welcomes your comments, suggestions for improvement, and any insights you gain from applying it within your organization. The Center would like to see the Process Classification Framework receive wide distribution, discussion, and use. Therefore, it grants permission for copying the framework, as long as acknowledgement is made to the American Productivity and Quality Center.*

## Process modelling

More than 25 years ago, the US Airforce adopted 'Integration Definition Function Modelling' (IDEF-0), as part of its Integrated Computer-Aided Manufacturing (ICAM) architecture. The IDEF-0 modelling language, now described in a Federal Information Processing Standards Publication (FIPS PUBS), provides a useful structured graphical

* Please direct your comments, suggestions, and questions to: APQC International Benchmarking Clearinghouse Information Services Department, 123 North Post Oak Lane, 3rd Floor, Houston, Texas 77024-7797, 713-681-4020 (phone), 713-681-8578 (fax), Internet: apqcinfo@apqc.org, *For updates, visit the website at http://www.apqc.org*

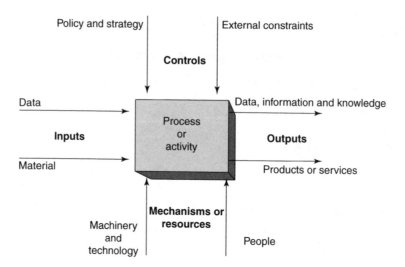

**Figure 10.3**
IDEF-0 process language. Federal Information Procession Standard Publication 183 (December 1993), National Institute of Standards and Technology (NIST)

framework for describing and improving business processes. The associated 'Integration Definition for Information Modelling' (IDEF-1X) language allows the development of a logical model of data associated with processes, such as measurement.

These techniques are widely used in business process re-engineering (BPR) and business process improvement (BPI) projects, and to integrate process information. A range of specialist software (including Windows/PC based) is also available to support the applications. IDEF-0 may be used to model a wide variety of new and existing processes, define the requirements, and design an implementation to meet the requirements.

An IDEF-0 model consists of a hierarchical series of diagrams, text and glossary, cross-referenced to each other through boxes (process components) and arrows (data and objects). The method is expressive and comprehensive and is capable of representing a wide variety of business, service and manufacturing processes. The relatively simple language allows coherent, rigorous and precise process expression, and promotes consistency. Figure 10.3 shows the basis of the approach.

> The ADePT method for design management presented in Chapter 6 gives an illustration of the IDEF-0 modelling technique being used to analyse the design process, this then leads to the Dependency Structure Matrix – the core of the ADePT process for the management of complex problems.

For a full description of the IDEF-0 methodology, it is necessary to consult the FIPS PUBS standard (Federal Information Processing Standard Publication 183 (December 1993), National Institute of Standards and Technology (NIST)). It should be possible, however, from the simple description given here, to begin process modelling (or mapping) using the technique.

Processes can be any combination of things, including people, information, software, equipment, systems, products or materials. The IDEF-0 model describes what a process

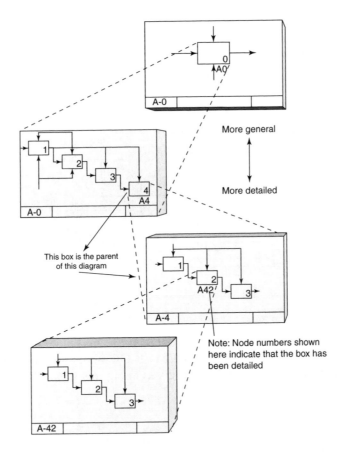

**Figure 10.4**
Decomposition structure – subprocesses

does, what controls it, what things it works on, what means it uses to perform its functions, and what it produces. The combined graphics and text comprise:

- *Boxes* – which provide a description of what happens in the form of an active verb or verb phrase.
- *Arrows* – which convey data or objects related to the processes to be performed (they do not represent flow or sequence as in the traditional process flow model).

Each side of the process box has a standard meaning in terms of box/arrow relationships. Arrows on the left side of the box are *inputs*, which are transformed or consumed by the process to produce *output* arrows on the right side. Arrows entering the top of the box are *controls* which specify the conditions required for the process to generate the correct outputs. Arrows connected to the bottom of the box represent '*mechanisms*' or *resources*. The abbreviation ICOR (inputs, controls, outputs, resources) is sometimes used.

Using these relationships, process diagrams are broken down or decomposed into more detailed diagrams, the top-level diagram providing a description of the highest level process. This is followed by a series of child diagrams providing details of the subprocesses (see Figure 10.4).

Each process model has a top-level diagram on which the process is represented by a single box with its surrounding arrows (e.g. Figure 10.5). Each subprocess is modelled

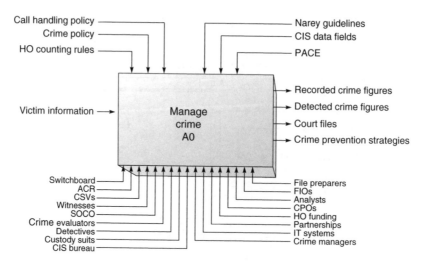

**Figure 10.5**
A0 crime management

individually by a box, with parent boxes detailed by child diagrams at the next lower level. An example of the application of IDEF or ICOR modelling in Crime Management and Reporting in West Yorkshire Police is given in Figures 10.6 to 10.8.

## Text and glossary

An IDEF-0 diagram may have associated structured text to give an overview of the process model. This may also be used to highlight features, flows, and interbox connections and to clarify significant patterns. A glossary may be used to define acronyms, key words and phrases used in the diagrams.

## Arrows

Arrows on high-level IDEF-0 diagrams represent data or objects as constraints. Only at low levels of detail can arrows represent flow or sequence. These high-level arrows may usefully be thought of as pipelines or conduits with general labels. An arrow may branch, fork, or join, indicating that the same kind of data or object may be needed or produced by more than one process or subprocess.

## IDEF-0 process modelling, improvement and teamwork

The IDEF-0 methodology includes procedures for developing and critiquing process models by a group or team of people. The creation of an IDEF-0 process model provides a disciplined teamwork procedure for process understanding and improvement. As the group works on the process following the discipline, the diagrams are changed to reflect corrections and improvements. More detail can be added by creating more diagrams, which in turn can be reviewed and altered. The final model represents an agreement on the process for a given purpose and from a given viewpoint, and can be the basis of new process or system improvement projects.

In the West Yorkshire Police example (Figures 10.5–10.8), teams involved in constructing these models gained appreciation of the role everyone plays in the overall

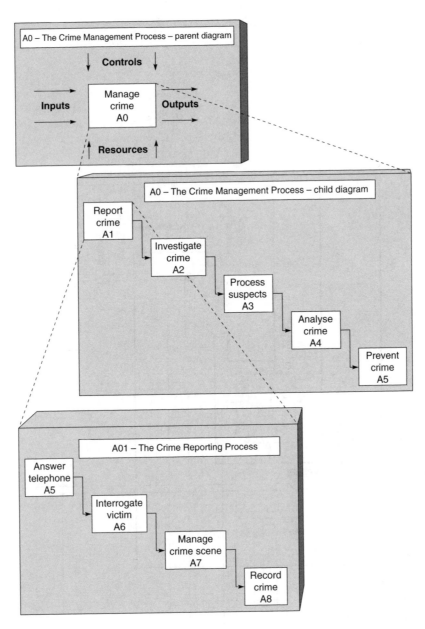

**Figure 10.6**
IDEF-0 decomposition structure – subprocesses – for crime management

process. This encouraged ownership of the process and acted as a spur to making improvements.

Initially the overall crime management process was considered at the macro level, identifying its various input, output, control and resource factors through a combination of brainstorming and consultation with managers. This allowed a high-level model to be compiled, as shown at Figure 10.8. Although the model reveals little in terms of the process interactions, it is helpful in illustrating the importance of crime management to the police force.

The outputs of the crime management process arguably represent the most important outputs of any police force. They include the levels of recorded crime, as well as the

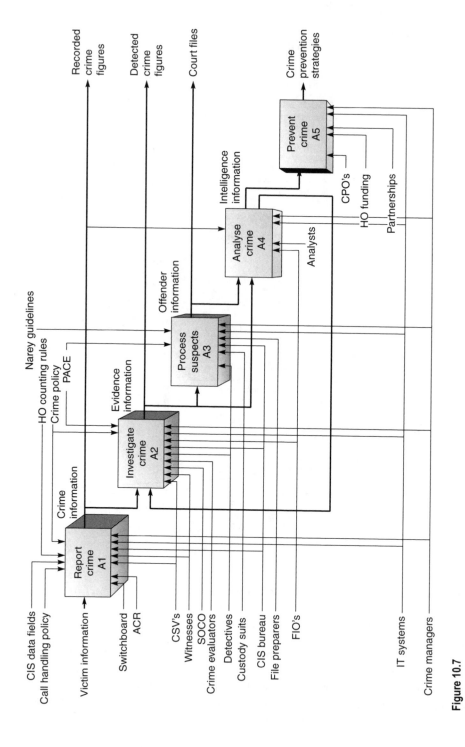

**Figure 10.7**
A0 crime management child diagram

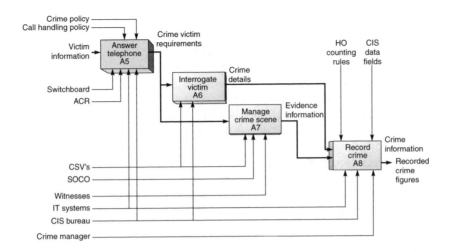

**Figure 10.8**
A01 – report crime

proportion of these offences that have been successfully detected. The publication of this information is of great interest to the public, the force and its political stakeholders, forming perceptions of safety and assessment of organizational performance. Also produced by the process are prosecution files for the courts and crime prevention strategies, often part of community safety partnerships. These outputs, though often less publicized, are measures of activities vital for the medium- and long-term reduction of crime.

The range of resources applied by West Yorkshire Police to its crime management process covers a large proportion of its functional areas. It includes uniform patrol, CID, Intelligence, Communication and Case Preparation. It gives an indication of the complexity, but also the appropriateness of using a process approach to manage improvement across the departmental structures of the force.

Having established the process at its macro level, the process was broken down into the main subprocesses of which it was composed. It was not possible to be prescriptive as to how many subprocesses this entailed. However, they needed to be sufficient to cover the range of the crime management process, and ensure that all the original input, output, control and resource factors could be reassigned on the child diagram.

Crime reporting was the initial process providing the interface with the victim and included the telephone crime reporting system. Crime investigation was the next process involving the greatest concentration of organizational resources, particularly those of an operational nature. The processing of suspects similarly attracted a heavy concentration of resources, reflecting the sometime onerous and even bureaucratic demands of its main output – court files. The crime analysis process would not have featured in a similar process model years ago, but is now an indispensable process in tackling crime, made possible through advances in IT. Finally, the crime prevention process has risen in prominence through community safety partnerships.

The IDEF-0/ICOR methodology allows processes to be broken down to as many levels as are required by the process under investigation. In this study, one further level was required in order to examine the crime reporting process in sufficient detail. This involved taking the crime reporting subprocess A1 from Figure 10.7 and decomposing its own constituent subprocesses.

This final level was composed of four subprocesses: telephone answering where the victim's call was handled between the switchboard, ACR and the CIS Bureau; victim interrogation where information pertaining to the crime was obtained directly through dialogue with the victim, either in person or by phone; crime scene management, which is the obtaining of forensic evidence or identifying potential witnesses at the crime scene; crime recording, where a detailed bundle of crime information is entered onto the CIS database as a recorded crime. The detailed subprocess model is shown in Figure 10.8.

Such an approach would be ideal, for example, to investigate well-known problem areas such as the causes of defective work in construction projects, variations or poor documentation within an organization. The IDEF-0/ICOR methodology could be used to model current processes in terms of inputs, controls, resources and outputs, and then to develop improved systems.

In using such techniques in process management, there can be a propensity for maps to assume disproportionate importance. This can result in participants becoming distracted in pursuit of accuracy or even the overall purpose of process improvement being supplanted by the modelling process itself.

## IDEF-1X

This is used to produce structural graphical information models for processes, which may support the management of data, the integration of information systems, and the building of computer databases. It is described in detail in the FIPS PUB 184 (December 1993, NIST). Its use is facilitated by the introduction of IDEF-0 modelling for process understanding and improvement.

A number of commercial software packages, which support IDEF-0 and IDEF-1X implementation, are available.

# Process flowcharting

In the systematic planning or detailed examination of any process, whether that be an administrative service delivery, manufacturing, or managerial activity, it is necessary to record the series of events and activities, stages and decisions in a form that can be easily understood and communicated to all. If improvements are to be made, the facts relating to the existing method must be recorded first. The statements defining the process should lead to its understanding and will provide the basis of any critical examination necessary for the development of improvements. It is essential therefore that the descriptions of processes are accurate, clear and concise.

The usual method of recording facts is to write them down, but this is not suitable for recording the complicated processes that exist in any organization, particularly when an exact record is required of a long process, and its written description would cover several pages requiring careful study to elicit every detail. To overcome this difficulty, certain methods of recording have been developed, and the most powerful of these is flowcharting. This method of describing a process owes much to computer programming, where the technique is used to arrange the sequence of steps required for the operation of the program. It has a much wider application, however, than computing.

Certain standard symbols are used on the chart, and these are shown in Figure 10.9. The starting point of the process is indicated by a circle. Each processing step, indicated by a rectangle, contains a description of the relevant operation, and where the process ends is indicated by an oval. A point where the process branches because of a decision is shown by a diamond. A parallelogram relates to process information but is not a processing step. The arrowed lines are used to connect symbols and to indicate direction of flow. For a complete description of the process, all operation steps (rectangles) and decisions (diamonds) should be connected by pathways to the start circle and end oval. If the flowchart cannot be drawn in this way, the process is not fully understood.

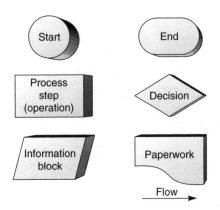

**Figure 10.9**
Flowcharting symbols

It is a salutary experience for most people to sit down and try to draw the flowchart for a process in which they take part every working day. It is often found that:

■ The process flow is not fully understood.
■ A single person is unable to complete the flowchart without help from others.

The very act of flowcharting will improve knowledge of the process, and will begin to develop the teamwork necessary to find improvements. In many cases, the convoluted flow and octopus-like appearance of the chart will highlight unnecessary movements of people and materials and lead to common-sense suggestions for waste elimination.

## Example of flowcharting in use – improving a travel procedure

We start by describing the original process for a male employee, though clearly it applies equally to females.

The process starts with the employee explaining his travel plans to his secretary. The secretary then calls the travel agent to enquire about the possibilities and gives feedback to the employee. The employee decides if the travel arrangements, e.g. flight numbers and dates, are acceptable and informs his secretary, who calls the agent to make the necessary bookings or examine alternatives. The administrative procedure, which starts as soon as the bookings have been made, is as follows:

1. The employee's secretary prepares the travel request (which is in four parts, A, B, C and D), and gives it to the employee. The request is then sent to the employee's manager, who approves it. The manager's secretary sends it back to the employee's secretary.
2. The employee's secretary sends copies A, B and C to the agent and gives copy D to the employee. The travel agent delivers the ticket to the employee's secretary, together with copy B of the travel request. The secretary endorses copy B for receipt of the ticket, sends it to Accounting, and gives ticket to employee.
3. The travel agent bills the credit-card company, and sends Accounting a pro-forma invoice with copy C of the travel request. Accounting matches copies B and C, and charges the employee's 181 account.
4. Accounting receives the monthly bill from the credit-card company, matches it against the travel request, and then pays the credit-card company.

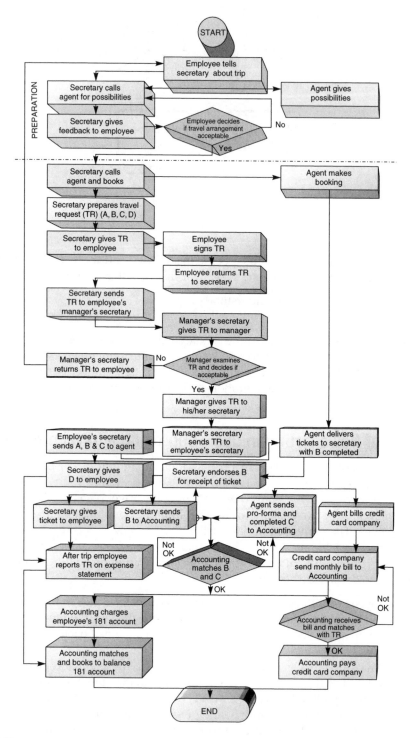

**Figure 10.10**
Original process for travel procedure

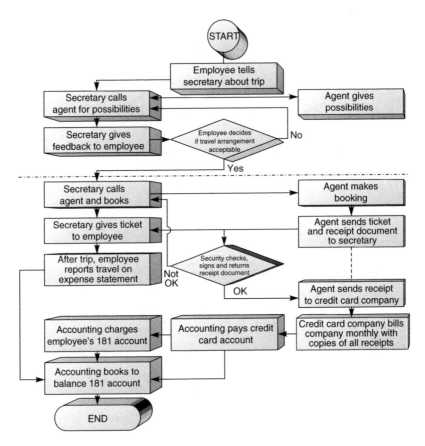

**Figure 10.11**
Improved process for travel procedure

5. The employee reports the travel request on his expense statement. Accounting matches and balances the employee's 181 account.

The total time taken for the administrative procedure, excluding the correction of errors and the preparation of overview reports, is 23 minutes per travel request.

The flowchart for the process is drawn in Figure 10.10. A quality-improvement team was set up to analyse the process and make recommendations for improvement, using brainstorming and questioning techniques. They made the following proposal to change the procedure. The preparation for the trip remained the same but the administrative steps, following the bookings being made, became:

1. The travel agent sends the ticket to the secretary, along with a receipt document, which is returned to the agent with the secretary's signature.
2. The agent sends the receipt to the credit-card company, which bills the company on a monthly basis with a copy of all the receipts. Accounting pays the credit-card company and charges the employee's 181 account.
3. The employee reports the travel on his expense statement, and Accounting balances the employee's 181 account.

The flowchart for the improved process is shown in Figure 10.11. The proposal reduced the total administrative effort per travel request (or per travel arrangement, because the travel request was eliminated) from 23 minutes to 5 minutes.

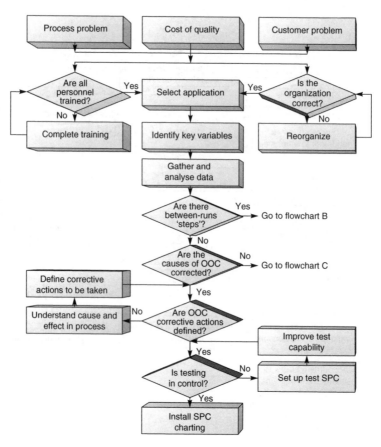

**Figure 10.12**
Flowchart (A) for installation of SPC charting systems. (The authors are grateful to Exxon Chemical International for permission to use and modify this chart.)

The details that appear on a flowchart for an existing process must be obtained from direct observation of the process, not by imagining what is done or what should be done. The latter may be useful, however, in the planning phase, or for outlining the stages in the introduction of a new concept. Such an application is illustrated in Figure 10.12 for the installation of statistical process control charting systems (see Chapter 13). Similar charts may be used in the planning of quality management systems.

It is surprisingly difficult to draw flowcharts for even the simplest processes, particularly managerial ones, and following the first attempt it is useful to ask whether:

- The facts have been correctly recorded.
- Any oversimplifying assumptions have been made.
- All the factors concerning the process have been recorded.

The authors have seen too many process flowcharts that are so incomplete as to be grossly inaccurate.

Summarizing, then, a flowchart is a picture of the steps used in performing an activity or task. Lines connect the steps to show the flow of the various tasks or steps. Flowcharts provide excellent documentation and are useful troubleshooting tools to

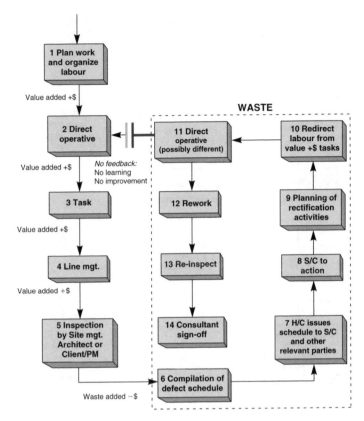

**Figure 10.13**
The current wasteful system of defect rectification

determine how each step is related to the others. By reviewing the flowchart, it is often possible to discover inconsistencies and determine potential sources of variation and problems. For this reason, flowcharts are very useful in process improvement when examining an existing process to highlight the problem areas. A group of people, with the knowledge about the process, should take the following simple steps:

1. Draw a flowchart of existing process.
2. Draw a second chart of the flow the process could or should follow.
3. Compare the two to highlight the changes necessary.

A number of commercial software packages which support process flowcharting are available.

The following flowchart was developed by one of the author's research group to depict the process typically found for defect identification and rectification on Sydney high-rise residential construction sites. Figure 10.13 depicts the current system and Figure 10.14 the proposed system. The difference between the two systems is stark; therefore, this is a very effective way of communicating the waste in the existing process and demonstrating the logic of change. Of course, to make the change requires many detailed changes in management processes, organizational relationships and skills.

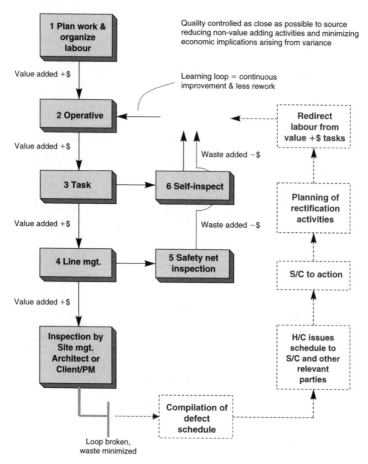

**Figure 10.14**
The proposed system of defect rectification showing the benefits of immediate action identification and correction

# Leadership, people and implementation aspects of process management

There are many top executives who have famously used process management to great effect, including Richard J. Leo, President and General Manager of Xerox, who suggested that process orientation is one of the key factors that delivered the remarkable turnaround in the performance of that company:

> Business processes are designed to be customer driven, cross functional and value based. They create knowledge, eliminate waste and abandon unproductive work yielding world-class productivity and higher perceived service levels for customers.

Alan Jones, Group Managing Director of multiple award-winning TNT Express, operates a process driven company and he believes this approach helps provide a logical framework for people in the organization to satisfy customers. The processes provide awareness for each person of his or her role in the business and this leads to superior performance:

> Publishing the process on the wall helps people understand their place in the big picture ... the continuously improving profits earned by TNT Express are a consequence

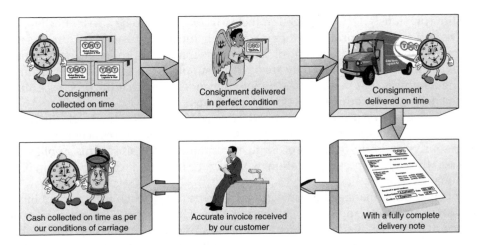

**Figure 10.15**
TNT Express Delivery Services: the perfect transaction process

of superlative performance that is derived from well thought out processes, ongoing measurement of a few carefully selected key indicators, good communications and the full involvement everywhere of all people working in the company.

The approach manifests itself in complete understanding by everyone of the end-to-end TNT Express process and this is described in the 'perfect transaction' (Figure 10.15).

Perhaps the most visible difference between a process management enterprise and a more traditional, functionally based one is the existence of process owners – top management with end-to-end responsibilities for individual processes. They have real responsibility for and authority over the process design, operation and measurement of performance. This requires working, at least initially, through functional heads, which usually leads to a major cultural challenge for the organization. Process owners cannot simply instruct workers in the process to do as they say, so style and the ability to influence is at least as important to process management as structure. High-level process owners rarely have any line responsibility.

This is a change in management structure with which organizations in the construction sector struggle. In part, this is driven by the project focus of the industry. Output is generally conceptualized as a series of projects through which value is created. Senior managers, in general contracting organizations where nearly all the actual production is outsourced, see the reduction of site-based *prelims* or *project overheads* as one of their key challenges. Head office functions are primarily oriented towards marketing and administration and the ability of head office to add value to project delivery is questioned. Senior head office managers find it difficult to integrate process-based thinking into their organizational structure. The lessons learned in other sectors are particularly relevant in construction as the major opportunities for improved efficiency and effectiveness are process based.

Managing the people who work in the processes demands attention to:

- designing, developing and delivering training programmes;
- setting performance targets;
- regular communication, preferably face-to-face:
  - keeping them informed of changing customer needs;
  - listening to concerns and ideas;
- negotiation and collaboration.

Celestica (previously D2D*) realized in the early 1980s that to be a cost-effective, competitive and indeed world-class organization, it must ensure that all processes are understood, measured and in control. Everyone is trained in process management and improvement and shown how they are part of a supplier–process–customer chain. The training reinforces that these chains are interdependent and that the processes all support the delivery of products or services to customers.

Operators of every process are properly trained, have necessary work instructions available, and have the appropriate tools (such as statistical process control), facilities and resources to perform the process to its optimum capability. This applies to all processes throughout the organization, whatever the outputs, including those in finance and human resources.

In many process-managed organizations this type of approach has changed the way they assign and train employees, emphasizing the whole process rather than narrowly focused tasks. It has made fundamental changes to cultures, stressing process-based teamwork and customers rather than functionally driven command and control. Creativity and innovation in process improvement are recognized as core competencies, and the annual performance reviews and personal development plans are linked to these.

---

Among the case studies, Sekisui Heim (p. 469), with its origins in manufacturing, is an interesting contrast to the more traditionally structured construction sector organizations; though, in all the case study companies, the dedication of people to critical process functions is a key strategy for implementing change.

---

The first thing that top management must recognize is that moving to process management requires much more than redrawing the organizational chart or structure. The changes needed are fundamental, leading to new ways of working and managing, and they will challenge any company or public service organization.

Many organizations today are facing a large number of changes and initiatives, often driven by public–private transitions, customer or government demands, technology, and so on. Before implementing process management, therefore, a senior management team needs to examine closely all its current change initiatives – using a simple framework, such as the Total Organizational Excellence Model 3 – to prune those that are not relevant to a process-managed business and combining/rationalizing those that are.

The introduction of process management is often driven or directly connected to a strategic initiative, such as reducing cycle times, increasing customer satisfaction, reducing working capital (perhaps tied up in work-in-progress), an enterprise resource planning (ERP) implementation, changes in technology, or introduction of e-business. The application of new enabling technologies is an ideal time to review the design and configuration of key processes. Failure to do so could lead to missed opportunities to extract maximum benefit from the technology.

Implementing process management like many change initiatives cannot be a quick fix and it will not happen overnight. Top management need the resolution and commitment for major changes in the way things are structured, carried out and measured. As with all such implementation, it needs careful planning and an understanding of what needs to be done first. Things high on the list will be the establishment of a core process framework, aligned to the needs of the business, and the appointment of key process owners. A process-based performance measurement framework then needs to be set up to track

---

* Celestica acquired Design to Distribution (D2D) Limited in 1997.

progress. As with all change initiatives, delivering some tangible measurable benefits early on will help overcome the inevitable resistance. In one pharmaceutical company, for example, the success of the work on the product development and product promotion processes helped significantly the cause of process management and the company extended its approach into the supply-chain management and other processes.

As companies and public service organizations move inexorably towards the wider introduction of e-commerce to do business, this will place a premium on rapid and fault-free execution of business processes. Putting a website in front of an inefficient, ineffectual or even broken process will soon bring it to its knees, together with everyone working in it and around it. This will also bring 'back-office' mistakes to the attention of the marketplace. Some of these processes will, of course, need to be redesigned – from customer order fulfilment to procurement. They will need to change 'shape' as demands, technology and market change. Without good process management in place this is going to be very difficult for functionally driven organizations.

# Acknowledgement

The authors are grateful to Mark Milsom of West Yorkshire Police for permission to use the material related to crime reporting in this chapter.

# References

1. 'The X-factor, winning performance through business excellence', European Centre for Business Excellence/British Quality Foundation, London 1999.
2. 'The Model in Practice' and 'The Model in Practice 2', British Quality Foundation, London 2000 and 2002 (prepared by European Centre for Business Excellence).
3. Oakland J. S., *Total Organizational Excellence*, Butterworth-Heinemann, Oxford 1999 and paperback 2001.

## *Chapter highlights*

### Process management vision

- Everything organizations do to create value for customers of their products or services is a process. Process management is key to improving performance.
- Process-managed organizations see things from a customer perspective – as a series of interconnected work and information flows that cut horizontally across the business functions.
- The key or core business processes are well-defined and developed sequences of steps with clear rationale, which add value by producing required outputs from a variety of inputs.
- Deployment of a common high-level process framework throughout the organization gives many benefits, including reduced costs and increased flexibility.
- Process management best practices include: identifying the key business processes, managing processes systematically, reviewing processes and setting improvement targets, using innovation and creativity to improve processes, changing processes, and evaluating the benefits.

### Process Classification Framework and process modelling

- The APQC's Process Classification Framework creates a high-level generic, cross-functional process view of an enterprise – a taxonomy of business processes.

- The IDEF (Integrated Definition Function Modelling) language provides a useful structured graphical framework for describing and improving business processes. It consists of a hierarchical series of diagrams and text, cross-referenced to each other through boxes. The processes are described in terms of inputs, controls, outputs and resources (ICOR).

### Process flowcharting
- Flowcharting is a method of describing a process in pictures, using symbols – rectangles for operation steps, diamonds for decisions, parallelograms for information, and circles/ovals for the start/end points. Arrow lines connect the symbols to show the 'flow'.
- Flowcharting improves knowledge of the process and helps to develop the team of people involved.
- Flowcharts document processes and are useful as troubleshooting tools and in process improvement. An improvement team would flowchart the existing process and the improved or desired process, comparing the two to highlight the changes necessary.

### Leadership, people and implementation
- Top management who have used process management to great effect recognize its contribution in creating knowledge and eliminating waste, yet they understand the importance of involving people, measurement and good communications.
- Process owners are key to effective process management. They have responsibility for and authority over process design, operation and measurement of performance.
- Managing the people who work in the processes requires attention to training programmes, performance targets, communicating changing customer needs, negotiation and collaboration.
- Moving to process management requires some challenging fundamental changes, leading to new ways of working and managing. Current initiatives should be carefully examined to ensure good planning and an understanding of what needs to be done first.
- As with all change initiatives, delivering some tangible measurable benefits early on will help overcome the inevitable resistance.
- With the wider introduction of e-commerce systems, there will be greater pressure to run rapid, fault-free business processes. Some of the processes will need to change 'shape' as demands, technologies and markets change.

# Process redesign

# Process redesign

Two of the main movements that have influenced process redesign in the past 20 years are 'lean production' and 'process re-engineering'. The first of these found its roots in a review of the causes of inefficiency in mass production while the latter was born out of the potential provided by information technology for traditional processes to be fundamentally redesigned.

# Lean construction

The starting point for the research that has become known as *lean production* was the International Motor Vehicle Programme (IMVP) research programme at MIT; first reported in Reference 1. However, as with many management terms, lean production is often used loosely. Recently, to redefine the term, Jim Womack sent a message to the Lean Enterprise Institute email list titled 'Deconstructing the Tower of Babel'.

He described how in 1987, working with a group of colleagues, they listed the performance attributes of a Toyota-style production system compared with traditional mass production. The Toyota-style production system:

- needed less human effort to design, make, and service products;
- required less investment for a given amount of production capacity;
- created products with fewer delivered defects and fewer in-process turnbacks;
- utilized fewer suppliers with higher skills;
- went from concept to launch, order to delivery, and problem to repair in less time with less human effort;
- could cost-effectively produce products in lower volume with wider variety to sustain pricing in the market while growing share;
- needed less inventory at every step from order to delivery and in the service system;
- caused fewer employee injuries, etc.

The group very quickly ascertained that the system needed less of everything to create a given amount of value, so they called it lean. And the term was born. In the intervening period, the term has become loosely applied to a great variety of things and so to set the record straight, Jim Womack wrote:

*… here's what lean means to me:*

- It always begins with the customer;
- The customer wants value: the right good or service at the right time, place and price with perfect quality;
- Value in any activity – goods, services or some combination – is always the end result of a process (design, manufacture, and service for external customers, and business processes for internal customers);
- Every process consists of a series of steps that need be taken properly in the proper sequence at the proper time;
- To maximize customer value, these steps must be taken with zero waste;
- To achieve zero waste, every step in a value-creating process must be valuable, capable, available, adequate and flexible, and the steps must flow smoothly and quickly from one to the next at the pull of the downstream customer;

- A truly lean process is a perfect process: perfectly satisfying the customer's desire for value with zero waste;
- None of us have ever seen a perfect process nor will most of us ever see one. But lean thinkers still believe in perfection, the never-ending journey toward the truly lean process.

Note that identifying the steps in the process, getting them to flow, letting the customer pull, etc., are not the objectives of lean practitioners. These are simply necessary steps to reach the goal of perfect value with zero waste.

Essentially what Womack defined was simply *a focus on the customer, on creating value and on eliminating waste* – the ideal of any production process. The wastes he referred to were those defined by Engineer Ohno of Toyota:

- overproduction;
- waiting;
- excess conveyance;
- extra processing;
- excessive inventory;
- unnecessary motion; and
- defects requiring rework or scrap.

Koskela and Howell[2] added the following for the construction sector:

- reduce the proportion of non-value adding activities;
- reduce lead time;
- reduce variability;
- simplify processes;
- increase flexibility; and
- increase transparency.

Over the past 12 years, the International Group for Lean Construction (IGLC)* has met annually and through their conferences they have gradually defined the lean construction agenda. It has been broadened to include all the causes of waste and the strategies for their elimination. The agenda has an increasing amount in common with TQM. The main difference being that the lean focus is on production processes whereas the TQM focus is on organizations as a whole. The IGLC agenda is structured around four themes: theory, project delivery, minimizing waste and maximizing value.

There has been a strong interest in the development of the underlying theoretical basis for project management,[2] though more recently the group is widening its interest to include the broader framework of social and management theory.

The scope of interest in the remaining three areas is:

# Project delivery

## *Management of people and teams*

Waste is created through the lack of alignment between the parties working side by side on construction projects. This translates into dysfunctional teams, poor levels of co-operation

---

* http://cic.vtt.fi/lean/

and lost opportunities for the optimum use of resources. The lean community is increasingly interested in research and practice in creating alignment, in creating trusting, open working relations, in developing effective teams and in effective collaboration between companies and their personnel delivering value through construction projects. These issues are addressed in Chapters 14, 15 and 16.

## Lean supply chain management

The application of the concepts of supply chain and value stream mapping to production analysis across the supply chain is the basis for improving the efficiency of procurement and production and the development of more closely integrated supply. This issue is given some consideration in Chapters 5 and 15. Analyses presented in research[3,4,5] document unproductive resources in engineered-to-order electrical and mechanical supply chains.

## Safety, quality and environment management systems

New management and collaboration ideas and tools are changing planning, implementation, feedback and process improvement in relation to safety, quality and environmental management in the construction phases of project delivery. These are providing the opportunity for industry leaders to radically improve their outcomes. These issues are considered in Chapters 7, 9 and 12.

## Performance measurement

One of the main drivers of continuous improvement or learning is performance measurement and benchmarking. Organizations that focus on key performance drivers measure their progress towards their goals continuously. Research and lean practice eliminates waste and improves efficiency at every level of the process. Chapters 7 and 9 consider performance measurement and benchmarking. Papers exploring these issues include Marosszeky *et al.* (2004).[6]

# Minimizing waste
## Production planning and control

New management and collaboration ideas and tools are being implemented; they are changing production and cost planning as well as controlling both the design and construction phases of project delivery, leading to substantial improvements in performance. These issues are considered in Chapter 10. Papers exploring these issues include Marosszeky *et al.* (2002)[7] and Saurin *et al.* (2002).[8]

## Information technology (IT) support for lean construction

IT through its rapidly developing potential for visualization and collaboration is supporting the transformation of processes within design, planning and production. Interorganizational collaboration integrating IT across functions in the supply chain offers opportunities for process re-engineering. These issues are considered in this chapter.

Process redesign

### Buffer management and work structuring

The efficiency of processes depends on the design of the overall production system, batch design and the structure of specific work packages. The IGLC group is looking at research and practice in the area of production systems design and tools that support more efficient process design. Some of these issues are considered in Chapter 6, although much of this topic is beyond the scope of this book.

### Prefabrication, assembly and open building

Production ideas such as modular production, tolerance mapping, dimensional co-ordination and prefabrication in lean production strategies are explored under this theme. Some of these issues are considered in Chapter 6, although much of this topic is beyond the scope of this book.

## Maximizing value
### Product development

This theme explores issues throughout the product development process from client briefing, design management, target costing, standardization, mass customization, procurement of design services and whole-life approach, including building in use disassembly and recycling. Some of these issues are considered in Chapter 6 although much of this topic is beyond the scope of this book.

### Strategy and implementation

Projects that achieve the ideals of lean production are set up to succeed from the outset. Lean practitioners are interested in research and practice in establishing procurement frameworks that are conducive to lean delivery, strategies for initiating lean-oriented projects in both the design and construction phase, in improving our understanding of the elements of contracting strategies for alignment. These issues are considered in Chapters 4 and 5.

# Re-engineering the organization?

When it has been recognized that a major business process requires radical reassessment, business process re-engineering or redesign (BPR) methods are appropriate. In their book *Re-Engineering the Corporation*, Hammer and Champy talked about reinventing the nature of work, 'starting again – reinventing our corporations from top to bottom'. BPR was launched on a wave of organizations needing to completely rethink how and why they do what they do in order to cope with the ever-changing world, particularly the development of technology-based solutions.

The reality of course, is, that many processes in many organizations are very good and do not need re-engineering, redesigning or reinventing, not for a while anyway. These processes should be subjected to a regime of continuous improvement (Chapter 12) at least until we have dealt with the very poorly performing processes that clearly do need radical review.

Some businesses and industries more than others have been through some pretty hefty changes – technological, political, financial and/or cultural. Customers of these organizations may be changing and demanding certain new relationships. Companies are finding leaner competitors encroaching into their marketplace, increased competition from other countries where costs are lower, and start-up competitors that do not share the same high bureaucracy and formal structures.

Enabling an organization, whether in the public or private sector, to be capable of meeting these changes is not a case of working harder but working differently. There have been many publicized BPR success stories and, equally, there have been some abject failures. In some cases, radical changes to major business processes have brought corresponding radical improvements in productivity. However, knowing how to reap such benefit, or indeed knowing if and how to apply BPR, has proved difficult for some organizations.

The concept of BPR was introduced to the world via two articles that described the radical changes to business processes being performed by a handful of western businesses. These were also among the first to embark on TQM initiatives in the 1980s and included Xerox, Ford, AT&T, Baxter Healthcare, and Hewlett-Packard.

Many companies adopted TQM initiatives in the 1980s hoping to win back business lost to Japanese competition. When Ford benchmarked Mazda's accounts payable department, however, they discovered a business process being run by five people, compared to Ford's 500. Even with the difference in scale of the two companies, this still demonstrated the relative inefficiency of Ford's accounts payable process. At Xerox, taking a customer's perspective of the company identified the need to develop systems rather than stand-alone products, which highlighted Xerox's own inefficient office systems.

Both Ford and Xerox realized that incremental improvement alone was not enough. They had developed high infrastructure costs and bureaucracies that made them relatively unresponsive to customer service. Focusing on internal customer/supplier interfaces improved quality, but preserved the current process structure and they could not hope to achieve in a few years what had taken the Japanese 30 years. To achieve the necessary improvements required a radical rethink and redesign of these processes.

What was being applied by organizations such as Ford and Xerox was *discontinuous improvement*. In order to respond to the competitive threats of Canon and Honda, Xerox and Ford needed TQM to catch up, but to get ahead they felt they required radical breakthroughs in performance. Central to these breakthrough improvements was IT.

# Information technology as a driver for BPR

BPR is often based on new possibilities for breakthrough performance provided by the emergence of new enabling technologies. The most important of these, the one that is the nominal ingredient in many BPR recipes, is IT.

Explosive advances in IT have enabled the dissemination, analysis, and use of information from and to customers and suppliers and within enterprises, in new ways and in timeframes that impact processes, organization designs and strategic competencies. Computer networks, open systems, client/server architecture, groupware, and electronic data interchange have opened up the possibilities for the integrated automation of business processes. Neural networks, enterprise analyser approaches, computer-assisted software engineering, and object-oriented programming now facilitate systems design around office processes.

The pace of change has, of course, been enormous and IT systems unavailable just ten to fifteen years ago have enabled sweeping changes in business process improvement, particularly in office systems. Just as statistical process control (SPC) enabled manufacturing processes to be improved by controlling variation and improving efficiency, so IT is enabling non-manufacturing processes to be fundamentally restructured.

IT in itself, however, did not offer all the answers, automation frequently being claimed not to produce the gains expected. Many companies putting in major new computer systems have achieved only the automation of existing processes. Frequently, different functions within the same organization have systems that are incompatible with each other. Locked into traditional functional structures, managers have spent large amounts on IT systems that have not been used cross-functionally. Yet it is in this cross-functional area that the big improvement gains through IT are to be made. Once a process view is taken to designing and installing an IT system, it becomes possible to automate cross-functional, cross-divisional, even cross-company processes.

The construction sector faces the same issues within, as well as across, organizations. A recent study of IT use within the Australian construction sector[9] found that while most of the opportunities for process automation had been taken up within the various specialist organizations in the industry, the challenges of process re-engineering and genuine interorganization collaboration have not. The study identified these areas to be substantial opportunities for overall productivity enhancement. The observation initially made in 2000, that IT offers much greater potential for the industry as a whole than it does for individual organizations and that this potential has been largely unexploited, is still true.

In summary, the study found that:

- automation of existing work practices has brought a range of generic benefits: productivity gains, increased business turnover, shorter cycle time, systems to manage larger and more complex projects and improved accuracy and consistency of documentation;
- there was a general perception that the use of IT brings with it an expectation of faster cycle and response times;
- among all the organizations interviewed, relatively few had re-engineered their business processes along with the adoption of IT; however, those that had done so had experienced significant gains in productivity and a commensurate gain in competitive advantage;
- for some firms IT had enabled expansion into new markets and had positioned them to compete internationally;
- among larger organizations IT use was widespread in all sectors and electronic intra- and interorganizational communication via e-mail was common;
- among smaller organizations there was a spread from those who were highly IT literate and were therefore able to use IT effectively across their operations to those who were just commencing to use IT;
- advanced users – both small and large – were creating networks based on IT across organizational, national and international boundaries;
- views regarding the ease of data exchange among different CAD documentation software were varied; though it was generally agreed that software developers were now addressing issues of interoperability.

The dominant concern among the more sophisticated users of IT was to further exploit communications and Internet technology to achieve more efficient and effective linkage across sectors.

## The supply chain opportunity

The potential of IT when used across industry sectors is best described relative to its use within the sector. It was found that most applications of IT have been confined to a single

sector. Consequently, benefits to date have been restricted. The greatest potential for trans-
forming the industry beyond another round of cost reduction lies in the re-engineering of
the supply chain to deliver increased value for the client, this requires cross-sectoral interorganizational collaboration. This potential can be perceived at three levels.

- Level 1 – IT can be and, typically, is used to improve the efficiency, speed and
  quality of communication across sectors; thereby, reducing cycle times and making
  a small gain in quality for the whole supply chain.
- Level 2 – IT can be but, as yet, is not used to facilitate the creation of a transformed
  supply chain. By taking a different approach to cross-sectoral relationships (e.g. by
  encouraging greater concurrency between tasks conducted by firms in different
  sectors through greater sharing of information) it may be possible to achieve
  substantial savings in time and money for the client.
- Level 3 – in a supply chain characterized by the sharing of information and
  knowledge, the potential exists to increase the total value to the end client (the
  developer/operator of the building or plant) by improving performance on
  multiple dimensions, including operational manageability and return on the asset.
  For example, if architects, engineers, contractors and clients would start to share
  information when a design is first conceived, through appropriately rich
  communication channels, it may become possible to design and build more
  efficiently. There would be fewer difficulties for the designers and builders and far
  greater benefits to the customer because new kinds of solution would be developed
  collaboratively. These solutions would be safe, more aesthetic, easier to build, and
  would perform better for the client. When Frank Gehry designed the Guggenheim
  Museum in Bilbao, he was able to create a totally innovative landmark because his
  design process was tightly linked through IT to his suppliers. This meant that he
  was able to ensure the feasibility of his design as he developed it.

Such potential is enabled by IT but requires more than mere adoption of the technology.
Successful implementation requires 'buy-in' by those who will use the technology. The
achievement of competitive benefits typically requires organizational change. For example, the productivity gains from CAD will become maximized for architects only when
drafting is integrated with the design task. Successful transformation of businesses is
achieved incrementally over several years through a cycle of learning and organizational
change.

---

**Case study 11**
**Flower & Samios Architects – business benefits from continuous innovation**

Flower & Samios is a Sydney-based firm of architects, who have gained significant benefits from their mastery of IT.
In 1987, they made their first tentative steps by leasing two personal computers with the intention of using them primarily for presentations of designs to clients. Within a few years, however, IT had become the engine room of the
practice, with all design work done on computers. Through an incremental process of project-by-project adoption, all
architects were given a computer and learned to design using CAD software. By 1992, drawing boards were no
longer used, and no draftsmen remained in the firm. Staff became sophisticated leading edge users of CAD, modelling and multimedia software. In ten years, the firm has invested about $500 000 in IT. Annual turnover has multi-
plied many times since 1987, with staff numbers growing only from 16 to 20.

   The adoption of IT has led to greater speed, flexibility and accuracy in design, and greater integration across
the business, with benefits extending to contract administration and project management. The outcomes have
included high client satisfaction, an industry reputation for IT leadership in architecture, lower costs and a competitive

edge that is difficult to imitate. With a reputation for expertise in 3D modelling, new opportunities have emerged in urban design and local government planning.

Leading the gradual transformation, partner John Flower developed three basic rules for successful management of IT in the practice. First, he determined that IT had to be mastered by the professional architects, starting with himself. Designing projects from the start in 3D on PCs became standard practice, which forced rapid learning. Second, for IT to provide real benefits, it had to be integrated into the business, which meant not employing IT specialists or CAD operators. This required a strong commitment to individual and organizational learning in the practice, to ensure that the professionals continued to advance their IT skills. Third, to enable this, Flower & Samios committed to using commercially available hardware and software because it was relatively easy to use and required no specialist IT capability. The competitive benefits were not obtained directly from the IT systems, which anyone could buy, but from the focused learning and mastery over time by the professionals of a range of IT applications, and from the integration of IT with the business.

Essentially this integration has meant that Flower & Samios has re-engineered its business processes to obtain maximum benefits from the technology. For example, by developing and maintaining a 3D model of the design from the outset, relevant drawings and documentation can be 'peeled off' the model at different stages without further work. Designs are easily amended, with instantaneous updating of elements of the model. Integrated use of multimedia packages has meant ease of communication of designs to customers. Furthermore, the combination of land modelling, measurement and costing capabilities in their systems has meant that they are not only faster, more efficient and more accurate, but that they have enhanced capabilities to manage and control the contracting and project administration process.

A significant aspect of change has come as a result of the development of complementary technologies that have enhanced modelling accuracy and immediacy of communication with clients. While these developments have not changed the way the practice operates internally in any fundamental sense, they have significantly enhanced the benefits that the practice can offer its clients as a result of their mastery of 3D-based design technology and benefit from significantly improved decision-making from clients and other stakeholders.

Communications with suppliers and consultants is increasingly electronic, with emailed CAD drawings and documents now the norm. Client access to their project has been enhanced because of the ongoing growth of the web and web-based technologies. Now clients can check in from anywhere in the world to look at the latest developments on their project website, helping them to keep in touch with the latest developments in their project. Furthermore clients are now emailed video presentations of their design to help them respond to design developments, and the availability of the 3D simulation of the project on their own computer greatly increases their access and input into the design process.

The parallel development of digital photography has also increased the speed with which realistic 3D simulations can be produced and hence the interaction of all stakeholders with the design as it develops. Existing features of the site including surrounding buildings, trees and aspects such as the access of sunlight can be photographed and the digital model of the design can very quickly be embedded into a realistic simulation of the real setting. Once again this has the affect of enabling engagement with the design by clients and other stakeholders. These techniques are accelerating client and stakeholder decision-making and dispute resolution where there is controversy around the impact that a proposal will have on the existing surroundings. The overall affect of this is better design, better capture of stakeholder interests and inputs and faster resolution.

John Flower foresees that in the near future, those external consultants involved in a project will access the 3D model of the project on the firm's server, and will undertake their specialist component of the design work on-line, while having read-only access to the rest of the model. 'When all the parties are working concurrently on a project in this way', John comments, 'they will transform the industry.' However, before that becomes a reality there is a need to resolve some new organizational issues, issues such as who is the model integrator, how is IP protected and what protocols are needed to manage the integrated design process. The technology also creates the opportunity for an integrated design team to meet and, in a very short space of time, resolve most of the key design decisions in a design. This has the potential both to accelerate design and at the same time to improve design integration.

For organizations within the supply chain to capture benefits available as a result of interorganizational integration, they need to go through a shared cycle of learning and change. For example, through the shared use of 3D CAD models, the collaborating organizations need to develop protocols for co-operative work practices as well as go through a period of shared learning. Only then can information be passed seamlessly throughout the supply chain and the potential of efficient computer integrated manufacture be realized in the sector.

# What is BPR and what does it do?

There are almost as many definitions of BPR as there are of TQM! However, most of them boil down to the same substance – the fundamental rethink and radical redesign of a business process, its structure and associated management systems, to deliver major or step improvements in performance (which may be in process, customer, or business performance terms).

Of course, BPR and TQM programmes are complementary under the umbrella of process management. The continuous and step change improvements must live side by side – when does continuous change become a step change anyway? There has been over the years much debate, including some involving the author, about this issue. Whether it gets resolved is not usually the concern of the organization facing today's uncertainties with the realization that 'business as usual' will not do and some major changes in the ways things are done are required.

Put into a strategic context, BPR is a means of aligning work processes with customer requirements in a dynamic, flexible way, in order to achieve long-term corporate objectives. This requires the involvement of customers and suppliers and thinking about future requirements. Indeed the secrets to redesigning a process successfully lie in thinking about how to reshape it for the future.

BPR then challenges managers to rethink their traditional methods of doing work and to commit to customer-focused processes. Many outstanding organizations have achieved and/or maintained their leadership through process re-engineering, especially where they found processes which were not customer focused. Companies using these techniques have reported significant bottom-line results, including better customer relations, reductions in cycle time to market, increased productivity, fewer defects/errors and increased profitability. BPR uses recognized methods for improving business results and questions the effectiveness of the traditional organizational structure. Defining, measuring, analysing and re-engineering work processes to improve customer satisfaction can pay off in many different ways.

For example, Motorola had set stretch goals of ten-fold improvement in defects and two-fold improvement in cycle time within five years. The time period was subsequently revised to three years and the now famous six sigma goal of 3.4 defects per million became a slogan for the company and probably one of the real drivers (see also Chapter 12). These stretch goals represent a focus on discontinuous improvement and there are many examples of other companies that have made dramatic improvements following major organizational and process redesign as part of TQM initiatives, including approaches such as the 'clean sheet' design of a 'green field' plant around work cells and self-managed teams.

Process redesign

A truly remarkable project is described in the IDC case study at the end of the book (p. 409). The silicone wafer plant (FAB) builder IDC realized in 1998 that it had to prepare for the inevitable demands of the high pressure computer industry, where time-to-market with new products can mean the difference between growth and stagnation for the client. IDC created a set of supereffective teams, which embraced all members of the subcontractor supply chain and cut across organizational boundaries. They prepared themselves for the challenge by developing independent multi-functional workgroups, adopted the latest IT-based technologies to integrate their efforts and developed their own management tools so that, when Intel asked the impossible of them for their Albuquerque plant, IDC and its team were able to meet the challenge and reduce time and cost by more than 25 percent compared to its previous similar project.

Most organizations have vertical functions: experts of similar backgrounds grouped together in a pool of knowledge and skills capable of completing any task in that discipline. This focus, however, fosters a vertical view and limits the organization's ability to operate effectively. Barriers to customer satisfaction evolve, resulting in unnecessary work, restricted sharing of resources, limited synergy between functions, delayed development time and no clear understanding of how one department's activities affect the total process of attaining customer satisfaction. Managers remain tied to managing singular functions with rewards and incentives for their narrow missions, inhibiting a shared external customer perspective.

BPR breaks down these internal barriers and encourages the organization to work in cross-functional teams with a shared horizontal view of the business. As we have seen in earlier chapters this requires shifting the work focus from managing functions to managing processes. Process owners, accountable for the success of major cross-functional processes, are charged with ensuring that employees understand how their individual work processes affect customer satisfaction. The interdependence between one group's work and the next becomes quickly apparent when everyone understands who the customer is and the value they add to the entire process of satisfying that customer.

# Processes for redesign

IT provided the means to achieve the breakthrough in process performance in some organizations. The inspiration, however, came from understanding both the current and potential processes. This required a more holistic view than that taken in traditional quality programmes, involving wholesale redesigns of the processes concerned.

Ford estimated a 20 percent reduction in head count if it automated the existing processes in accounts payable. Taking an overall process perspective, Ford achieved a 75 percent reduction in one department. Xerox took an organizational view and concentrated on the cross-functional processes to be re-engineered, radically changing the relationship between supplier and external customer.

Clearly, the larger the scope of the process, the greater and farther reaching are the consequences of the redesign. At a macro level, turning raw materials into a produce used by a delighted customer is a process made up of subsets of smaller processes. The aim of the overall process is to add value to the raw materials. Taking a holistic view of the process makes it possible to identify non-value-adding elements and remove them. It enables people to question why things are done, and to determine what should be done.

---

The case study (p. 487) of the re-engineering of timber floor construction for dwellings on sloping sites in the Australian residential sector describes process re-engineering across the supply chain. Such changes are more difficult to achieve than process changes which are entirely within a single organization. However, some areas of construction innovation require such change. It may be best achieved through partnerships among the organizations within the supply chain or by the major players who have the intellectual and financial capacity to make the change and to capture the benefits from them.

---

Some of the re-engineering literature advised starting with a bank sheet of paper and redesigning the process anew. The problems inherent in this approach are:

- the danger of designing another inefficient system; and
- not appreciating the scope of the problem.

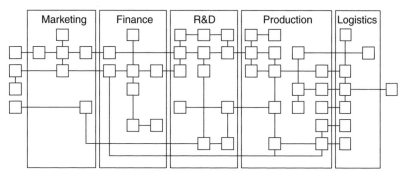

Each box is a process. Lines identify linked processes

**Figure 11.1**
Simplified process map

Therefore, the authors and their colleagues recommend a thorough understanding of current processes before embarking on a re-engineering project.

Current processes can be understood and documented by process mapping and flowcharting. As processes are documented, their interrelationships become clear and a map of the organization emerges. Figure 11.1 shows a much simplified process map. As the aim of BPR is to make discontinuous, major improvements, this invariably means organizational change, the extent of which depends on the scope of the process re-engineered.

Taking the organization depicted in Figure 11.1 as an example, if the decision is made to redesign the processes in finance, the effect may be that in Figure 11.2a: eight individual processes have become three. There has been no organizational effect on the processes in the other functions, but finance has been completely restructured. In Figure 11.2b, a chain of processes crossing all the functions has been re-engineered. The effect has been the loss of redundant processes and possibly many heads but much of the organization has been unaffected. Figure 11.2 shows the organization after a thorough re-engineering of all its processes. Some elements may remain the same, but the effect is organization-wide.

Whatever the scope of the redesign, head count is not the only change. When work processes are altered, the way people work alters. Figures 11.1 and 11.2 show an organization's functional departments with process running through them. These are the handful of core processes that make up what an organization does (see Figure 11.3) and in many organizations these would benefit from re-engineering to improve added value output and efficiency.

## Focus on results

BPR is not intended to preserve the status quo, but to fundamentally and radically change what is done; it is *dynamic*. Therefore, it is essential for a BPR effort to focus on required customers which will determine the scope of the BPR exercise. A simple requirement may be a 30 percent reduction in costs or a reduction in delivery time of two days. These would imply projects with relatively narrow scope, which are essentially inwardly focused and probably involve only one department; for example, the finance department in Figure 11.2a.

When Wal-Mart focused on satisfying customer needs as an outcome it started a redesign that not only totally changed the way it replenished inventory, but also made this

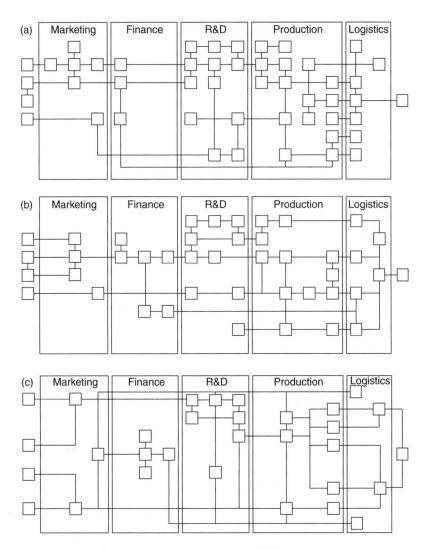

**Figure 11.2**
(a) Process redesign in finance, (b) cross-functional process design, (c) organizational process redesign

the centrepiece of its competitive strategy. The system put in place was radical, and required tremendous vision. In ten years, Wal-Mart grew from being a small niche retailer to the largest and most profitable retailer in the world.

Focusing on results rather than just activities can make the difference between success and failure in change projects. The measures used, however, are crucial. At every level of redesign and re-engineering, a focus on results gives direction and measurability; whether it be cost reduction, head count reduction, increase in efficiency, customer focus, identification of core processes and non-value-adding components, or strategic alignment of business processes. Benchmarking is a powerful tool for BPR and is the trigger for many BPR projects, as in Ford's accounts payable process. As shown in Chapter 9, the value of benchmarking does not lie in what can be copied, but in its ability to identify goals. If used well, benchmarking can shape strategy and identify potential competitive advantage.

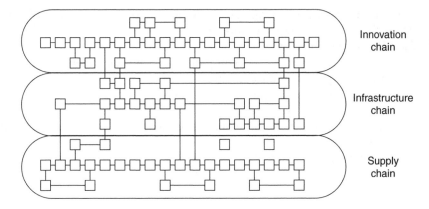

Figure 11.3
Process organization

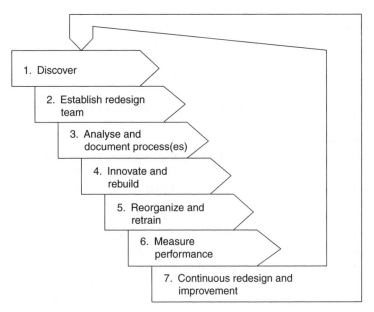

Figure 11.4
The seven phases of BPR

# The redesign process

Central to BPR is an objective overview of the processes to be redesigned. Whereas information needs to be obtained from the people directly involved in those processes, it is never initiated by them. Even at its lowest level, BPR has a top-down approach and most BPR efforts, therefore, take the form of a project. There are numerous methodologies proposed, but all share common elements. Typically, the project takes the form of seven phases, shown in Figure 11.4.

1. **Discover**

    This involved first identifying a problem or unacceptable outcome, followed by determining the desired outcome. This usually requires an assessment of the business need and will certainly include determining the processes involved,

217

including the scope, identifying process customers and their requirements, and establishing effectiveness measurements.

2. **Establish redesign team**

   Any organization, even a small company, is a complex system. There are customers, suppliers, employees, functions, processes, resources, partnerships, finances, etc., and many large organizations are incomprehensible – no one person can easily get a clear picture of all the separate components. Critical to the success of the redesign is the make-up of a redesign team. The team should comprise as a minimum the following:

   - senior manager as sponsor;
   - steering committee of senior managers to oversee overall re-engineering strategy;
   - process owner;
   - team leader;
   - redesign team members.

   It is generally recommended that the redesign team have between five and ten people; represent the scope of the process (that is, if the process to be re-engineered is a cross-functional, so is the team); only work on one redesign at a time; and is made up of both insiders and outsiders. Insiders are people currently working within the process concerned who help gain credibility with co-workers. Outsiders are people from outside the organization who bring objectivity and can ask the searching questions necessary for the creative aspects of the redesign. Many companies use consultants for this purpose.

3. **Analyse and document process(es)**

   Making visible the invisible, documenting the process(es) through mapping and/or flowcharting is the first crucial step that helps an organization see the way work really is done and not the way one thinks or believes it is done. Seeing the process as it is provides a baseline from which to measure, analyse, test and improve.

   Collecting supporting process data, including benchmarking information and IT possibilities, allows people to weigh the value each task adds to the total process, to rank and select areas for the greatest improvement, and to spot unnecessary work and points of unclear responsibility. Clarifying the root causes of problems, particularly those that cross department lines, safeguards against quick-fix remedies and assures proper corrective action, including the establishment of the right control systems.

4. **Innovate and rebuild**

   In this phase the team rethink and redesign the new process, using the same process mapping technique, in an iterative approach involving all the stakeholders, including senior management. A powerful method for challenging existing practices and generating breakthrough ideas is 'assumption busting' – see later section.

5. **Reorganize and retrain**

   This phase includes piloting the changes and validating their effectiveness. The new process structure and operation/system will probably lead to some reorganization, which may be necessary for reinforcement of the process strategy and to achieve the new levels of performance.

   Training and/or retraining for the new technology and roles play a vital part in successful implementation. People need to be equipped to assess, re-engineer, and support – with the appropriate technology – the key processes that contribute to customer satisfaction and corporate objectives. Therefore, BPR efforts can involve substantial investment in training but they also require considerable top management support and commitment.

6. **Measure performance**

It is necessary to develop appropriate metrics for measuring the performance of the new process(es), subprocesses, activities, and tasks. These must be meaningful in terms of the inputs and outputs of the processes, and in terms of the customers of and suppliers to the process(es) (see Chapter 7).

7. **Continuous redesign and improvement**

The project approach to BPR suggests a one-off approach. When the project is over, the team is disbanded, and business returns to normal, albeit a radically different normal. It is generally recommended that an organization does not attempt to re-engineer more than one major process at a time, because of the disruption and stress caused. Therefore, in major re-engineering efforts of more than one process, as one team is disbanded, another is formed to redesign yet another process. Considering that Ford took five years to redesign its accounts payable process, BPR on a large scale is clearly a long-term commitment.

In a rapidly changing, ever more competitive business environment, it is becoming more likely that companies will re-engineer one process after another. Once a process has been redesigned, continuous improvement of the new process by the team of people working in the process should become the norm.

# Assumption busting

Within BPR is a powerful method for challenging existing practices and generating break-through ideas for improvement. 'Assumption busting', as it was named by Hammer and Champy, aims to identify the rules that govern the way we do business and then uncover the real underlying *assumptions* behind the adoption of these rules. Business processes are governed by a number of rules that determine the way the process is designed, how it inter-faces with other activities within the organization, and how it is operated. These rules can exist in the form of explicit policies and guidelines or, what is more often the case, in the mind of the people who operate the process. These unwritten rules are the product of assumptions about the process environment that have been developed over a number of years and often emerge from uncertainties surrounding trading relationships, capabilities, resources, authorities, etc. Once these underlying assumptions are uncovered they can be challenged for relevance and, in many cases, can be found to be false. This opens up new opportunities for process redesign and, as a consequence, the creation of new value and improved performance.

For example, BBC Resources Ltd, a supplier of TV and radio studio and outside broad-cast resource services, faced the requirement to improve business performance. The busi-ness was losing money and faced competition from independent providers. They needed to improve the efficiency of their processes while retaining their core capability that created competitive advantage.

A team was commissioned to review the core value adding processes, setting challeng-ing targets for improvements in performance in order to stimulate breakthrough thinking. The team decided to take a more radical approach by using assumption busting and pre-pared a six week programme of work. Within that timeframe they used an established eight-step method (shown in Table 11.1) to redesign the core end-to-end service delivery processes for two major business units – studios and outside broadcasts. The work involved identifying the key areas of cost consumption, challenging the rules and assump-tions that governed the existing process and generating a set of improvement opportuni-ties. When they had evaluated their findings, the team presented ideas to deliver an improvement in excess of 15 percent in process efficiency.

| Table 11.1 | Assumption busting – a proven technique |
| --- | --- |
| Step 1 – Identify the core value that must be delivered to the customer, the business and the key stakeholders. |
| Step 2 – Map the process or processes to be improved at a high level only, identifying key problems. |
| Step 3 – Select a particular problem to resolve (e.g. process cost efficiency, process quality, process speed) and collect supporting performance data. |
| Step 4 – Brainstorm the rules that have an effect on the problem being resolved. Test the rule statements for validity and prioritize them for further analysis. |
| Step 5 – Undertake a rigorous review of each rule, uncovering the underlying assumption behind each. |
| Step 6 – Identify the modified assumptions and in turn a modified set of process rules. |
| Step 7 – Identify the impact of these rules on the process and construct a new set of process design principles. |
| Step 8 – Develop a revised process design and test for validity. |

One of the process rules concerned the use of a highly technically qualified member of staff for the planning and delivery of all of the programmes supported. The core underlying assumption was that all of the programmes were complex in nature. When this assumption was challenged the team in BBC Resources realized that, as not all programmes were so complex in nature, less technically qualified members of staff could be utilized at lower cost to the business.

Similarly the mail order company Grattan was presented with a different challenge. With the catalogue shopping market facing intense competition from other retailing methods, the Grattan team realized the need to look at their value chain processes to find ways of improving the service they provide to their agent community. One target for service improvement was the reduction in the value of query calls coming into the service centres.

Once the relevant processes had been defined, a cross-functional team decided to employ the assumption busting method to identify areas for improvement. One rule that quickly came to light concerned the issue of order acknowledgements, which, in the form of account statements, are posted out to agents on the same day that the order is processed. However, the current rule stated that statements should be sent via third class mail. This rule was created out of an assumption that this was the most cost-efficient way of communicating order details to the agent. Third class mail can take up to ten days, which is beyond the current expected delivery time. So, when consignments did not arrive as expected, the agent called to progress the order. As a consequence of changing the process, statements are now being sent out within the expected delivery time leading to a reduction in query calls and improved service.

## Application of the technique

In practice the authors' colleagues have found this technique to be of greatest value when applied by a cross-functional group of process operators and supervisors who are given a specific problem to fix. In using the technique, care must be exercised in the use of terms such as rule and assumption. They often cause initial confusion and there can be real difficulty in uncovering the core underlying assumptions. Rules should be clearly stated and tested for validity before proceeding down what eventually could become a bind alley. Furthermore, a rigorous approach to the identification of the core assumption is vital to uncovering the real opportunities for improvement. An assumption by definition 'is a statement/belief that is accepted or supposed to be true without proof or demonstration'. In some cases, rules are created from specific knowledge about the business and its environs and not based on assumptions. Assumptions spring from our beliefs about the environment and not our specific knowledge.

## Other applications of assumption busting

Assumption busting is of particular benefit when applied by partners within a supply chain. The trading relationships and practices that exist in a modern supply chain, such as a supermarket and its multiple tiers of suppliers, are the product of a number of assumptions made by the supply chain partners about what is possible. Once teams from each of the partnering businesses work collaboratively to uncover the rules and assumptions that govern their trading relationships the door is unlocked to new methods and economies.

The method can also be immensely powerful when companies are introducing new technology. Breakthrough technologies can lead to breakthrough performance as new technologies make possible what is considered impossible today. Hence, a number of current rules and assumptions are there to be challenged as processes are redesigned to take advantage of the new technology. We are often just as constrained by our lack of imagination regarding the possibilities of tomorrow as we are by our knowledge of what is possible today. One example of this was in BBC World Service where the introduction of digital technology to replace analogue was accompanied by assumption busting-led process redesign to take advantage of the new technological capability.

While assumption busting has been primarily applied to the generation of new process designs, it exists in its own right as a method for developing more 'lateral' solutions to problems. In the early 1970s, Dr Edward do Bono introduced the concept of lateral thinking as an alternative method of generating ideas to that of the more traditional logical or 'vertical' thinking. Dr de Bono argued that our thinking is constrained by patterns that form in our minds over time and channel our future thoughts. Assumption busting helps people break out of this 'channelled thinking' to develop creative ideas. Managers could benefit from applying assumption busting to a number of problems or opportunities in their businesses – assumptions constrain us everywhere, not just within our business processes.

Whether it is in response to specific customer requirements, new technology, or in the quest for competitive advantage, assumption busting provides a simple but effective method for breaking into new areas of adding value. World-class performance will not be achieved by effort alone; creativity and innovation are cornerstones of future success. Innovative ways of delivering new value will be rewarded. Assumption busting provides a powerful method for generating new ideas from looking at today and tomorrow in a different way.

# BPR – the people and the leaders

For an organization to focus on its core processes almost certainly requires an understanding of its core competencies. Moreover, core process redesign can channel an organization's competencies into an outcome that gives it strategic competitive advantage and the key element is visioning that outcome. Visioning the outcome may not be enough, however, since many companies' 'vision' desires results without simultaneously 'visioning' the systems that are required to generate them. Without a clear vision of the systems, processes, methods, and approaches that will allow achievement of the desired results, dramatic improvement is frequently not obtained as the organization fails to align around a common tactical strategy. Such an 'operational' vision is lacking in many organizations.

The fallout from BPR has profound impacts on the employees in any enterprise at every level – from executives to operators. In order for BPR to be successful, therefore, significant changes in organization design and enterprise culture are also often required. Unless the leaders of the enterprise are committed to undertake these changes, the BPR initiative will flounder. The point is, of course, that organization design and culture change are much more difficult than modifying processes to take advantage of new IT.

While the enabling IT is often necessary and is clearly going to play a role in many BPR exercises, it is by no means sufficient, nor is it the most difficult hurdle on the path to success. Thanks to IT we can radically change the processes an organization operates and, hopefully, achieve dramatic improvements in performance. However, in any BPR project there will be considerable risk attached to building the information system that will support the new, redesigned processes. Information systems should be but rarely are described so that they are easy for people to understand.

While BPR may be a distinct, short-term activity for a specific business function, the record indicates that BPR activities are most successful when they occur within the framework of a long-term thrust for excellence. Within a TQM culture a BPR effort is more likely to find the process focus supportive workforce, organization design, and mindset changes needed for its success.

Process improvement is sometimes positioned as a bottom-up activity. In some contrast, TQM involves setting longer-term goals at the top and modifying the business as necessary to achieve the goals. Often, the modifications to the business required to achieve the goals are extensive and groundbreaking. The history of successful TQM thrusts in award-winning companies in Europe and the United States is replete with new organization designs, with flattened structures, and with empowered employees in the service of end customers. In many successful organizations, BPR has been an integral part of the culture – a process-driven change dedicated to the ideals and concepts of TQM. That change must create something that did not exist before, namely a 'learning organization' capable of adapting to a changing competitive environment. When processes, or even the whole business, needs to be re-engineered, the radical change may not, probably will not, be readily accepted.

# Acknowledgement

The authors are grateful to the contribution made by their colleagues Ken Gadd and Mike Turner in the preparation of this chapter.

# References

1. Womack, J.D., Jones, D.T. and Roos, D. (1991) *The Machine that Changed the World*. Harper-Collins.
2. Koskela, L. and Howell, G. (2002) 'The Theory of Project Management: explanation to novel methods', Proceedings IGLC-10, August, Gramado, Brazil, http://www.cpgec.ufrgs.br/norie/iglc10/
3. Arbulu, R.J. and Tommelein, I.D. (2002) 'Alternative Supply-Chain Configurations for Engineered or Catalogued Made-to-Order Components: case study on pipe supports used in power plants', Proceedings IGLC-10, August, Gramado, Brazil, http://www.cpgec.ufrgs.br/norie/iglc10/

4. Elfving, J., Tommelein, I.D. and Ballard, G. (2003) 'An International Comparison of the Delivery Process of Power Distribution Equipment', Proceedings IGLC-11, August, Blacksburg, USA, http://strobos.cee.vt.edu/IGLC11/

5. Tommelein, I.D., Akel, N.G. and Boyers, J.C. (2004) 'Application of Lean Supply Chain Concepts to a Vertically-Intergrated Company: A Case Study', Proceedings IGLC-12, August, Ellsinore, Denmark, http://www.iglc2004.dk/13729

6. Marosszeky, M., Thomas, R., Karim, K., Davis, S. and McGeorge, D. (2004) 'Quality Management Tools for Lean Production: moving from enforcement to empowerment', Proceedings IGLC-10, August, Gramado, Brazil (p. 87), http://www.cpgec.ufrgs.br/norie/iglc10/frame_proceedings.htm

7. Marosszeky, M., Karim, K., Davis, S. and Naik, N. (2004) 'Lessons Learnt in Developing Effective Performance Measures for Construction Safety Management, A Case Study', Proceedings IGLC-12, August, Ellsinore, Denmark, http://www.iglc2004.dk/13728

8. Saurin, T., Formoso, C., Guimarães, L.M. and Soares, A. (2002) 'Safety and Production: an integrated planning and control model', Proceedings IGLC-10, August, Gramado, Brazil (p. 61), http://www.cpgec.ufrgs.br/norie/iglc10/frame_proceedings.htm

9. Marosszeky, M., Sauer, C., Johnson, K., Karim, K. and Yetton, P. (2000) 'Information Technology in the Building and Construction Industry: The Australian Experience', *INCITE 2000 – Implementing IT to Obtain a Competitive Advantage in the 21st Century*, The Hong Kong Polytechnique University, Hong Kong.

## *Chapter highlights*

### Lean construction

- Lean focuses on the client, the elimination of waste and the maximization of value.
- The primary forms of waste are overproduction, waiting, excess conveyance, extra processing, excessive inventory, unnecessary motion, and defects requiring rework or scrap. To these, Koskela added for the construction sector reducing the proportion of non-value-adding activities, reduction in lead time, reduction in variability, simplification of processes, increasing flexibility, and increasing transparency.

### Re-engineering the organization?

- When a major business process requires radical reassessment, perhaps through the introduction of new technology, discontinuous methods of business process re-engineering or redesign (BPR) are appropriate.
- The opportunity for radical change in construction processes may involve collaboration across the supply chain and this might be best achieved through partnerships among organizations.
- Drives for process change include information technology (IT), political, financial, cultural and competitive aspects. These often require a change of thinking about the ways processes are and could be operated.
- IT often creates opportunities for breakthrough performance but BPR is needed to deliver it. Successful practitioners of BPR have made striking improvements in customer satisfaction and productivity in short periods of time.
- Interorganizational integration of IT is one of the greatest opportunities and challenges facing the sector – it has the potential to unlock significant value.

### What is BPR and what does it do?

- There are many definitions of BPR but the basic elements involve a fundamental rethink and radical redesign of a business process, its structure and associated management systems to deliver step improvements in performance.

Process redesign

- BPR and TQM are complementary under the umbrella of process management – the continuous and discontinuous improvements living side by side. Both require the involvement of customers and suppliers and their future requirements.
- BPR challenges managers to rethink their traditional methods of doing work and to commit to customer-focused processes. This breaks down organizational barriers and encourages cross-functional teams.

### Processes for redesign/focus on results
- Much larger savings and head count reductions are possible through properly applied BPR than simply automating existing processes. The larger the scope of the process the greater and farther reaching the consequences of the redesign.
- A thorough understanding of the current process is needed before embarking on a re-engineering project. Documentation of processes through mapping and flowcharting allows interrelationships to be clarified.
- Focusing on results rather than activities can make the difference between success and failure in BPR and other change projects, but the measures used are critical. Benchmarking is a powerful tool for BPR and often the trigger for many projects.

### The redesign process/assumption busting
- BPR has a top-down approach and needs an objective overview of the process to be redesigned to drive the project.
- Typically a BPR project will have seven phases: discover – identifying the problem or unacceptable outcome; establish redesign team; analyse and document process(es); innovate and rebuild; reorganize and retrain; measure performance; continuous redesign and improvement.
- Assumption busting is a useful eight-step BPR method which aims to identify and challenge the 'rules' and assumptions that govern and underlie the way business is done. A team is formed to: identify the core value to be delivered to customers and stakeholders; map the process at high level; select problems to resolve and collect performance data; brainstorm and test the rules; rigorously review each rule to uncover underlying assumptions; identify modified assumptions and process rules; identify impact and construct a new set of process principles; develop revised process and test validity.

### BPR – the people and the leaders
- For an organization to focus on its core processes requires an understanding of its core competencies, and the channelling of these into outcomes that deliver strategic competitive advantage.
- BPR has profound impacts on employees from the top to the bottom of an organization. In order to be successful significant changes in organization design and enterprise culture are also often required. This requires commitment from the leaders to undertake these changes.
- TQM ideals and concepts provide a perfect platform for BPR projects and the creation of a 'learning organization' capable of adapting to a radically changing environment.

# 12

# Quality management systems

# Why a quality management system?

In the construction sector, worldwide, the value of formal quality management systems is still widely questioned and it is small wonder. In the UK, HK and Australia government policy has been to simply mandate ISO 9000 for all government suppliers. Some company managements saw the need to obtain certification as simply the need to get onto government tender lists and, hence, primarily as a marketing problem. In cases where senior management gave the issue scant attention, marketing and/or other managers hired a consultant to develop a compliant system.

In the early 1990s, quality consultants generally had backgrounds either in power generation or manufacturing and they understood little or nothing about construction. However, they knew how to put together a top-down ISO 9000 compliant management system and that satisfied the needs of construction marketing managers. Hence, there was little or no workplace involvement in early management system development in the construction sector. Usually, system manuals were very thick and they did not reflect the business goals or management needs of the enterprise. Because of this difficult start, to this day there are few construction organizations worldwide that have grasped the strategic significance of the total quality philosophy for their businesses. The case studies in this book bring together the stories of companies that have understood and have succeeded in meeting this challenge.

In earlier chapters we have seen how the keystone of quality management is the concept of customer and supplier working together for their mutual advantage. For any particular organization this becomes 'total' quality management if the supplier/customer interfaces extend beyond the immediate customers, back inside the organization, and beyond the immediate suppliers. In order to achieve this, a company must organize itself in such a way that the human, administrative and technical factors affecting quality will be under control. This leads to the requirement for the development and implementation of a quality management system that enables the objectives set out in the quality policy to be accomplished. Clearly, for maximum effectiveness and to meet individual customer requirements, the management system in use must be appropriate to the type of activity and product or service being offered.

It may be useful to reflect on why such a device is necessary to achieve control of processes. The author still remembers being at a table in a restaurant with eight people who all ordered the 'chef's special individual soufflé'. All eight soufflés arrived together at the table, magnificent in their appearance and consistency, each one exhibiting an almost identical size and shape – a truly remarkable demonstration of culinary skill. How had this been achieved? The chef had *managed* such consistency by making sure that, for each soufflé, he used the same ingredients (materials), the same equipment (plant), the same method (procedure) in exactly the same way every time. The process was under control. This is the aim of a good quality management system, to provide the 'operator' of the process with consistency and satisfaction in terms of methods, materials, equipment, etc. (Figure 12.1). Two feedback loops are also required: the 'voice' of the customer (marketing activities) and the 'voice' of the process (measurement activities).

The chef's soufflés – they were not British Standard, NIST Standard, Australian Standard, or ISO Standard soufflés – were the 'chef's special soufflés'. It is not conceivable that the chef sat down with a blank piece of paper to invent a soufflé recipe. Why reinvent wheels? He probably used a standard formula and changed it slightly to make it his own. This is exactly the way in which successful organizations use the international standards on quality management systems that are available. The 'wheel' has been invented but it must be built in a way that meets the specific organizational and product or service requirements. The international family of standards ISO 9000:2000 'Quality

Management Systems' specifies systems which can be implemented in an organization to ensure that all the product/service performance requirements and needs of the customer are fully met.

Let us return to the chef in the restaurant and propose that his success leads to a desire to open eight restaurants in which are served his special soufflés. Clearly he cannot rush from each one of these establishments to another every evening making soufflés. The only course open to him to ensure consistency of output, in all eight restaurants, is for him to write down in some detail the system he uses, and then make sure that it is used on all sites, every time a soufflé is produced. Moreover, he must periodically visit the different sites to ensure that:

1. The people involved are operating according to the designed system (a system audit).
2. The soufflé system still meets the requirements (a system review).

If in his system audits and reviews he discovers that an even better product or less waste can be achieved by changing the method or one of the materials, then he may wish to effect a change. To maintain consistency, he must ensure that the appropriate changes are made to the management system, *and* that everyone concerned is issued with the revision and begins to operate accordingly.

A good quality management system will ensure that two important requirements are met:

- *The customer's requirements* – for confidence in the ability of the organization to deliver the desired product or service consistently.
- *The organization's requirements* – both internally and externally including regulatory, and at an optimum cost, with efficient utilization of the resources available – material, human, technological, and information.

The requirements can be truly met only if objective evidence is provided, in the form of information and data, which supports the system activities, from the ultimate supplier through to the ultimate customer.

A *quality management system* may be defined, then, as an assembly of components, such as the management, responsibilities, processes and resources for implementing total quality management. These components interact and are affected by being in the system, so the isolation and study of each one in detail will not necessarily lead to an understanding of the system as a whole. Often the interactions between the

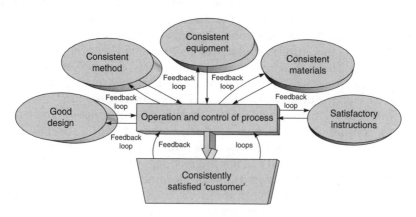

**Figure 12.1**
The systematic approach to process management

components – such as materials and processes, people and responsibilities – are just as important as the components themselves, and problems can arise from these interactions as much as from the components. Clearly, if one of the components is removed from the system, the whole thing will change.

The adoption of a quality management system is, of course, a strategic decision and its design should be influenced by the organization's objectives, structure and size, the products or services offered, and its processes.

# Quality management system design

The quality management system should apply to and interact with all processes in the organization. It begins with the identification of the customer requirements and ends with their satisfaction, at every transaction interface. The activities may be classified in several ways – generally as processing, communicating and controlling, but more usefully and specifically as shown in the quality management process model described in ISO 9001:2000, Figure 12.2. This reflects graphically the integration of four major areas:

- Management responsibility.
- Resource management.
- Product realization.
- Measurement, analysis and improvement.

The management system requirements under these headings are specified in the international standard. Table 12.1 lists the ISO 9000:2000 family which together forms a

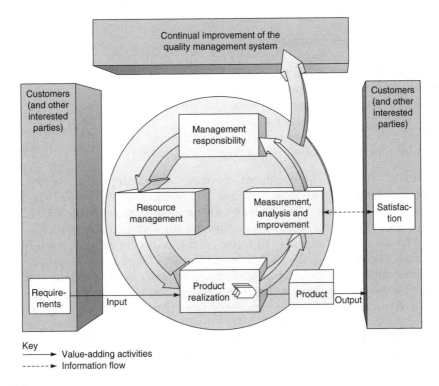

Key
——▶ Value-adding activities
------▶ Information flow

**Figure 12.2**
Model of a process-based quality management system

**Table 12.1** The ISO 9000:2000 family of standards on quality management systems

| BS EN ISO | Name | Purpose |
|---|---|---|
| 9000:2000 | Quality management systems – Fundamentals and vocabulary | Describes the fundamentals and specifies the terminology for QMS |
| 9001:2000 | Quality management systems – Requirements | Specifies the requirement for a QMS where an organization needs to demonstrate its ability to provide products that fulfil customer and applicable regulatory requirements and aims to enhance customer satisfaction |
| 9004:2000 | Quality management systems – Guidelines for performance improvements | Provides guidelines that consider both the effectiveness and efficiency of the QMS, with the aim of improving the performance of the organization and satisfaction of customers and other interested parties |

Note: ISO 19011 provides guidance on auditing quality and environmental management systems – see Chapter 8.

coherent set of quality management system standards to hopefully facilitate mutual understanding across national and international trade.

ISO 9000:2000 – the fundamentals and vocabulary – sets down the principles behind quality management which formed the basis for the quality management system standards in the ISO 9000 family.

Eight principles were identified to be used by top management as they lead their organizations and improve performance.

1. Customer focus – see Chapter 1.
2. Leadership – see Chapter 3.
3. Involvement of people – see Chapters 14–15.
4. Process approach – see Chapter 10.
5. System approach to management – see Chapters 4 and 10.
6. Continual improvement – see Chapter 13.
7. Factual approach to decision-making – see Chapters 7 and 13.
8. Mutually beneficial supplier relationships – see Chapter 5.

ISO 9001:2000 'Quality Management Systems – Requirements' is the current International Standard and it supersedes ISO 9001:1994, ISO 9002:1994 and ISO 9003:1994. ISO 9002:1994 and ISO 9003:1994 are now obsolete. A new ISO 9004:2000 provides a set of guidelines for continuous improvement. Detailed information on the ISO 9000:2000 family of standards may be found on the following website: www.iso.org.

A fundamental change in the structure and focus of ISO 9000:2000 compared to the previous version of the standard has made it much more effective and this difference is important for the construction sector. The 1994 version of the standard was management system focused so that compliance could be obtained by a company on the basis of its management system, regardless of the quality of the actual product and services they provided. The new version of the standard links product and service quality to system quality through a focus on customer satisfaction and builds into the process the demonstration of continuous improvement. The new standard is a far more solid basis for a quality system; and compliance can no longer be achieved by the rather cynical purchase of a *quality system*.

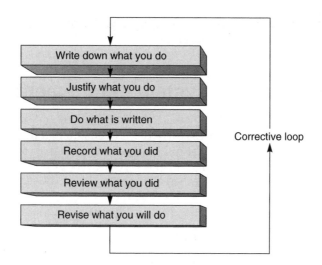

**Figure 12.3**
The quality system and never-ending improvement

It is interesting to bring together the concept of Deming's cycle of continuous improvement – Plan, Do, Check, Act – and quality management systems. A simplification of what a good management system is trying to do is given in Figure 12.3, which follows the improvement cycle.

In many organizations established methods of working already exist around identified processes, and all that is required is the *documenting of what is currently done*. In some instances companies may not have procedures to satisfy the requirements of a good standard, and they may have to begin to devise them. Alternatively, it may be found that two people, supposedly performing the same task, are working in different ways, and there is a need to standardize procedures. Some organizations use the effective slogan 'If it isn't written down, it doesn't exist'. This can be a useful discipline, provided it doesn't lead to bureaucracy.

*Justify* that the *system* as it is designed *meets the requirements of a good international standard*, such as ISO 9001. There are other excellent standards that are used, and these provide similar checklists of things to consider in the establishment of the quality system.

One person alone cannot document a quality management system; the task is the job of all personnel who have responsibility for any part of it. This means that a quality system, by definition, has to be built from the operational level up and cannot be imposed by external consultants. The quality system must be a *practical working one* – that way it ensures that consistency of operation is maintained and it may be used as a training aid.

In the operation of any process, a useful guide is:

- No process without data collection (measurement).
- No data collection without analysis.
- No analysis without decisions.
- No decisions without actions (improvement) – which can include doing nothing.

This excellent discipline is built into any good quality management system, primarily through the audit and review mechanism. The requirement to *audit or 'check'* that the system is functioning according to plan, and to *review* possible system improvements, utilizing audit results, should ensure that the *improvement* cycle is engaged through the *corrective action* procedures. The overriding requirement is that the systems must reflect the established practices of the organization, improved where necessary to bring them into line with current and future requirements.

# Quality management system requirements

The quality management system that needs to be documented and implemented will be determined by the nature of the process carried out to ensure that the product or service conforms to customer requirements. Certain fundamental principles are applicable, however, throughout industry, commerce, and the services. These fall into generally well-defined categories which are detailed in ISO 9001:2000.

## 1. Management responsibility

### Customer needs/requirements (see Chapter 1)

The organization must focus on customer needs and specify them as defined requirements for the organization. The aim of this is to achieve customer confidence in the products and/or services provided. It is also necessary to ensure that the defined requirements are understood and fully met.

### Quality policy (see Chapter 3)

The organization should define and publish its quality policy, which forms one element of the corporate policy. Full commitment is required from the most senior management to ensure that the policy is communicated, understood, implemented and maintained at all levels in the organization. For every project, the quality plans must fully reflect the company quality policy and the project leadership must be responsible for implementing company quality policy and goals within the supply chain at the project level. The company quality policy should be authorized by top management and signed by the chief executive, or equivalent, who must also ensure that it:

(a) is suitable for the needs/requirements of the customers and the purpose of the organization;
(b) includes commitment to meeting requirements and continual improvement for all levels of the organization;
(c) provides a framework for establishing and reviewing quality objectives;
(d) is regularly reviewed for its suitability and objectiveness.

### Quality objectives and planning

Organizations should establish written quality objectives and define the responsibilities of each function and level in the organization.

One manager reporting to top management, with the necessary authority, resources, support, and ability, should be given the responsibility to co-ordinate, implement, and maintain the quality management system, resolve any problems and ensure prompt and effective corrective action. This includes responsibility for ensuring proper handling of the system and reporting on needs for improvement. Those who control sales, service operations, warehousing, delivery, and reworking of non-conforming product or service processes should also be identified.

At the project level, responsibility for conformance with the quality policy and its implementation lies fully with line management and should not be separated out as a special responsibility. Doing so would only lead to a conflict within the project organization.

## Management review

Management reviews of the system must be carried out, by top management at defined intervals, with records to indicate the actions decided upon. The effectiveness of these actions should be considered during subsequent reviews. Reviews typically include data on the internal quality audits, customer feedback, product conformance analysis, process performance, and the status of preventive, corrective and improvement actions.

## Quality manual

The organization should prepare a 'quality manual' that is appropriate. It should include but not necessarily be limited to:

(a) the quality policy;
(b) definition of the quality management system – scope, exclusions, etc.;
(c) description of the interaction between the processes of the quality management system;
(d) documented procedures required by the QMS, or reference to them.

In the quality manual for a large organization it may be convenient to indicate simply the existence and contents of other manuals, those containing the details of procedures and practices in operation in specific areas of the system.

Before an organization can agree to supply to a specification, it must ensure that:

(a) the processes and equipment (including any that are subcontracted) are capable of meeting the requirements;
(b) the operators have the necessary skills and training;
(c) the operating procedures are documented and not simply passed on verbally;
(d) the plant and equipment instrumentation is capable (e.g. measuring the process variables with the appropriate accuracy and precision);
(e) the quality-control procedures and any inspection, check, or test methods available provide results to the required accuracy and precision, and are documented;
(f) any subjective phrases in the specification, such as 'finely ground', 'low moisture content', 'in good time', are understood, and procedures to establish the exact customer requirement exist.

## Control of documents

The organization needs to establish procedures for controlling the new and revised documents required for the operation of the quality management system. Documents of external origin must also be controlled. These procedures should be designed to ensure that:

(a) documents are approved;
(b) documents are periodically reviewed, and revised as necessary;

(c) the current versions of relevant documents are available at all locations where activities essential to the effective functioning of the processes are performed;

(d) obsolete documents are promptly removed from all points of issue and use, or otherwise controlled to prevent unplanned use;

(e) any obsolete documents retained for legal or knowledge-preservation purposes are suitably identified.

Documentation needs to be legible, revision controlled, readily identifiable and maintained in an orderly manner. Of course, the documentation may be in any form or any type of media.

### Control of quality records

Quality records are needed to demonstrate conformance to requirements and effective operation of the quality management system. Quality records from suppliers also need to be controlled. This aspect includes record identification, collection, indexing, access, filing, storage and disposition. In addition, the retention times of quality records need to be established.

## 2. Resource management

The organization should determine and provide the necessary resources to establish and improve the quality management system, including all processes and projects. The general infrastructure needed to achieve conformity to the product or service requirements needs to be provided and maintained, including buildings, equipment, any supporting services, and the work environment.

### Human resources

The organization needs to select and assign people who are competent, on the basis of applicable education, training, skills and experience, to those activities which impact the conformity of product and/or service. On construction sites, where most of the work is actually undertaken by subcontractors, this includes workers and supervisors across the supply chain.

The organization also needs to:

(a) determine the training needed to achieve conformity of product and/or service;

(b) provide the necessary training to address these needs;

(c) evaluate the effectiveness of the training on a continual basis.

Individuals clearly need to be educated and trained to qualify them for the activities they perform. Competence, including qualification levels achieved, needs to be demonstrated and documented. It may be beneficial to conduct joint training of supervisors and managers on a project to ensure that they are working within a consistent framework of values and expectations.

Information is ever increasingly a vital resource and any organization needs to define and maintain the current information and the infrastructure necessary to achieve conformity of products and/or services. The management of information, including access and protection of information to ensure integrity and availability, needs also to be considered. Once again, the seamless use of real-time information across the supply chain is essential to the efficient operation of the delivery process as a whole.

# 3. Product realization

As we have seen in Figure 12.2, any organization needs to determine the processes required to convert customer requirements into customer satisfaction, by providing the required product and/or service. In determining such processes the organization needs to consider the outputs from the quality planning process.

The sequence and interaction of these processes need to be determined, planned and controlled to ensure they operate effectively, and there is a need to assign responsibilities for the operation and monitoring of the product/service generating processes.

These processes clearly need to be operated under controlled conditions and produce outputs which are consistent with the organization's quality policy and objective and it is necessary to:

(a) determine how each process influences the ability to meet product and/or service requirements;
(b) establish methods and practices relevant to process activities, to the extent necessary to achieve consistent operation of the process;
(c) verify processes can be operated to achieve product and/or service conformity;
(d) determine and implement the criteria and methods to control processes related to the achievement of product and/or service conformity;
(e) determine and implement arrangements for measurement, monitoring and follow-up actions, to ensure processes operate effectively and the resultant product/service meets the requirements;
(f) ensure availability of process documentation and records which provide operating criteria and information, to support the effective operation and monitoring of the processes. (This documentation needs to be in a format to suit the operating practices, including written quality plans);
(g) provide the necessary resources for the effective operation of the processes.

## Customer-related processes

One of the first processes to be established is the one for identifying customer requirements: both internal and external customers and immediate and end user customers. This needs to consider the:

(a) extent to which customers have specified the product/service requirements;
(b) requirements not specified by the customer but necessary for fitness for purpose;
(c) obligations related to the product/service, including regulatory and legal requirements;
(d) other customer requirements, e.g. for availability, delivery and support of product and/or service.

The identified customer requirements need also to be reviewed before a commitment to supply a product/service is given to the customer (e.g. submission of a tender, acceptance of a contract or order). This should determine that:

(a) identified customer requirements are clearly defined for the product and/or service;
(b) the order requirements are confirmed before acceptance, particularly where the customer provides no written statement of requirements;

(c) the contract or order requirements differing from those in any tender or quotation are resolved.

This should also apply to amended customer contracts or orders. Moreover, each commitment to supply a product/service, including amendment to a contract or order, needs to be reviewed to ensure the organization will have the ability to meet the requirements.

Any successful organization needs to implement effective communication and liaison with customers, particularly regarding:

(a) product and/or service information;
(b) enquiry and order handling, including amendments;
(c) customer complaints and other reports relating to non-conformities;
(d) recall processes, where applicable;
(e) customer responses relating to conformity of product/service.

Where an organization is supervising or using customer property, care needs to be exercised to ensure verification, storage and maintenance. Any customer product or property that is lost, damaged or otherwise found to be unsuitable for use should, of course, be recorded and reported to the customer. Customer property may, of course, include intellectual property, e.g. information provided in confidence.

## Design and development

The organization needs to plan and control design and development of products and/or services, including:

(a) stages of the design and development process;
(b) required review, verification and validation activities;
(c) responsibilities for design and development activities.

Interfaces between different groups involved in design and development need to be managed to ensure effective communication and clarity of responsibilities, and any plans and associated documentation should be:

(a) made available to personnel that need them to perform their work;
(b) reviewed and updated as design and development evolves.

The requirements to be met by the product/service need to be defined and recorded, including identified customer or market requirements, applicable regulatory and legal requirements, requirements derived from previous similar designs, and any other requirements essential for design and development. Incomplete, ambiguous or conflicting requirements must be resolved.

The outputs of the design and development process need to be recorded in a format that allows verification against the input requirements. So, the design and development output should:

(a) meet the design and development input requirements;
(b) contain or make reference to design and development acceptance criteria;
(c) determine characteristics of the design essential to safe and proper use, and application of the product or service;
(d) output documents should also be reviewed and approved before release.

Validation needs to be performed to confirm that the resultant product/service is capable of meeting the needs of the customers or users under the planned conditions. Wherever possible, validation should be defined, planned and completed prior to the delivery or implementation of the product or service. Partial validation of the design or development output may be necessary at various stages to provide confidence in their correctness, using such methods as:

(a) reviews involving other interested parties;
(b) modelling and simulation studies;
(c) pilot production, construction or delivery trials of key aspects of the product and/or service.

Design and development changes or modifications need to be determined as early as possible, recorded, reviewed and approved, before implementation. At this stage, the effect of changes on compatibility requirements and the usability of the product or service throughout its planned life need to be considered.

Because of the size and complexity of constructed products in-process evaluation may not provide sufficient feedback regarding design quality. Evaluations of design effectiveness are often best conducted by interviewing end users and owners, and undertaking an expert evaluation some years after a project has been completed. Only in this way can we ensure that the cycle of learning about design quality is continuous and that the lessons learned are fed back into the initiation and development process. The responsibility for initiating this kind of improvement cycle lies both with the designers (who need the information to enable them to improve their services) and the developers (who need to ensure that future projects are informed by lessons from current practice).

## Purchasing

Purchasing processes need to be controlled to ensure purchased products/services conform to the organization's requirements. The type and extent of methods to do this are dependent on the effect of the purchased product/service upon the final product/service. Clearly suppliers need to be evaluated and selected on their ability to supply the product or service in accordance with the organization's requirements. Supplier evaluations, supplier audit records and evidence of previously demonstrated ability should be considered when selecting suppliers and when determining the type and extent of supervision applicable to the purchased materials/services.

The purchasing documentation should contain information clearly describing the product/service ordered, including:

(a) requirements for approval or qualification of product and/or service, procedures, processes, equipment and personnel;
(b) any management system requirements.

Review and approval of purchasing documents, for adequacy of the specification of requirements prior to release, is also necessary.

Any purchased products/services need some form of verification. Where this is to be carried out at the supplier's premises, the organization needs to specify the arrangements and methods for product/service release in the purchasing documentation.

## Production and service delivery processes

The organization needs to control production and service delivery processes through:

(a) information describing the product/service characteristics;
(b) clearly understandable work standards or instructions;
(c) suitable production, installation and service provision equipment;
(d) suitable working environments;
(e) suitable inspection, measuring and test equipment, capable of the necessary accuracy and precision;
(f) the implementation of suitable monitoring, inspection or testing activities;
(g) provision for identifying the status of the product/service, with respect to required measurement and verification activities;
(h) suitable methods for release and delivery of products and/or services.

Where applicable, the organization needs to identify the product/service by suitable means throughout all processes. Where traceability is a requirement for the organization, there is a need to control the identification of the product/service. There is also a need to ensure that, during internal processing and final delivery of the product/service, the identification, packaging, storage, preservation, and handling do not adversely affect conformity with the requirements. This applies equally to parts or components of a product and elements of a service.

Where the resulting output cannot be easily or economically verified by monitoring, inspection or testing, including where processing deficiencies may become apparent only after the product is in use or the service has been delivered, the organization needs to validate the production and service delivery processes to demonstrate their effectiveness and acceptability.

The arrangements for validation might include:

(a) processes being qualified prior to use;
(b) qualification of equipment or personnel;
(c) use of specific procedures or records.

Evidence of validated processes, equipment and personnel needs to be recorded and maintained, of course.

## Post-delivery services

Where there is a requirement for the organization to provide support services, after delivery of the product or service, this needs to be both planned and in line with the customer requirement.

Note the Sekisui Heim case study (p. 469) where the company builds into its contractual arrangements inspections at six months, one and two years, and requests permission to inspect at five and ten years. Furthermore, the company seeks to develop a lifetime relationship involving regular five yearly inspections and maintenance. This ongoing relationship helps to separate construction defects from maintenance neglect and provides direct feedback into product improvement. A further welcomed by-product of this contract is that it enables a company to build long-term relationships with buyers, it attracts customer recommendations and improves market perceptions of the company's products and service.

## Monitoring and measuring devices

There is a need to control, calibrate, maintain, handle and store the applicable measuring, inspection and test equipment to specified requirements. In construction this applies in the main to survey measuring equipment, quality assessment measuring equipment and precision manufacturing equipment. Measuring, inspection, and test equipment should be used in a way which ensures that any measurement uncertainty, including accuracy and precision, is known and is consistent with the required measurement capability. Any test equipment software should meet the applicable requirements for the design and development of the product (see above).

The organization certainly needs to:

(a) calibrate and adjust measuring, inspection and test equipment at specified intervals or prior to use, against equipment traceable to international or national standards. Where no standards exist, the basis used for calibration needs to be recorded;
(b) identify measuring, inspection and test equipment with a suitable indicator or approved identification record to show its calibration status;
(c) record the process for calibration of measuring, inspection and test equipment;
(d) ensure the environmental conditions are suitable for any calibrations, measurements, inspections and tests;
(e) safeguard measuring, inspection and test equipment from adjustments which would invalidate the calibration;
(f) verify validity of previous inspection and test results when equipment is found to be out of calibration;
(g) establish the action to be initiated when calibration verification results are unsatisfactory.

# 4. Measurement, analysis and improvement

Any organization needs to define and implement measurement, analysis and improvement processes to demonstrate that the products, services and processes conform to the specified requirements. The type, location and timing of these measurements need to be determined and the results recorded based on their importance. The results of data analysis and improvement activities should be an input to the management review process, of course.

---

This aspect is best described in the Sekisui Heim and Woh Hup case studies (pp. 469, 499) where the process of analysing errors and developing ways to improve product and services is described.

---

## Measurement and monitoring

There is a need to determine and establish processes for measurement of the quality management system performance. Customer satisfaction must be a primary measure of system output and the internal audits should be used as a primary tool for evaluating ongoing system compliance.

The organization needs to establish a process for obtaining and monitoring information and data on both immediate and end user customer satisfaction for all essential processes. The methods and measures for obtaining customer satisfaction information and data and the nature and frequency of reviews need to be defined to demonstrate the level of customer confidence in the delivery of conforming product and/or service supplied by the organization. Suitable measures for establishing internal improvement need to be implemented and the effectiveness of the measures periodically evaluated.

The organization must establish a process for performing internal audits of the quality management system and related processes. The purpose of the internal audit is to determine whether:

(a) the quality management system established by the organization conforms to the requirements of the International Standard; and
(b) the quality management system has been effectively implemented and maintained.

The internal audit process should be based on the status and importance of the activities, areas or items to be audited, and the results of previous audits.

The internal audit process should include:

(a) planning and scheduling the specific activities, areas or items to be audited;
(b) assigning trained personnel independent of those performing the work being audited;
(c) assuring that a consistent basis for conducting audits is defined.

The results of internal audits should be recorded including:

(a) activities, areas, and processes audited;
(b) non-conformities or deficiencies found;
(c) status of commitments made as the result of previous audit, such as corrective actions or product audits;
(d) recommendations for improvement.

The last point is a critical one for construction projects. The authors have observed that on many construction projects even though errors are detected and rectified on a regular basis they are repeated again and again. Although the project organization may have set processes in place for error identification, they are not avoided. This suggests that the focus is on detection rather than avoidance.

The results of the internal audits should be communicated to the area audited and the management personnel responsible need to take timely corrective action on the non-conformities recorded.

Suitable methods for the measurement of processes necessary to meet customer requirements need to be applied, including monitoring the output of the processes that control conformity of the product or service provided to customers. The measurement results then need to be used to determine opportunities for improvements.

The organization needs also to apply suitable methods for the measurement of the product or service to verify that the requirements have been met. Evidence from any inspection and testing activities and the acceptance criteria used need to be recorded. If there is an authority responsible for release of the product and/or service, this should also be recorded.

Products or services should not be dispatched until all the specified activities have been satisfactorily completed and the related documentation is available and authorized. The only exception to this is when the product or service is released under positive recall procedures.

## Control of non-conforming products

Products and services which do not conform to requirements need to be controlled to prevent unplanned use, application or installation, and the organization needs to identify, record and review the nature and extent of the problem encountered, and determine the action to be taken. This needs to include how non-conforming service will be:

(a) corrected or adjusted to conform to requirements; or
(b) accepted under concession, with or without correction; or
(c) reassigned for an alternative, valid application; or
(d) rejected as unsuitable.

The responsibility and authority for the review and resolving of non-conformities needs to be defined, of course.

When required by the contract, the proposed use or repair of non-conforming product or a modified service needs to be reported for concession to the customer. The description of any corrections or adjustments, accepted non-conformities, product repairs or service modifications also need to be recorded. Where it is necessary to repair or rework a product or modify a service, verification requirements need to be determined and implemented.

## Analysis of data

Analysis of data needs to be established as a means of determining where system improvements can be made. Data needs to be collected from relevant sources, including internal audits, corrective and preventive action, non-conforming product service, customer complaints and customer satisfaction results.

The organization should then analyse the data to provide information on:

(a) the effectiveness of the quality management system;
(b) process operation trends;
(c) customer satisfaction;
(d) conformance to customer requirements of the product/service;
(e) suppliers.

There is also a need to determine the statistical techniques to be used for analysing data, including verifying process operations and product service characteristics. Of course the statistical techniques selected should be suitable and their use controlled and monitored.

## Improvement

The organization needs to establish a process for eliminating the causes of non-conformity and preventing recurrence. Non-conformity reports, customer complaints and other suitable quality management system records are useful as inputs to the corrective action process. Responsibilities for corrective action need to be established together with the procedures for the corrective action process, which should include:

(a) identification of non-conformities of the products, services, processes, the quality management system, and customer complaints;
(b) investigation of causes of non-conformities, and recording results of investigations;

(c) determination of corrective actions needed to eliminate causes of non-conformities;
(d) implementation of corrective action;
(e) follow-up to ensure corrective action taken is effective and recorded.

Corrective actions also need to be implemented for products or services already delivered, but subsequently discovered to be non-conforming and customers need to be notified where possible.

The organization needs to establish a process for eliminating the causes of potential non-conformities to prevent their occurrence. Quality management system records and results from the analysis of data should be used as inputs for this and responsibilities for preventive action established. The process should include:

(a) identification of potential product, service and process non-conformities;
(b) investigation of the causes of potential non-conformities of products/services, processes and the quality management system, and recording the results;
(c) determination of preventive action needed to eliminate causes of potential non-conformities;
(d) implementation of preventive action needed;
(e) follow-up to ensure preventive action taken is effective, recorded and submitted for management review.

Processes for the continual improvement of the quality management system need to be established including methods and measures suitable for the products/services.

# Other management systems and models

Organizations of all kinds are increasingly concerned to achieve and demonstrate sound environmental performance. Many have undertaken environmental audits and reviews to assess this. To be effective these need to be conducted within a structured management system, which in turn is integrated with the overall management activities dealing with all aspects of desired environmental performance.

Such a system should establish processes for setting environmental policy and objectives, and achieving compliance to them. It should be designed to place emphasis on the prevention of adverse environmental effects, rather than on detection after occurrence. It should also identify and assess the environmental effects arising from the organization's existing or proposed activities, products, or services and from incidents, accidents, and potential emergency situations. The system must identify the relevant regulatory requirements, the priorities, and pertinent environmental objective and targets. It needs also to facilitate planning, control, monitoring, auditing and review activities to ensure that the policy is complied with, that it remains relevant, and is capable of evolution to suit changing circumstances.

The International Standard ISO 14001 contains a specification for environmental management systems for ensuring and demonstrating compliance with stated policies and objectives. The standard is designed to enable any organization to establish an effective management system, as a foundation for both sound environmental performance and participation and environmental auditing schemes.

ISO 14001 shares common management system principles with the ISO 9001:2000 standard and organizations may elect to use an existing management system, developed in conformity with that standard, as a basis for environmental management. The ISO 14001 standard defines environmental policy, objectives, targets, effect, management, systems, manuals, evaluation, audits and reviews. It mirrors the ISO 9001:2000 standard requirements in many of its own requirements, and it includes a guide to these in an informative annex. EN ISO 9001:2000 also gives a useful annex of 'correspondence between ISO 9001:2000 and ISO 14001:1996' under each of the main headings of both standards. This shows that the quality management system standard has been aligned with requirements of the environmental management system standard in order to enhance the compatibility of the two for the benefit of the users. As such ISO 9001:2000 enables an organization to align and integrate its own quality management systems with other related management system requirements. In this way, it should be possible for any organization to adapt its existing management system(s) in order to establish one that complies with ISO 9001:2000.

ISO 9000:2000 also makes comments on the relationship between quality management systems and excellence models. It claims that the approaches set down in the ISO 9000 family of standards and in the various excellence models (see Chapter 2) are based on common principles and that the only differences are in the scope of application. Excellence models and ISO 9000:

(a)  help organizations identify strengths and weaknesses;
(b)  aid the evaluation of organizations;
(c)  establish a basis for continuous improvement;
(d)  allow and support external recognition.

It is also recognized that excellence models add value to the QMS approach by providing criteria that allow comparative evaluation of organizational performance with other organizations.

Management systems are needed in all areas of activity, whether large or small businesses, manufacturing, service or public sector. The advantages of systems in manufacturing are obvious, but they are just as applicable in areas such as marketing, sales, personnel, finance, research and development, as well as in the service industries and public sectors. No matter where it is implemented a good management system will improve process control, reduce wastage, lower costs, increase market share (or funding), facilitate training, involve staff, and raise morale.

# Reflections on the changes and improvements made to quality management systems

One measure of the changes made between 1994 and 2000 in the ISO 9000 family of quality management system standards is in the language used. Table 12.2 shows a simple count of the number of times certain words are to be found in each of the versions of ISO 9001. These reflect the main thrusts of the new standard:

■  increased customer focus;
■  process approach (at the expense of procedural documentation);
■  continuous improvement;
■  skills-based approach to human resource management.

The adoption of a process approach reflects the changes in management thinking in recent years and supports the authors' 'campaign' for improved process management

**Table 12.2** Change in vocabulary in ISO 9000:2000

**Change in ISO 9001 vocabulary (text and notes)**

| Word | Occurrence | |
|------|------|------|
| | 1994 | 2000 |
| Improvement | 0 | 16 |
| Processes | 21 | 39 |
| Procedures | 36 | 9 |
| Customer | 15 | 32 |
| Competency | 0 | 4 |
| Capability | 5 | 1 |
| Analyse(is) | 0 | 5 |

to deliver better performance in most organizations. The standard now explicitly recognizes the process-oriented nature of many organizations and addresses all sorts of important areas – risk assessment, relationship and interactions between functions/departments – the internal customer/supplier chains of Chapter 1 – and the vital one of capability.

Hence, the revised standard is a major step forward from the earlier versions. It pays a lot of attention to the management of people, an area which many organizations have neglected in their past 'quality initiatives', especially those concerned with documenting quality systems.

Organizations making the transition from the 1994 version to ISO 9002:2001 are finding how vital are the management of skills, including those of the auditors – external as well as internal! A mindset change is required away from the procedures-based earlier approaches which led to changes of mechanistic and even bureaucratic systems. With understanding, commitment and the right attitude, particularly in the senior management, however, a quality management system can support a dynamic organization and help it achieve its aims and objectives ('whats') through a thorough understanding of the 'hows'. The ISO 9000:2000 family of standards is now much more aligned with the EFQM Excellence Model in terms of the intuitively appealing direction–process–people–performance structure.

# Reference

1. Dissanayaka, S.M., Kumaraswamy, M.M., Karim, K. and Marosszeky, M. (2000) 'Evaluating Outcomes from IS0 9000 Certified Quality Systems of Hong Kong Constructors', *Total Quality Management Journal*, Vol. 12, No. 1.

## Chapter highlights

**Why a quality management system?**
- While formal quality systems have a poor name in construction due to the cynical way in which they first implemented, construction industry leaders the world over have demonstrated the benefits they have derived from them.
- An appropriate quality management system will enable the objectives set out in the quality policy to be accomplished.

- The International Organization for Standardization (ISO) 9000:2000 series set out methods by which a system can be implemented to ensure that the specified customer requirements are met.
- A quality system may be defined as an assembly of components, such as the management responsibilities, process, and resources.

### Quality management system design
- ISO 9000:2000 links management system quality to product and process quality through a focus on demonstrating customer satisfaction and continuous improvement.
- Quality management systems should apply to and interact with processes in the organization. The activities are generally processing, communicating, and controlling. These should be documented in the form of a quality manual.
- The system should follow the Plan, Do, Check, Act cycle, through documentation, implementation, audit and review.
- The ISO 9000:2000 family together forms a coherent set of quality management system standards to facilitate mutual understanding across national and international trade.

### Quality management system requirements
- The general categories of the ISO 9001:2000 standard on quality management systems include: management responsibility, resource management, product realization, measurement analysis and improvement, which are detailed in the standard.

### Other management systems and models
- The International Standard ISO 14001 contains specifications for environmental management systems for ensuring and demonstrating compliance with the stated policies and objectives, and acting as a base for auditing and review schemes.
- ISO 14001 shares common principles with the ISO 9001:2000 standard on quality management systems. The latter shows, by 'correspondence' between the two, under the main headings of both standards, that the quality standard has been aligned with the requirements of the environmental standard.
- ISO 9000:2000 also makes comments on the relationship between quality management systems and excellence models. The two are based on the common principles of identifying strengths and weaknesses, evaluation, continuous improvement, and external recognition.

### Improvements made to quality management systems
- The language used in the 1994 and 2000 versions of the ISO 9000 family shows changes in emphasis to increased customer focus, process rather than procedural approach, continuous improvement, and skills-based approach to people management.
- The ISO 9000:2000 standard is a major step forward from the earlier versions, paying much more attention to human resource management and being in tune with the EFQM Excellence Model in terms of the direction–process–people–performance alignment.

# 13

# Continuous improvement

# A systematic approach

The most obvious feature of the management practices described in the case studies was the use of performance measurement to drive continuous improvement. In the never-ending quest for improvement in business processes and outcomes, numbers and information should always form the basis for understanding, decisions and actions; and a thorough data gathering, recording and presentation system is essential:

*Record* data      – all processes can and should be measured.
                         – all measurements should be recorded.
*Use* data            – if data is recorded and not used it will be abused.
*Analyse* data     – data analysis should be carried out by means of some basic systematic tools.
*Act* on the results – recording and analysis of data without action leads to frustration.

Within the construction sector there is a general perception that as the sector deals with one-of-a-kind production, numerical processes are either of very limited or of no value. They have been developed by manufacturing and therefore have very limited application in construction. This may seem to be logical while the industry is viewed as delivering an unrelated sequence of projects. However, as soon as the industry is regarded as the producer of generically similar though superficially different products through a set of largely repetitive processes, an entirely different rationale must be applied. Suddenly the tools for presenting and analysing data are just as critical to any process improvement exercise in construction as in any other sector of the economy. The companies in the case studies all use data analysis extensively to help achieve excellent outcomes. The challenge for every senior manager in a construction industry organization is to discover how he/she can use the management tools and numerical techniques presented in this chapter to improve his/her processes.

Readers who are new to TQM and structured process improvement should go to the section on DRIVE (p. 273) – a process developed by John Oakland for implementing an improvement plan. It will provide you with a framework for action and place the other tools into context.

In addition to the basic elements of a quality management system that provide a framework for recording, there exists a set of methods the Japanese quality guru Ishikawa has called the 'seven basic tools'. These should be used to interpret and derive the maximum use from data. The simple methods listed below, of which there are clearly more than seven, will offer any organization a means of collecting, presenting, and analysing most of its data:

- Process flowcharting – what is done?
- Check sheets/tally charts – how often is it done?
- Histograms – what do overall variations look like?
- Scatter diagrams – what are the relationships between factors?
- Stratification – how is the data made up?
- Pareto analysis – which are the big problems?
- Cause and effect analysis and brainstorming, including CEDAC, NGT, and the five whys – what causes the problems?
- Force-field analysis – what will obstruct or help the change or solution?
- Emphasis curve – which are the most important factors?
- Control charts – which variations to control and how?

Sometimes more sophisticated techniques, such as analysis of variance, regression analysis, and design of experiments, need to be employed.

The effective use of the tools requires their application by the people who actually work on the processes. Their commitment to this will be possible only if they are assured that management cares about improving quality. Managers must show they are serious by establishing a systematic approach and providing the training and implementation support required.

Improvements cannot be achieved without specific opportunities, commonly called problems, being identified or recognized. A focus on improvement opportunities leads to the creation of teams whose membership is determined by their work on and detailed knowledge of the process, and their ability to take improvement action. The teams must then be provided with good leadership and the right tools to tackle the job.

The systematic approach (Figure 13.1) should lead to the use of factual information, collected and presented by means of proven techniques, to open a channel of communications

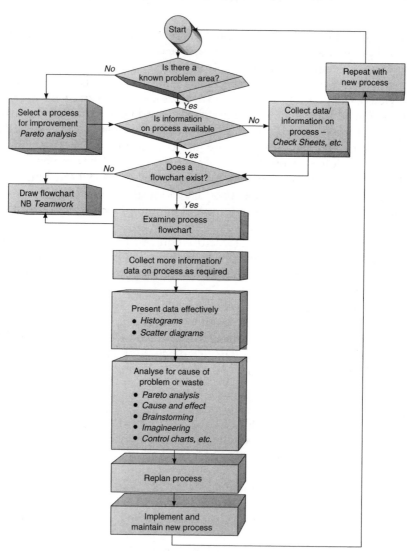

**Figure 13.1**
Strategy for process improvement

not available to the many organizations that do not follow this or a similar structured approach to problem solving and improvement. Continuous improvements in the quality of products, services, and processes can often be obtained without major capital investment, if an organization marshals its resources, through an understanding and breakdown of its processes in this way.

By using reliable methods, creating a favourable environment for team-based problem solving, and continuing to improve using systematic techniques, the never-ending improvement helix (see Chapter 3) will be engaged. This approach demands the real time management of data, and actions focused on processes and inputs rather than outputs. It will require a change in the language of many organizations from percentage defects, percentage 'prime' product, and number of errors, to *process capability*. The climate must change from the traditional approach of 'If it meets the specification, there are no problems and no further improvements are necessary.' The driving force for this will be the need for better internal and external customer satisfaction levels, which will lead to the continuous improvement question, 'Could we do the job better?'

# Some basic tools and techniques

Understanding processes so that they can be improved by means of the systematic approach requires knowledge of a simple kit of tools or techniques. What follows is a brief description of each technique, but a full description and further examples of some of them may be found in Reference 1.

## Process flowcharting

The use of this technique, which is described in Chapter 10, ensures a full understanding of the inputs, outputs and flow of the process. Without that understanding, it is not possible to draw the correct flowchart of the process. In flowcharting it is important to remember that in all but the smallest tasks no single person is able to complete a chart without help from others. This makes flowcharting a powerful team forming exercise.

## Check sheets or tally charts

A check sheet is a tool for data gathering, and a logical point to start in most process control or problem-solving efforts. It is particularly useful for recording direct observations and helping to gather in facts rather than opinions about the process. In the recording process it is essential to understand the difference between data and numbers.

Data are pieces of information, including numerical, that are useful in solving problems, or provide knowledge about the state of a process. Numbers alone often represent meaningless measurements or counts, which tend to confuse rather than to enlighten. Numerical data on quality will arise either from counting or measurement.

The use of simple check sheets or tally charts aids the collection of data of the right type, in the right form, at the right time. The objectives of the data collection will determine the design of the record sheet used.

## Histograms

Histograms show, in a very clear pictorial way, the frequency with which a certain value or group of values occurs. They can be used to display both attribute and variable data,

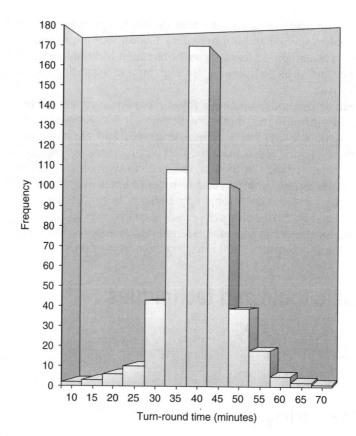

**Figure 13.2**
Frequency distribution for track turn-around times (histogram)

and are an effective means of letting the people who operate the process know the results of their efforts. Data gathered on truck turn-round times is drawn as a histogram in Figure 13.2.

## Scatter diagrams

Depending on the technology, it is frequently useful to establish the association, if any, between two parameters or factors. A technique to begin such an analysis is a simple X-Y plot of the two sets of data. The resulting grouping of points on scatter diagrams (e.g. Figure 13.3) will reveal whether or not a strong or weak, positive or negative, correlation exists between the parameters. The diagrams are simple to construct and easy to interpret, and the absence of correlation can be as revealing as finding that a relationship exists.

## Stratification

Stratification is simply dividing a set of data into meaningful groups. It can be used to great effect in combination with other techniques, including histograms and scatter diagrams. If, for example, three shift teams are responsible for a certain product output 'stratifying' the data into the shift groups might produce histograms that indicate 'process adjustments' were taking place at shift changeovers.

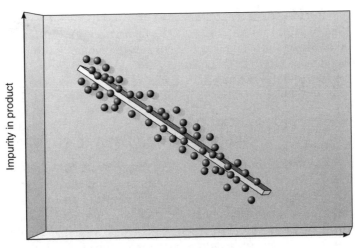

**Figure 13.3**
Scatter diagram showing a negative correlation between two variables

## Pareto analysis

If the symptoms or causes of defective output or some other 'effect' are identified and recorded, it will be possible to determine what percentage can be attributed to any cause, and the probable results will be that the bulk (typically 80 percent) of the errors, waste, or 'effects', derive from a few of the causes (typically 20 percent). For example, Figure 13.4 shows a *ranked frequency distribution* of incidents in the distribution of a certain product. To improve the performance of the distribution process, therefore, the major incidents (broken bags/drums, truck scheduling, and temperature problems) should be tackled first. An analysis of data to identify the major problems is known as *Pareto analysis*, after the Italian economist who realized that approximately 90 percent of the wealth in his country was owned by approximately 10 percent of the people. Without an analysis of this sort, it is far too easy to devote resources to addressing one symptom only because its cause seems immediately apparent.

## Cause and effect analysis and brainstorming

A useful way of mapping the inputs that affect quality is the *cause and effect diagram*, also known as the Ishikawa diagram (after its originator) or the fishbone diagram (after its appearance, Figure 13.5). The effect or incident being investigated is shown at the end of a horizontal arrow. Potential causes are then shown as labelled arrows entering the main cause arrow. Each arrow may have other arrows entering it as the principal factors or causes are reduced to their subcauses and subsubcauses by *brainstorming*.

Brainstorming is a technique used to generate a large number of ideas quickly, and may be used in a variety of situations. Each member of a group, in turn, may be invited to put forward ideas concerning a problem under consideration. Wild ideas are safe to offer, as criticism or ridicule is not permitted during a brainstorming session. The people taking part do so with equal status to ensure this. The main objective is to create an atmosphere of enthusiasm and originality. All ideas offered are recorded for subsequent analysis. The process is continued until all the conceivable causes have been included. The proportion of

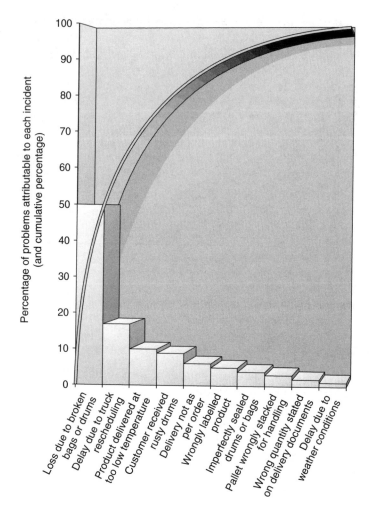

**Figure 13.4**
Incidents in the distribution of a chemical product

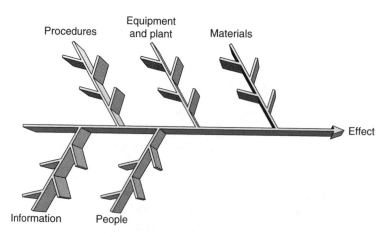

**Figure 13.5**
The cause and effect, Ishikawa or fishbone diagram

non-conforming output attributable to each cause, for example, is then measured or estimated, and a simple Pareto analysis identifies the causes that are most worth investigating.

A useful variant on the technique is negative brainstorming. Here the group brainstorms all the things that would need to be done to ensure a negative outcome. For example, in the implementation of TQM, it might be useful for the senior management team to brainstorm what would be needed to make sure TQM *was not* implemented. Having identified in this way the potential roadblocks, it is easier to dismantle them.

## CEDAC

A variation on the cause and effect approach, which was developed at Sumitomo Electric and is now used by many major corporations across the world, is the cause and effect diagram with addition of cards (CEDAC).

The effect side of a CEDAC chart is a quantified description of the problem, with an agreed and visual quantified target and continually updated results on the progress of achieving it. The cause side of the CEDAC chart uses two different coloured cards for writing facts and ideas. This ensures that the facts are collected and organized before solutions are devised. The basic diagram for CEDAC has the classic fishbone appearance.

## Nominal group technique

The nominal group technique (NGT) is a particular form of team brainstorming used to prevent domination by particular individuals. It has specific application for multi-level, multi-disciplined teams, where communication boundaries are potentially problematic.

In NGT a carefully prepared written statement of the problem to be tackled is read out by the facilitator (F). Clarification is obtained by questions and answers and then the individual participants (P) are asked to restate the problem in their own words. The group then discusses the problem until its formulation can be satisfactorily expressed by the team (T). The method is set out in Figure 13.6. NGT results in a set of ranked

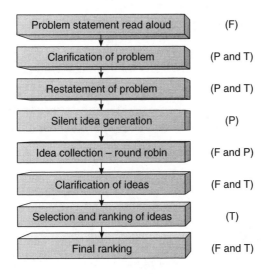

**Figure 13.6**
Nominal group technique

255

ideas that are close to a team consensus view obtained without domination by one or two individuals.

Even greater discipline may be brought to brainstorming by the use of 'soft systems methodology (SSM)', developed by Peter Checkland.[1] The component stages of SSM are gaining a 'rich understanding' through 'finding out', input/output diagrams, root definition (which includes the so-called CATWOE analysis: customers, 'actors', transformations, 'world-view', owners, environment), conceptualization, comparison, and recommendation.

## Force field analyses

Force field analysis is a technique used to identify the forces that either obstruct or help a change that needs to be made. It is similar to negative brainstorming and cause/effect analysis and helps to plan how to overcome the barriers to change or improvement. It may also provide a measure of the difficulty in achieving the change.

The process begins with a team describing the desired change or improvement, and defining and objectives or solution. Having prepared the basic force field diagram, it identifies the favourable/positive/driving forces and the unfavourable/negative/restraining forces, by brainstorming. These forces are placed in opposition on the diagram and, if possible, rated for their potential influence on the ease of implementation. The results are evaluated. Then comes the preparation of an action plan to overcome some of the restraining forces, and increase the driving forces. Figure 13.7 shows a force field analysis produced by a senior management team considering the implementation of TQM in its organization.

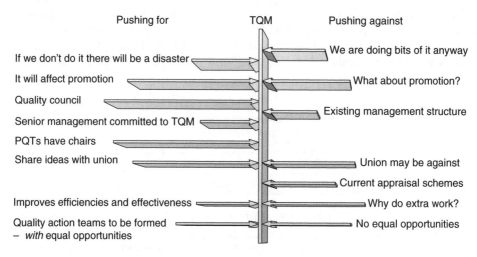

**Figure 13.7**
Force field analysis

## The emphasis curve

This is a technique for ranking a number of factors, each of which cannot be readily quantified in terms of cost, frequency of occurrence, etc., in priority order. It is almost impossible for the human brain to make a judgement of the relative importance of more than three or four non-quantifiable factors. It is, however, relatively easy to judge which is the most important of two factors, using some predetermined criteria. The emphasis curve technique uses this fact by comparing only two factors at any one time.

The procedural steps for using the 'emphasis curve chart' are as follows:

1. List the factors for ranking under a heading 'Scope'.
2. Compare factor 1 with factor 2 and rank the most important. To assist in judging the relative importance of two factors, it may help to use weightings, e.g. degree of seriousness, capital investment, speed of completion, etc., on a scale of 1 to 10.
3. Compare factor 1 with 3, 1 with 4, 1 with 5 and so on – ringing the most important number in the matrix.
4. Having compared factor 1 against the total scope, proceed to compare factor 2 with 3, 2 with 4 and so on.
5. Count the number of 'ringed' number 1s in the matrix and put the total in a right-hand column against Number 1. Next count the total number of 2s in the matrix and put total in column against Number 2 and so on.
6. Add up the numbers in the column and check the total, using the formula $\{n(n-1)\}/2$, where $n$ is the number of entries in the column. This check ensures that all numbers have been 'ringed' in the matrix.
7. Proceed to rank the factors using the numbers in the column.
8. Generally the length of time to make a judgement between two factors does not significantly affect the outcome; therefore the rule is 'accept the first decision, record it and move quickly onto the next pair'.

# Control charts

A control chart is a form of traffic signal whose operation is based on evidence from the small samples taken at random during a process. A green light is given when the process should be allowed to run. All too often processes are 'adjusted' on the basis of a single measurement, check or inspection, a practice that can make a process much more variable than it is already. The equivalent of an amber light appears when trouble is possibly imminent. The red light shows that there is practically no doubt that the process has changed in some way and that it must be investigated and corrected to prevent production of defective material or information. Clearly, such a scheme can be introduced only when the process is 'in control'. Since samples taken are usually small, there are risks of errors, but these are small, calculated risks and not blind ones. The risk calculations are based on various frequency distributions.

These charts should be made easy to understand and interpret and they can become, with experience, sensitive diagnostic tools to be used by operating staff and first-line supervision to prevent errors or defective output being produced. Time and effort spent to explain the working of the charts to all concerned are never wasted.

The most frequently used control charts are simple run charts, where the data is plotted on a graph against time or sample number. There are different types of control charts for variables and attribute data: for variables mean ($\bar{X}$) and range ($R$) charts are used together; number defective or **np** charts and proportion defective or **p** charts are the most common ones for attributes. Other charts found in use are moving average and range charts, numbers of defects (**c** and **u**) charts, and cumulative sum (cusum) charts. The latter offer very powerful management tools for the detection of trends or changes in attributes and variable data.

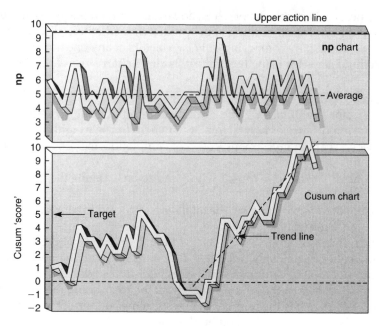

**Figure 13.8**
Comparison of cusum and **np** charts for the same data

The cusum chart is a graph that takes a little longer to draw than the conventional control chart, but gives a lot more information. It is particularly useful for plotting the evolution of processes, because it presents data in a way that enables the eye to separate true changes from a background of random variation. Cusum charts can detect small changes in data very quickly, and may be used for the control of variables and attributes. In essence, a reference or 'target value' is subtracted from each successive sample observation, and the result accumulated. Values of this cumulative sum are plotted, and 'trend lines' may be drawn on the resulting graphs. If they are approximately horizontal, the value of the variable is about the same as the target value. A downward slope shows a value less than the target and an upward slope a value greater. The technique is very useful, for example, in comparing sales forecast with actual sales figures.

Figure 13.8 shows a comparison of an ordinary run chart and a cusum chart that have been plotted from the same data – errors in samples of 100 invoices. The change, which is immediately obvious on the cusum chart, is difficult to detect on the conventional control chart.

The range of type and use of control charts is now very wide, and within the present text it is not possible to indicate more than the basic principles underlying such charts.[1] All of them can be generated electronically using the various software tools available.

## Statistical process control

The responsibility for quality in any transformation process must lie with the operators of that process. To fulfil this responsibility, however, people must be provided with the tools necessary to:

- Know whether the process is capable of meeting the requirements.
- Know whether the process is meeting the requirements at any point in time.

- Make correct adjustment to the process or its inputs when it is not meeting the requirements.

The techniques of statistical process control (SPC) will greatly assist in these stages. To begin to monitor and analyse any process, it is necessary first of all to identify what the process is, and what the inputs and outputs are. Many processes are easily understood and relate to known procedures, e.g. drilling a hole, compressing tablets, filling cans with paint, polymerizing a chemical using catalysts. Others are less easily identifiable, e.g. servicing a customer, delivering a lecture, storing a product in a warehouse, inputting to a computer. In many situations it can be extremely difficult to define the process. For example, if the process is inputting data into a computer terminal, it is vital to know if the scope of the process includes obtaining and refining the data, as well as inputting. Process definition is so important because the inputs and outputs change with the scope of the process.

All processes can be monitored and brought 'under control' by gathering and using data – to measure the performance of the process and provide the feedback required for corrective action, where necessary. SPC methods, backed by management commitment and good organization, provide objective means of *controlling* quality in any transformation process, whether used in the manufacture of artefacts, the provision of services, or the transfer of information.

SPC is not only a tool kit, it is a strategy for reducing variability, the cause of most quality problems: variation in products, in times of deliveries, in ways of doing things, in materials, in people's attitudes, in equipment and its use, in maintenance practices, in everything. Control by itself is not sufficient. Total quality management requires that the processes should be improved continually by reducing variability. This is brought about by studying all aspects of the process, using the basic question: 'Could we do this job more consistently and on target?' The answer drives the search for improvements. This significant feature of SPC means that it is not constrained to measuring conformance, and that it is intended to lead to action on processes that are operating within the 'specification' to minimize variability.

Process control is essential, and SPC forms a vital part of the TQM strategy. Incapable and inconsistent processes render the best design impotent and make supplier quality assurance irrelevant. Whatever process is being operated, it must be reliable and consistent. SPC can be used to achieve this objective.

In the application of SPC there is often an emphasis on techniques rather than on the implied wider managerial strategies. It is worth repeating that SPC is not only about plotting charts on the walls of a plant or office, it must become part of the company-wide adoption of TQM and act as the focal point of never-ending improvement. Changing an organization's environment into one in which SPC can operate properly may take several years rather than months. For many companies SPC will bring a new approach, a new 'philosophy', but the importance of the statistical techniques should not be disguised. Simple presentation of data using diagrams, graphs, and charts should become the means of communication concerning the state of control of processes. It is on this understanding that improvements will be based.

## The SPC system

A systematic study of any process through answering the questions:

- Are we capable of doing the job correctly?
- Do we continue to do the job correctly?

- Have we done the job correctly?
- Could we do the job more consistently and on target?[2]

provides knowledge of the *process capability* and the sources of non-conforming outputs. This information can then be fed back quickly to marketing, design, and the 'technology' functions. Knowledge of the current state of a process also enables a more balanced judgement of equipment, with regard to the tasks within its capability and its rational utilization.

Statistical process control procedures exist because there is variation in the characteristics of all material, articles, services, and people. The inherent variability in each transformation process causes the output from it to vary over a period of time. If this variability is considerable, it is impossible to predict the value of a characteristic of any single item or at any point in time. Using statistical methods, however, it is possible to take meagre knowledge of the output and turn it into meaningful statements that may then be used to describe the process itself. Hence, statistically based process control procedures are designed to divert attention from individual pieces of data and focus it on the process as a whole. SPC techniques may be used to measure and control the degree of variation of any purchased materials, services, processes, and products, and to compare this, if required, to previously agreed specifications. In essence, SPC techniques select a representative, simple, random sample from the 'population', which can be an input to or an output from a process. From an analysis of the sample it is possible to make decisions regarding the current performance of the process.

Organizations that embrace the TQM concepts should recognize the value of SPC techniques in areas such as sales, purchasing, invoicing, finance, distribution, training, and in the service sector generally. These are outside the traditional areas for SPC use, but it needs to be seen as an organization-wide approach to reducing variation with the specific techniques integrated into a programme of change throughout. A Pareto analysis, a histogram, a flowchart, or a control chart is a vehicle for communication. Data is data and, whether the numbers represent defects or invoice errors, weights or delivery times, or the information relates to machine settings, process variables, prices, quantities, discounts, sales or supply points, is irrelevant – the techniques can always be used.

In the authors' experience, some of the most exciting applications of SPC have emerged from organizations and departments which, when first introduced to the methods, could see little relevance in them to their own activities. Following appropriate training, however, they have learned how to, for example:

- *Pareto analyse* errors on invoices to customers and industry injury data.
- *Brainstorm and cause and effect analyse* reasons for late payment and poor purchase invoice matching.
- *Histogram* defects in invoice matching and arrival of trucks at certain times during the day.
- *Control chart* the weekly demand of a product.

Distribution staff have used control charts to monitor the proportion of late deliveries, and Pareto analysis and force field analysis to look at complaints about the distribution system. Bank operators have been seen using cause and effect analysis, NGT and histograms to represent errors in the output from their services. Moving average and cusum charts have immense potential for improving processes in the marketing area.

Those organizations that have made most progress in implementing continuous improvement have recognized at an early stage that SPC is for the whole organization. Restricting it to traditional manufacturing or operational activities means that a window of opportunity for improvement has been closed. Applying the methods and techniques outside manufacturing will make it easier, not harder, to gain maximum benefit from SPC.

# Some additional techniques for process design and improvement

Seven 'new' qualitative tools may be used as part of quality function deployment (see Chapter 6) or to improve processes. These do not replace the basic systematic tools described earlier in this chapter, neither are they extensions of these. The new tools are systems and documentation methods used to achieve success in design by identifying objectives and intermediate steps in the finest detail. The seven new tools are:

1. Affinity diagram.
2. Interrelationship diagraph.
3. Tree diagram.
4. Matrix diagram or quality table.
5. Matrix data analysis.
6. Process decision programme chart (PDPC).
7. Arrow diagram.

The tools are interrelated, as shown in Figure 13.9 and are summarized below. A fuller treatment of the use of these tools is given by Oakland in *Total Quality Management – the route to improved performance*, 2nd edition, 1994, Butterworth-Heinemann.

## 1. Affinity diagram

This is used to gather large amounts of language data (ideas, issues, opinions) and organizes them into groupings based on the natural relationship between the items. In other words, it is a form of brainstorming.

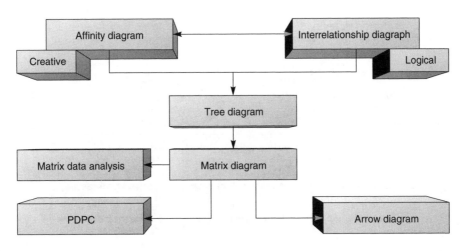

**Figure 13.9**
The seven new tools of quality design

The steps for generating an affinity diagram are as follows:

1. Assemble a group of people familiar with the problem of interest. Six to eight members in the group works best.
2. Phrase the issue to be considered. It should be vaguely stated so as not to prejudice the responses in a predetermined direction.
3. Give each member of the group a stack of cards and allow 5–10 minutes for everyone individually in the group to record ideas on the cards, writing down as many ideas as possible.
4. At the end of the 5–10 minutes each member of the group, in turn, reads out one of his/her ideas and places it on the table for everyone to see, without criticism or justification.
5. When all ideas are presented, members of the group place together all cards with related ideas repeating the process until the ideas are in a few groups.
6. Look for one card in each group that captures the meaning of that group.

The output of this exercise is a compilation of a maximum number of ideas under a limited number of major headings. This data can then be used with other tools to define areas for attack. One of these tools is the interrelationship diagram.

## 2. Interrelationship diagraph

This tool is designed to take a central idea, issue or problem, and map out the logical or sequential links among related factors. While this still requires a very creative process, the interrelationship diagraph begins to draw the logical connections that surface in the affinity diagram. In designing, planning, and problem solving it is obviously not enough just to create an explosion of ideas. The affinity diagram allows some organized creative patterns to emerge but the interrelationship diagraph lets *logical* patterns become apparent.

Figure 13.10 given an example of a simple interrelationship diagraph.

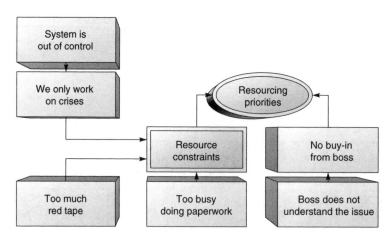

**Figure 13.10**
Example of the interrelationship diagraph

# 3.  Systems flow/tree diagram

The systems flow/tree diagram (usually referred to as a tree diagram) is used to systematically map out the full range of activities that must be accomplished in order to reach a desired goal. It may also be used to identify all the factors contributing to a problem under consideration. One of the strengths of this method is that it forces the user to examine the logical and chronological link between tasks.

Depending on the type of issue being addressed, the tree diagram will be similar to either a cause and effect diagram or a flowchart, although it may be easier to interpret because of its clear linear layout. If a problem is being considered, each branch of the tree diagram will be similar to a cause and effect diagram.

# 4.  Matrix diagrams

The matrix diagram is the heart of the seven new tools and the house of quality described in Chapter 6. The purpose of the matrix diagram is to outline the interrelationships and correlations between tasks, functions or characteristics, and to show their relative importance. There are many versions of the matrix diagram, but the most widely used is a simple L-shaped matrix known as the *quality table* in which customer demands (the whats) are analysed with respect to substitute quality characteristics (the hows). See Figure 13.11. Correlations between the two are categorized as strong, moderate and possible. The customer demands shown on the left of the matrix are determined in co-operation with the customer in a joint meeting if possible.

The right side of the chart is often used to compare current performance to competitors' performance, company plan, and potential sales points with reference to the customer demands. Weights are given to these items to obtain a 'relative quality weight', which can be used to identify the key customer demands. The relative quality weight is then used with the correlations identified on the matrix to determine the key quality characteristics.

## T-shaped matrix diagram

A T-shaped matrix is nothing more than the combination of two L-shaped matrix diagrams. Figure 13.12 shows one application, the relationship between a set of courses in a curriculum and two important sets of considerations: who should do the training for each course and which would be the most appropriate functions to attend each of the courses?

There are other matrices that deal with ideas such a product or service function, cost, failure modes, capabilities, etc., and there are at least 40 different types of matrix diagrams available.

Substitute quality characteristics

| | | MFR | Ash | Importance | Current | Best competitor | Plan | IR | SP | RQW |
|---|---|---|---|---|---|---|---|---|---|---|
| Customer demands | No film breaks | ◯ 17 | ▲ 6 | 4 | 4 | 4 | 4 | 1 | ◯ | 5.6 |
| | High rates | ◉ 23 | | 3 | 3 | 4 | 4 | 1.3 | | 4.6 |
| | Low gauge variability | ◉ 37 | ▲ 7 | 4 | 3 | 4 | 4 | 1.3 | ◯ | 7.3 |

◉  Strong correlation      IR   Improvement ratio
◯  Some correlation        SP   Sales point
▲  Possible correlation    RQW  Relative quality weight

**Figure 13.11**
An example of the matrix diagram (quality table)

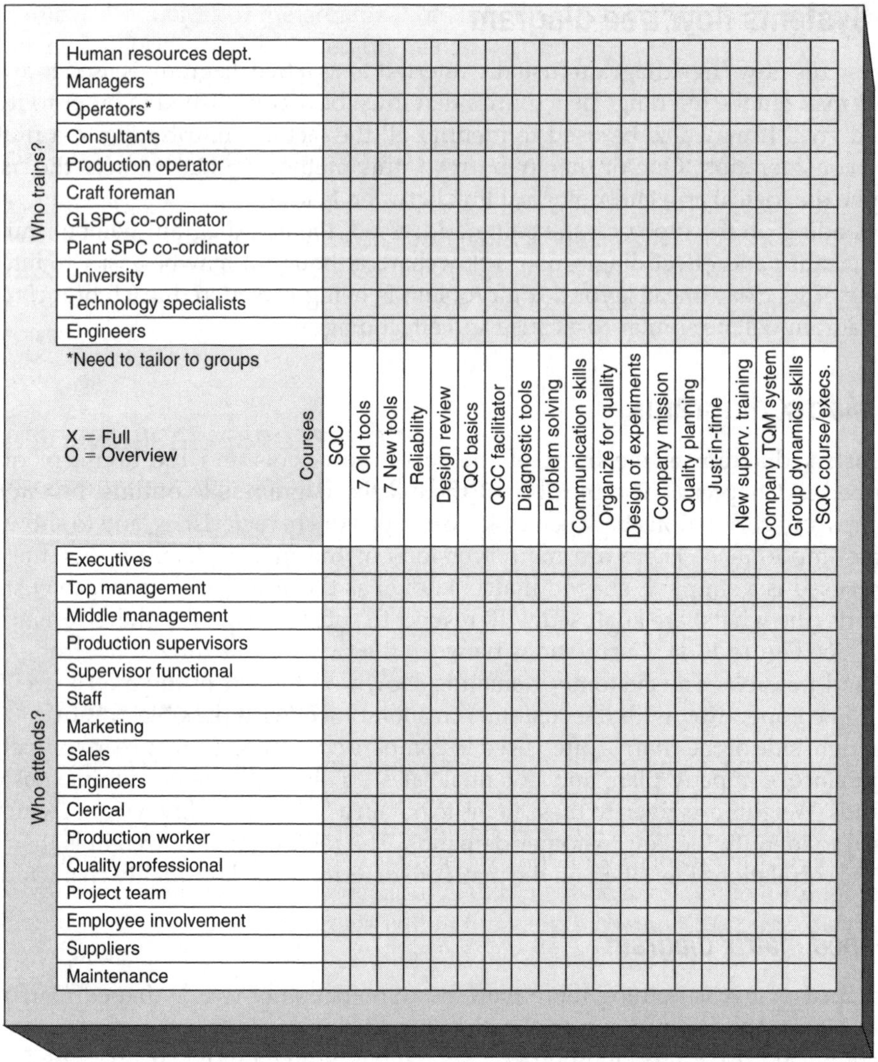

**Figure 13.12**
T-matrix diagram on company-wide training

The T-matrix diagram contains the following labels:

**Who trains?**
- Human resources dept.
- Managers
- Operators*
- Consultants
- Production operator
- Craft foreman
- GLSPC co-ordinator
- Plant SPC co-ordinator
- University
- Technology specialists
- Engineers

*Need to tailor to groups

X = Full
O = Overview

**Courses**
SQC; 7 Old tools; 7 New tools; Reliability; Design review; QC basics; QCC facilitator; Diagnostic tools; Problem solving; Communication skills; Organize for quality; Design of experiments; Company mission; Quality planning; Just-in-time; New superv. training; Company TQM system; Group dynamics skills; SQC course/execs.

**Who attends?**
- Executives
- Top management
- Middle management
- Production supervisors
- Supervisor functional
- Staff
- Marketing
- Sales
- Engineers
- Clerical
- Production worker
- Quality professional
- Project team
- Employee involvement
- Suppliers
- Maintenance

## 5.  Matrix data analysis

Matrix data analysis is used to take data displayed in a matrix diagram and arrange it so that it can be more easily viewed and show the strength of the relationship between variables. It is used most often in marketing and product research. The concept behind matrix data analysis is fairly simple, but its execution (including data gathering) is complex.

## 6.  Process decision programme chart

A PDPC is used to map out each event and contingency that can occur when progressing from a problem statement to its solution. The PDPC is used to anticipate the unexpected and plan for it. It includes plans for counter-measures on deviations. The PDPC is related to a failure mode and effect analysis and its structure is similar to that of a tree diagram.

## 7. Arrow diagram

The arrow diagram is used to plan or schedule a task. To use it, one must know the sub-task sequence and duration. This tool is essentially the same as the standard Gantt chart used in project planning. Although it is a simple and well-known tool for planning work, it is surprising how often it is ignored. The arrow diagram is useful in analysing a repetitive job in order to make it more efficient.

## Summary

What has been described in this section is a system for improving the design of products, processes, and services by means of seven 'new' qualitative tools. For the most part the seven tools are neither new nor revolutionary, but rather a compilation and modification of some methods that have been around for a long time. These tools do not replace statistical methods or other techniques, but they are meant to be used together as part of continuous process improvement.

The tools work best when representatives from all parts of an organization take part in their use and execution of the results. Besides the structure that the tools provide, the co-operation between functions or departments that is required will help break down barriers within the organizations.

While design, marketing and operations people will see the most direct applications for these tools, proper use of the 'philosophy' behind them requires participation from all parts of an organization. In addition, some of the seven new tools can be used in direct problem-solving activities.

# Taguchi methods for process improvement

Genichi Taguchi was a noted Japanese engineering specialist who advanced 'quality engineering' as a technology to reduce costs and improve quality simultaneously. The popularity of Taguchi methods today testifies to the merit of his philosophies on quality. The basic elements of Taguchi's ideas, which have been extended here to all aspects of product, service and process quality, may be considered under four main headings.

## 1. Total loss function

An important aspect of the quality of a product or service is the total loss to society that it generates. Taguchi's definition of product quality as 'the loss imparted to society from the time a product is shipped, is rather strange, since the word *loss* denotes the very opposite of what is normally conveyed by using the word *quality*. The essence of his definition is that the smaller the loss generated by a product or service from the time it is transferred to the customer, the more desirable it is.

The main advantage of this idea is that it encourages a new way of thinking about investment in quality improvement projects, which become attractive when the resulting savings to customers are greater than the cost of improvements.

Taguchi claims with some justification that any variation about a target value for a product or process parameter causes loss to the customer. The loss may be some simple inconvenience, but it can represent actual cash losses, owing to rework or badly fitting parts, and it may well appear as loss of customer goodwill and eventually market share.

The loss (or cost) increases exponentially as the parameter value moves away from the target, and is at a minimum when the product or service is at the target value.

## 2. Design of products, services and processes

In any product or service development three stages may be identified: product or service design, process design, and production or operations. Each of these overlapping stages has many steps, the output of one often being the input to others. The output/input transfer points between steps clearly affect the quality and cost of the final product or service. The complexity of many modern products and services demands that the crucial role of design be recognized. Indeed the performance of the quality products from the Japanese automotive, banking, camera, and machine tool industries can be traced to the robustness of their product and process designs.

The prevention of problems in using products or services under varying operations and environmental conditions must be built in and at the design stage. Equally, the costs during production or operation are determined very much by the actual manufacturing or operating process. Controls, including SPC methods, added to processes to reduce imperfections at the operational stage are expensive, and the need for controls *and* the production of non-conformance can be reduced by correct initial designs of the process itself.

Taguchi distinguishes between *off-line* and *on-line* quality control methods, 'quality control' being used here in the very broad sense to include quality planning, analysis and improvement. Off-line QC uses technical aids in the *design* of products and processes, whereas on-line methods are technical aids for controlling quality and costs in the *production* of products or services. Too often the off-line QC methods focus on evaluation rather than improvement. The belief by some people (often based on experience!) that it is unwise to buy a new model or a motor car 'until the problems have been sorted out' testifies to the fact that insufficient attention is given to improvement at the product and process design stages. In other words, the bugs should be removed *before* not after product launch. This may be achieved in some organizations by replacing detailed quality and reliability evaluation methods with approximate estimates, and using the liberated resources to make improvements.

## 3. Reduction of variation

The objective of a continuous quality improvement programme is to reduce the variation of key products performance characteristics about their target values. The widespread practice of setting specifications in terms of simple upper and lower limits conveys the wrong idea that the customer is satisfied with all values inside the specification band, but is suddenly not satisfied when a value slips outside one of the limits. The practice of stating specifications as tolerance intervals only can lead manufacturers to produce and despatch goods whose parameters are just inside the specification band. Owing to the interdependence of many parameters of component parts and assemblies, this is likely to lead to quality problems.

The target value should be stated and specified as the ideal, with known variability about the mean. For those performance characteristics that cannot be measured on the continuous scale, the next best thing is an ordered categorical scale such as excellent, very good, good, fair, unsatisfactory, very poor, rather than the binary classification of 'good' or 'bad' that provides meagre information with which the variation reduction process can operate.

Taguchi has introduced a three-step approach to assigning nominal values and tolerances for product and process parameters:

(a) *System design* – the application of scientific engineering and technical knowledge to produce a basic functional prototype design. This requires a fundamental understanding of the needs of the customer *and* the production environment.
(b) *Parameter design* – the identification of the settings of product or process parameters that reduce the sensitivity of the designs to sources of variation. This requires a study of the whole process system design to achieve the most robust operational settings, in terms of tolerance to ranges of the input variables.
(c) *Tolerance design* – the determination of tolerances around the nominal settings identified by parameter design. This requires a trade-off between the customer's loss due to performance variation and the increase in production or operational costs.

## 4.  Statistically planned experiments

Taguchi has pointed out that statistically planned experiments should be used to identify the settings of product and process parameters that will reduce variation in performance. He classifies the variables that affect the performance into two categories: design parameters and sources of 'noise'. As we have seen earlier, the nominal settings of the *design parameters* define the specification for the product or process. The *sources of noise* are all the variables that cause the performance characteristics to deviate from the target values. The *key* noise factors are those that represent the major sources of variability, and these should be identified and included in the experiments to design the parameters at which the effect of the noise factors on the performance is minimum. This is done by systematically varying the design parameter settings and comparing the effect of the noise factors for each experimental run.

Statistically planned experiments may be used to identify:

(a) The design parameters that have a large influence on the product or performance characteristic.
(b) The design parameters that have no influence on the performance characteristics (the tolerances of these parameters may be relaxed).
(c) The settings of design parameters at which the effect of the sources of noise on the performance characteristic is minimal.
(d) The settings of design parameters that will reduce cost without adversely affecting quality.[2]

Taguchi methods have stimulated a great deal of interest in the application of statistically planned experiments to product and process designs. The use of 'design of experiments' to improve industrial products and processes is not new – Tippett used these techniques in the textile industry more than 50 years ago. What Taguchi has done, however, is to acquaint us with the scope of these techniques in off-line quality control.

Taguchi's methods, like all others, should not be used in isolation, but be an integral part of continuous improvement.

# Six sigma

Since the early 1980s, most of the world has been in what the authors call a 'quality revolution'. Based on the simple premise that organizations of all kinds exist mainly to serve the needs of the customers of their products or services, good quality management has assumed great importance. Competitive pressures on companies and government demands on the public sector have driven the need to find more effective and efficient approaches to managing businesses and non-profit-making organizations.

In the early days of the realization that improved quality was vital to the survival of many companies, especially in manufacturing, senior managers were made aware, through national campaigns and award programmes, that the basic elements had to be right. They learned through adoption of quality management systems, the involvement of improvement teams and the use of quality tools that improved business performance could be achieved only through better planning, capable processes and the involvement of people. These are the basic elements of a TQM approach and this has not changed no matter how many sophisticated approaches and techniques come along.

The development of TQM has seen the introduction and adoption of many dialects and components, including quality circles, international systems and standards, statistical process control, business process re-engineering, lean manufacturing, continuous improvement, benchmarking and business excellence.

An approach finding favour in some companies is six sigma, most famously used in Motorola, General Electric, and Allied Signal. This operationalized TQM into a project-based system, based on delivering tangible business benefits, often directly to the bottom line. Strange combinations of the various approaches have led to lean sigma and other company-specific acronyms such as 'Statistically Based Continuous Improvement (SBCI)'.

## The six-sigma improvement model

There are five fundamental phases or stages in applying the six-sigma approach to improving performance in a process: Define, Measure, Analyse, Improve, and Control (DMAIC). These form an improvement cycle grounded in Deming's original Plan, Do, Check, Act (PDCA) (Figure 13.13). In the six-sigma approach, DMAIC provides a breakthrough strategy and disciplined methods of using rigorous data gathering and statistically based analysis to identify sources of errors and ways of eliminating them. It has become increasingly common in so-called 'six-sigma organizations', for people to refer to 'DMAIC projects'. These revolve around the three major strategies for processes to bring about rapid bottom-line achievements – design/redesign, management and improvement.

Table 13.1 shows the outline of the DMAIC steps.

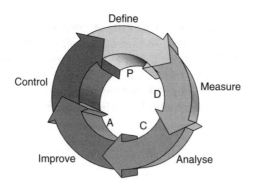

**Figure 13.13**
The six-sigma improvement model – DMAIC

**Table 13.1** The DMAIC steps

| | |
|---|---|
| D | Define the scope and goals of the improvement project in terms of customer requirements and the process that delivers these requirements – inputs, outputs, controls and resources. |
| M | Measure the current process performance – input, output and process – and calculate the short- and longer-term process capability – the sigma value. |
| A | Analyse the gap between the current and desired performance, prioritize problems and identify root causes of problems. Benchmarking the process outputs, products or services against recognized benchmark standards of performance may also be carried out. |
| I | Generate the improvement solutions to fix the problems and prevent them from recurring so that the required financial and other performance goals are met. |
| C | This phase involves implementing the improved process in a way that 'holds the gains'. Standards of operation will be documented in systems such as ISO 9000 and standards of performance will be established using techniques such as statistical process control (SPC). |

## Building a six-sigma organization and culture

Six-sigma approaches question many aspects of business, including its organization and the cultures created. The goal of most commercial organizations is to make money through the production of saleable goods or services and, in many, the traditional measures used are capacity or throughput based. As people tend to respond to the way they are being measured, the management of an organization tends to get what it measures. Hence, throughput measures may create work-in-progress and finished goods inventory thus draining the business of cash and working capital. Clearly, supreme care is needed when defining what and how to measure.

Six-sigma organizations focus on:

- understanding their customers' requirements;
- identifying and focusing on core-critical processes that add value to customers;
- driving continuous improvement by involving all employees;
- being very responsive to change;
- basing managing on factual data and appropriate metrics;
- obtaining outstanding results, both internally and externally.

The key is to identify and eliminate variation in processes. Every process can be viewed as a chain of independent events and, with each event subject to variation, variation accumulates in the finished product or service. Because of this, research suggests that most businesses operate somewhere between the three- and four-sigma level. At this level of performance, the real cost of quality is about 25–40 percent of sales revenue. Companies that adopt a six-sigma strategy can readily reach the five-sigma level and reduce the cost of quality to 10 percent of sales. They often reach a plateau here and to improve to six-sigma performance and 1 percent cost of quality takes a major rethink.

Properly implemented six-sigma strategies involve:

- leadership involvement and sponsorship;
- whole organization training;
- project selection tools and analysis;
- improvement methods and tools for implementation;
- measurement of financial benefits;

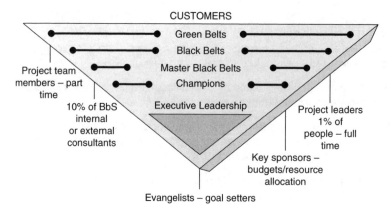

**Figure 13.14**
A six-sigma company

- communication;
- control and sustained improvement.

One highly publicized aspect of the six-sigma movement, especially in its application in companies such as General Electric (GE), Motorola, Allied Signal and GE Capital in Europe, is the establishment of process improvement experts, known variously as 'master black belts', 'black belts' and 'green belts'. In addition to these martial arts-related characters, who perform the training, lead teams and do the improvements, are other roles which the organization may consider, depending on the seriousness with which they adopt the six-sigma discipline. These include the:

- leadership group or council/steering committee;
- sponsors and/or champions/process owners;
- implementation leaders or directors – often master black belts;
- six-sigma coaches – master black belts or black belts;
- team leaders or project leaders – black belts or green belts;
- team members – usually green belts.

Many of these terms will be familiar from TQM and continuous improvement activities. The 'black belts' reflect the finely honed skill and discipline associated with the six-sigma approaches and techniques. The different levels of green, black and master black belts recognize the depth of training and expertise.

Mature six-sigma programmes, such as at GE, Johnson & Johnson and Allied Signal, have about 1 percent of the workforce as full-time black belts. There is typically one master black belt to every ten black belts or about one to every 1000 employees. A black belt typically oversees/completes five to seven projects per year, which are led by green belts who are not employed full time on six-sigma projects (Figure 13.14).

The means of achieving six-sigma capability are, of course, the key. At Motorola this included millions of dollars spent on a company-wide education programme, documented quality systems linked to quality goals, formal processes for planning and achieving continuous improvements, individual QA organizations acting as the customer's advocate in all areas of the business, a corporate quality council for co-ordination, promotion, rigorous measurement and review of the various quality systems/programmes to facilitate achievement of the policy.

# Ensuring the financial success of six-sigma projects

Six-sigma approaches are not looking for incremental or 'virtual' improvements, but breakthroughs. This is where six sigma has the potential to outperform other improvement initiatives. An intrinsic part of implementation is to connect improvement to bottom line benefits and projects should not be started unless they plan to deliver significantly to the bottom line.

Estimated cost savings vary from project to project, but reported average results range from £100 to 150 000 per project, which typically last four months. The average black belt will generate £500 000–1 000 000 benefits per annum, and large savings are claimed by the leading exponents of six sigma. For example, GE has claimed returns of $1.2 bn from its investment of $450 m.

Six-sigma project selection takes on different faces in different organizations. While the overall goal of any six-sigma project should be to improve customer results and business results, some projects will focus on production/service delivery processes, and others will focus on business/commercial processes. Whichever they are, all six-sigma projects must be linked to the highest levels of strategy in the organization and be in direct support of specific business objectives. The projects selected to improve business performance must be agreed upon by both the business and operational leadership, and someone must be assigned to 'own' or be accountable for the projects, as well as someone to execute them.

At the business level, projects should be selected based on the organization's strategic goals and direction. Specific projects should be aimed at improving such things as customer results, non-value add, growth, cost and cash flow. At the operations level, six-sigma projects should still tie to the overall strategic goals and direction but directly involve the process/operational management. Projects at this level then should focus on key operational and technical problems that link to strategic goals and objectives.

When it comes to selecting six-sigma projects, key questions such as the following should be asked:

- What is the nature of the projects being considered?
- What is the scope of the projects being considered?
- How many projects should be identified?
- What are the criteria for selecting projects?
- What types of results may be expected from six-sigma projects?

Project selection can rely on a 'top-down' or 'bottom-up' approach. The top-down approach considers a company's major business issues and objectives and then assigns a champion – a senior manager most affected by these business issues – to broadly define the improvement objectives, establish performance measures, and propose strategic improvement projects with specific and measurable goals that can be met in a given time period. Following this, teams identify processes and critical-to-quality characteristics, conduct process baselining, and identify opportunities for improvement. This is the favoured approach and the best way to align 'localized' business needs with corporate goals.

At the process level, six-sigma projects should focus on those processes and critical-to-quality characteristics that offer the greatest financial and customer results potential. Each project should address at least one element of the organization's key business objectives, and be properly planned.

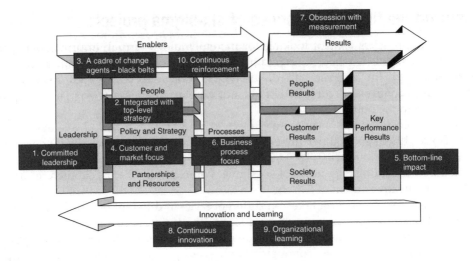

**Figure 13.15**
The Excellence Model and six sigma

## Concluding observations and links with TQM, SPC excellence, etc.

Six sigma is not a new technique, its roots can be found in total quality management (TQM) and statistical process control (SPC) but it is more than TQM or SPC rebadged. It is a framework within which powerful TQM and SPC tools can be allowed to flourish and reach their full improvement potential. With the TQM philosophy, many practitioners promised long-term benefits over 5–10 years, as the programmes began to change hearts and minds. Six sigma is about delivering breakthrough benefits in the short term and is distinguished from TQM by the intensity of the intervention and pace of change.

Excellence approaches such as the EFQM Excellence Model and six sigma are complementary vehicles for achieving better organizational performance. The Excellence Model can play a key role in the baselining phase of strategic improvement, while the six-sigma breakthrough strategy is a delivery vehicle for achieving excellence through:

1. Committed leadership.
2. Integration with top-level strategy.
3. A cadre of change agents – black belts.
4. Customer and market focus.
5. Bottom-line impact.
6. Business process focus.
7. Obsession with measurement.
8. Continuous innovation.
9. Organizational learning.
10. Continuous reinforcement.

These are 'mapped' onto the Excellence Model in Figure 13.15 (see also Reference 3).

There is a whole literature and many conferences have been held on the subject of six sigma and it is not possible here to do justice to the great deal of thought that has gone into the structure of these approaches. As with Taguchi methods the major contribution of six sigma has not been in the creation of new technology or methodologies, but in bringing to the attention of senior management the need for a disciplined structured approach and their commitment, if real performance and bottom-line improvements are to be achieved.

## Technical note

Sigma is a statistical unit of measurement that describes the distribution about the mean of any process or procedure. A process or procedure that can achieve plus or minus *six-sigma* capability can be expected to have a defect rate of no more than a few parts per million, even allowing for some shift in the mean. In statistical terms, this approaches *zero defects*.

In a process in which the characteristic of interest is a variable, defects are usually defined as the values which fall outside the specification limits. Assuming and using a normal distribution of the variable, the percentage and/or parts per million defects can be found. For example, in a centred process with a specification set average $\pm3\sigma$ there will be 0.27 percent or 2700 ppm defects. This may be referred to as 'an unshifted $\pm3$ sigma process' and the quality called '$\pm3$ sigma quality'. In an 'unshifted $\pm6$ sigma process', the specification range is average $6\sigma$ and it produces only 0.002 ppm defects.

It is difficult in the real world, however, to control a process so that the mean is always set at the nominal target value – in the centre of the specification. Some shift in the process mean is expected. The ppm defects produced by such a 'shifted process' are the sum of the ppm outside each specification limit, which can be obtained from the normal distribution.

# The 'DRIVE' model for continuous improvement

Oakland and his colleagues have developed a framework for a structured approach to problem solving in teams, the *DRIVE* model. The mnemonic provides landmarks to keep the team on track and in the right direction.

| | |
|---|---|
| *Define* | the problem. *Output:* written definition of the task and its success criteria. |
| *Review* | the information. *Output:* presentation of known data and action plan for further data. |
| *Investigate* | the problem. *Output:* documented proposals for improvement and action plans. |
| *Verify* | the solution. *Output:* proposed improvements that meet success criteria. |
| *Execute* | the change. *Output:* task achieved and improved process documented. |

The various stages are discussed in detail below. Some of the steps may be omitted if they have already been answered or are clearly not relevant to a particular situation.

## Define

At this stage the improvement team is concerned with gaining a common understanding and agreement within the groups of the task that it faces, in terms of the problem to be solved and the boundaries of the process or processes that contain it. It is necessary to generate at the outset a means of knowing when the team has succeeded. There is no concern at this stage with solutions. The key steps are:

1. *Look at the task*
   Typical questions:
   (a) What is the brief?
   (b) Is it understood?
   (c) Is there agreement with it?
   (d) Is it sufficiently explicit?
   (e) Is it achievable?
   There may be a need for clarification with the 'sponsor' at this stage, and possibly some redefinition of the task.
2. *Understand the process*
   (a) What processes 'contain' the problem?
   (b) What is wrong at present?
   (c) Brainstorm – ideas for improvement.
   (d) Perhaps draw a rough flowchart to focus thinking.
3. *Prioritize*
   (a) Set boundaries to the investigation.
   (b) Make use of ranking, Pareto, matrix analysis, etc., as appropriate.
   (c) Review and gain agreement in the team of what is 'do-able'.
4. *Define the task*
   (a) Produce a written description of the process or problem area that can be confirmed with the team's sponsor.
   (b) Confirm agreement in the team.
   (c) This step may generate further questions for clarification by the sponsor of the process.
5. *Agree success criteria*
   (a) List possible success criteria. How will the team know when it has succeeded?
   (b) Choose and agree success criteria in the team.
   (c) Discuss and agree time scales for the project.
   (d) Agree with 'sponsor'.
   (e) Document the task definition, success criteria and time scale for the complete project.

## Review

This stage is concerned with finding out what information is already available, gathering it together, structuring it, identifying what further information might be needed, and agreeing in the team WHAT is needed, HOW it is going to be obtained, and WHO is going to get it.

1. *Gather existing information*
   (a) Locate sources – verbal inputs, existing files, charts, quality records, etc.
   (b) Go and collect, ask, investigate.
2. *Structure information*
   Information may be available but not in the right format.
3. *Define gaps*
   (a) Is enough information available?
   (b) What further information is needed?

(c) What equipment is affected?

(d) Is the product/service from one plant or area?

(e) How is the product/service at fault?

4. *Plan further data collection*

(a) Use any data already being collected.

(b) Draw up check sheet(s).

(c) Agree data-collection tasks in the team – WHO, WHAT, HOW, WHEN.

(d) Seek to consult others, where appropriate. Who actually has the information? Who really understands the process?

(e) This is a good opportunity to start to 'extend the team' in preparation for the *Execute* stage later on.

## Investigate

This stage is concerned with analysing all the data, considering all possible improvements, and prioritizing these to come up with one or more solutions to the problem, or improvements to the process, which can be verified as being the answer which meets the success criteria.

1. *Implement data-collection action plan*

Check at an early stage that the plan is satisfying the requirements.

2. *Analyse data*

(a) What picture is the data painting?

(b) What conclusions can be drawn?

(c) Use all appropriate tools to give a clearer picture of the process.

3. *Generate potential improvements*

(a) Brainstorm improvements.

(b) Discuss all possible solutions.

(c) Write down all suggestions (have there been any from outside the team?).

4. *Agree proposed improvements*

(a) Prioritize possible proposals.

(b) Decide what is achievable in what time scales.

(c) Work out how to test proposed solution(s) or improvement(s).

(d) Design check sheets to collect all necessary data.

(e) Build a checking/verifying plan of action.

## Verify

This stage is concerned with testing the plans and proposals to make sure that they work before any commitment to major process changes. This may require a relatively short discussion round a table in a meeting or lengthy pilot trials in a laboratory, office or even a main operations area or production plant.

1. *Implement action plan*

Carry out the agreed tests on the proposals.

2. *Collect data*

(a) Consider the use of questionnaires, if appropriate.

(b) Make sure the check sheets are accumulating the data properly.

3. *Analyse data*

4. *Verify that success criteria are met*
   (a) Compare performance of new or changed process with success criteria from *Define* stage.
   (b) If success criteria are not met, return to appropriate stage in drive model (usually the *Investigate*) stage.
   (c) Continue until the success criteria have been met. For difficult problems, it may be necessary to go a number of times round this loop.

## Execute

This stage is concerned with selling the solution or process improvement to others, e.g. the process owner, who may not have taken part in the investigation but whose commitment is vital to ensure success. Part of this stage may well be the need to address the existing documented quality management system, especially in the case of BS 5750/ISO 9000-registered organizations.

1. *Develop implementation plan to gain commitment*
   (a) Is there commitment from others? Consider all possible impacts.
   (b) Actions?
   (c) Timing?
   (d) Selling required?
   (e) Training required for new or modified process?
2. *Review appropriate system paperwork/documentation*
   (a) Who should do this? The team? The activity/process owner?
   (b) What are the implications for other systems?
   (c) What controlled documents are affected?
3. *Gain agreement to all facets of the execution plan from the process owner*
4. *Implement the plan*
5. *Monitor success*
   (a) Extent of original team involvement? Initially perhaps and then at intervals?
   (b) 'Delegate' to process owner/department concerned? At what stage?
6. *Responsibility*
   Balance between team taking responsibility for meeting its agreed project success criteria and ownership within the organization of processes and continuous improvement. In the case of registered firms, responsibility for continued monitoring can be delegated to the quality management.

The problem-solving tools most likely to be used at each of the DRIVE stages are shown in Table 13.2.

## An example of the DRIVE model used in practice

The example below shows how the DRIVE model for a particular project worked out in practice. The sort of responses made by the team are given in quotes thus ' '.

*Stated task:* 'We lose orders because our response time in making quotations is too long. We must significantly reduce our response time.'

1. *Define stage*
   (a) Look at the task:
      (i) Can we accept the problem as stated? 'Yes, this is generally known to be a problem area.'

**Table 13.2** Likely tools in the DRIVE model

| | Define | Review | Investigate | Verify | Execute |
|---|:---:|:---:|:---:|:---:|:---:|
| Brainstorming | ■ | ■ | ■ | | |
| Cause and effect diagram | ■ | | | | |
| Pareto | ■ | ■ | | ■ | ■ |
| Matrix analysis | ■ | ■ | ■ | | |
| Check sheets | | ■ | ■ | ■ | ■ |
| Flowcharts | | ■ | ■ | | ■ |
| Force field | | ■ | | | ■ |
| Scatter diagram | | ■ | | ■ | |
| Histograms | | ■ | | | |
| Charts | | ■ | ■ | ■ | ■ |
| Project bar chart | | | | ■ | ■ |

(ii) Does this apply to all customers, or to specific product lines? 'All customers.'

(iii) Does anyone measure response time at the moment? 'There was a one-off assessment a long time ago but it is not routinely measured.'

(iv) Have we the right expertise in our team to tackle it? 'Not really sure until we understand the problem.'

(v) Can we succeed with this problem? 'Yes, if we don't let it get too big.'

(b) Understand the process:

(i) What processes 'contain' this problem? 'Customer visits by sales reps, telephone enquiries system, telex/fax enquiries, development department (technical vetting and provision of samples), pricing department', etc.

(ii) What is wrong at present? 'No real liaison between departments, labs have other priorities, visit reports are not explicit.'

(c) Prioritize:

(i) What should be the boundaries of our investigation? 'Enquiries arising from direct sales visits to customers compose about 70 percent of all enquiries. We will restrict our project to this area initially.'

(d) Define the task

(e) Agree success criteria

(f) In response to (d) and (e) the team finally documented:

'Customer quotations project:
Our task is to investigate and reduce delays in the handling of those customer enquiries which arise from direct visits by our sales force. The project will be conducted in three phases:

1. (a) To establish the average time (in days) between the salesperson's visit and receipt by the customer of our quotation.
   (b) To agree a target reduction in the average response time to enquiries.

2. To make recommendations that will enable the target reduction in response time to be achieved.

3. To implement the recommendations and monitor response times on a sample basis to demonstrate that the desired reduction has been achieved.

Milestones for the completion of each phase, measured from the formal go-ahead date for these proposals, are:

Phase 1, 1 month

Phase 2, 4 months

Phase 3, 8 months.'

2. *Review stage*

   The team was not able to locate the original study report, but did discover a memo that gate the following summary:

   ■ Average time to process a quotation request: 17 days

   ■ Average time of quotations judged 'too late': 20 days

   ■ Percentage of quotations 'too late': 30 percent

   From this, the team concluded that the average time would have to be reduced to about nine days to give a frequency of exceeding 20 days of only 1 in 100, i.e. only 1 percent of quotations 'too late'.

3. *Investigate stage*

   The team constructed flowcharts of the various stages of the process. Major 'grey areas' occurred in dealing with quotations for new(er) products or new customers where more technical letting by the laboratory was required, because (a) customer requirements were not clear, and (b) salesmen were not authorized to offer new specifications without checking with the technical department. It was in these areas that the sales visit reports were not sufficiently specific.

   The flowchart of the process was changed to include a path giving early warning to the technical department of requirements from 'new business areas', by identifying specifically named customers and product types. Quotations in these areas were treated with priority.

   A check sheet to measure quotation turn-round times was designed.

4. *Verify stage*

   ■ The modified process was implemented.

   ■ The check sheet was used to gather data.

   ■ A 'c chart' was used to monitor the average turn-round time in days for each week's orders, with the new procedure introduced at week 10. Action and warning lines for the chart were based on a target average of nine days. The ensuing chart is shown in Figure 13.16.

5. *Execute stage*

   ■ The above data was presented to a meeting of the sales and technical departments.

   ■ The changed procedures were agreed, documented and circulated, including the list of current customers and products for special vetting.

   ■ A procedure, which was documented, called for *all* quotations for new customers and products to be added to the special list and retained until familiarity enabled them to become 'standard'. They were then removed from the list for special attention.

   ■ Continued monitoring, using the chart in Figure 13.15, showed the average turn-round time reduce eventually to ten days. Only 2 percent of quotations were then taking longer than 20 days.

The DRIVE framework has been found invaluable in structuring the thinking and approach by numerous situations including large project management.

Like the six-sigma DMAIC model it helps groups of people cover all aspects of the continuous improvement cycle.

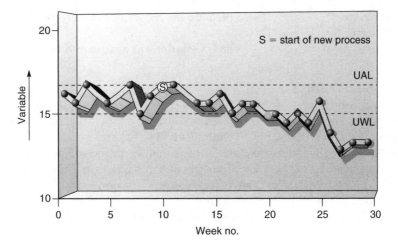

**Figure 13.16**
Charting the effects of the improvement through DRIVE

# References

1. Checkland, P. (1990) *Soft Systems Methodology in Action*, Wiley.
2. Oakland, J.S. (2002) *Statistical Process Control*, 5th edition, Butterworth-Heinemann, Oxford.
3. Porter, L. (2002) 'Six Sigma Excellence' by Les Porter, *Quality World*, April, pp. 12–15.

## *Chapter highlights*

### A systematic approach

- Numbers and information will form the basis for understanding, decisions, and actions in never-ending improvement – record data, use/analyse data, act on results.
- A set of simple tools is needed to interpret fully and derive maximum use from data. More sophisticated techniques may need to be employed occasionally.
- The effective use of the tools requires the commitment of the people who work on the processes. This in turn needs management support and the provision of training.

### Some basic tools and techniques

- The basic tools and the questions answered are:

| | |
|---|---|
| Process flowcharting | – what is done? |
| Check/tally charts | – how often is it done? |
| Histograms | – what do variations look like? |
| Scatter diagrams | – what are the relationships between factors? |
| Stratification | – how is the data made up? |
| Pareto analysis | – which are the big problems? |
| Cause and effect analysis and brainstorming (also CEDAC and NGT) | – what causes the problem? |
| Force-field analysis | – what will obstruct or help the change or solution? |

Continuous improvement

**279**

| Emphasis curve | – which are the most important factors? |
| Control charts | |
| (including cusum) | – which variations to control and how? |

### Statistical process control

- People operating a process must know whether it is capable of meeting the requirements, know whether it is actually doing so at any time, and make correct adjustments when it is not. SPC techniques will help here.
- Before using SPC, it is necessary to identify what the process is, what the inputs/outputs are, and how the suppliers and customers and their requirements are defined. The most difficult areas for this can be in non-manufacturing.
- All processes can be monitored and brought 'under control' by gathering and using data. SPC methods, with management commitment, provide objective means of controlling quality in any transformation process.
- SPC is not only a tool kit, it is a strategy for reducing variability, part of never-ending improvement. This is achieved by answering the following questions:
  - Are we capable of doing the job correctly?
  - Do we continue to do the job correctly?
  - Have we done the job correctly?
  - Could we do the job more consistently and on target?
- SPC provides knowledge and control of process capability.
- SPC techniques have value in the service sector and in the non-manufacturing areas, such as marketing and sales, purchasing, invoicing, finance, distribution, training and personnel.

### Some additional techniques for process design and improvement

- Seven 'new tools' may be used as part of quality function deployment (QFD, see Chapter 6) or to improve processes. These are systems and documentation methods for identifying objectives and intermediate steps in the finest detail.
- The seven new tools are: affinity diagram, interrelationship digraph, tree diagram, matrix diagrams or quality table, matrix data analysis, PDPC, and arrow diagram.
- The tools are interrelated and their promotion and use should lead to better designs in less time. They work best when people from all parts of an organization are using them. Some of the tools can be used in activities related to problem solving and design.

### Taguchi methods for process improvement

- Genichi Taguchi has advanced 'quality engineering' as a technology to reduce costs and make improvements.
- Taguchi's approach may be classified under four headings: total loss function; design of products, services and processes; reduction of variation; and statistically planned experiments.
- Taguchi methods, like all others, should not be used in isolation, but as an integral part of continuous improvement.

### Six sigma

- Six sigma is not a new technique – its origins may be found in TQM and SPC. It is a framework through which powerful TQM and SPC tools flourish and reach their full potential. It delivers breakthrough benefits in the short term through the intensity and speed of change. The Excellence Model is a useful framework for mapping the key six-sigma breakthrough strategies.

■ A process that can achieve six-sigma capability (where sigma is the statistical measure of variation) can be expected to have a defect rate of a few parts per million, even allowing for some drift in the process setting.

■ Six sigma is a disciplined approach for improving performance by focusing on enhancing value for the customer and eliminating costs which add no value.

■ There are five fundamental phases/stages in applying the six-sigma approach: Define, Measure, Analyse, Improve, and Control (DMAIC). These form an improvement cycle similar to Deming's Plan, Do, Check, Act (PDCA), to deliver the strategies of process design/redesign, management and improvement, leading to bottom-line achievements.

■ Six-sigma approaches question organizational cultures and the measures used. Six-sigma organizations, in addition to focusing on understanding customer requirements, identify core processes, involve all employees in continuous improvement, are responsive to change, base management on fact and metrics, and obtain outstanding results.

■ Properly implemented six-sigma strategies involve: leadership involvement and sponsorship, organization-wide training, project selection tools and analysis, improvement methods and tools for implementation, measurement of financial benefits, communication, control and sustained improvement.

■ Six-sigma process improvement experts, named after martial arts – master black belts, black belts and green belts – perform the training, lead teams and carry out the improvements. Mature six-sigma programmes have about 1 percent of the workforce as black belts.

### The DRIVE framework for continuous improvement

■ A structured approach to problem solving is provided by the DRIVE model: Define the problem, Review the information, Investigate the problem, Verify the solution, and Execute the change.

■ After initial problems are solved, others should be tackled – successful solutions motivating new teams. In all cases teams should follow a disciplined approach to problem solving, using proven techniques.

■ The DRIVE model helps structure – thinking and approaches and, like the six-sigma DMAIC framework, can be invaluable in all aspects of continuous improvement.

# Human resource management

# Introduction

The cyclic variation of demand for construction sector products and services is perhaps among the most extreme of all industry sectors – investment in the property sector varies widely. It is at the whim of investors – share market buoyancy, taxation policy and government interest rate policy are just three of its main drivers. This has led to extreme highs and lows in demand; the variation between peaks and troughs at times exceeding 100 percent. This naturally places a consequent strain on resources at the peaks, forcing many in the sector to look for work elsewhere during the slumps.

This relative instability and the consequent cyclic hiring and downsizing in many companies may have contributed to an attitude, among management, that staff can simply be brought in whenever needed. However, in recent years, the way in which people are managed and developed at work has come to be recognized as one of the primary keys to improved and sustained organizational performance. This is reflected by popular idioms such as 'people are our most important asset' or 'people make the difference'. Indeed, such axioms now appear in the media and on corporate public relations documents with such regularity, that the accuracy and integrity of such assertions has begun to be questioned (see, for example, References 1 and 2). This chapter draws on some of the research undertaken by the European Centre for Business Excellence (EC*for*BE), the research and education division of Oakland Consulting, which focused on world-class, successful and, in many cases, award-winning organizations. It describes the main people management activities that are currently being used in these leading edge organizations.

There is an overwhelming amount of evidence that successful organizations pay much more than lip service to the claim that people are their most important resource. This is consistent with the recognition in the past decade that intellectual capital reflects a significant part of any company's value; and that knowledge management (Chapter 16) is a key strategic activity, especially if the tacit knowledge within an organization is to be properly leveraged for the benefit of the company.

On a general level, successful organizations share a fundamental philosophy to value and invest in their employees. More specifically, world-class organizations value and invest in their people through the following activities:

- Strategic alignment of human resource management (HRM) policies.
- Effective communication.
- Employee empowerment and involvement.
- Motivation through recognition of excellence.
- Training and development.
- Teams and teamwork.
- Review and continuous improvement.
- The fostering of social cohesion.

GraniteRock, one of the case studies in the book (p. 375), is rated by *Fortune* magazine as the sixth preferred employer in the entire US economy. Examine the study carefully and gain some insight into the way in which its people are central to the development strategy of the company; particularly, the way in which individual attention is given to the development of every person.

# Strategic alignment of HRM policies

It is clear that leading edge organizations adopt a common approach or plan, illustrated in Figure 14.1, to align their HR policies to the overall business strategy.

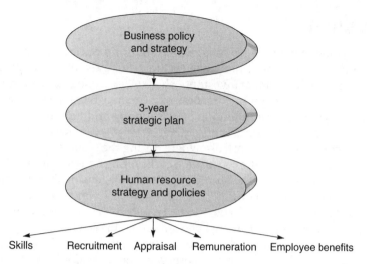

**Figure 14.1**
Strategic alignment of HRM policies

Key elements of the HR strategy (e.g. skills, recruitment and selection, health and safety, appraisal, employee benefits, remuneration, training, etc.) are first identified, usually by the HR director who then reports regularly to the board and the HR plan, typically spanning three years, is aligned with the overall business objectives and is an integral part of company strategy. For example, if a business objective is to expand at a particular site, then the HR plan provides the necessary additional manpower with the appropriate skills profile and training support. The HR plan is revised as part of the overall strategic planning process. Divisional boards then liaise with the HR director to ensure that the HR plan supports and is aligned with overall policy.

In addition, the HR director holds regular meetings with key personnel from employee relations, health and safety, training and recruitment, etc., to review and monitor the HR plan, drawing upon published data and benchmarking activities in all relevant areas of policy and practice. Divisional managing directors and the HR director report progress on how the HR plan is supporting the business to the quality committee or board. An overview of this human resource process is illustrated in Figure 14.2.

Although it is beyond the scope of this chapter to make a detailed examination of HR policy, it is prudent to outline briefly some of the common practices that emerged from the identified best practice relating to selection and recruitment, skills and competencies, appraisal, and employee reward, recognition and benefits.

## Selection and recruitment

The following practices are common among the organizations studied regarding selection and recruitment:

1. Ensure fairness by using standard tools and practices for job descriptions and job evaluations.
2. Enhance 'transparency' and communication through jargon-free booklets that provide detailed information to new recruits about performance, appraisal, job conditions and so on.
3. Ensure that job descriptions are responsibility rather than task oriented.

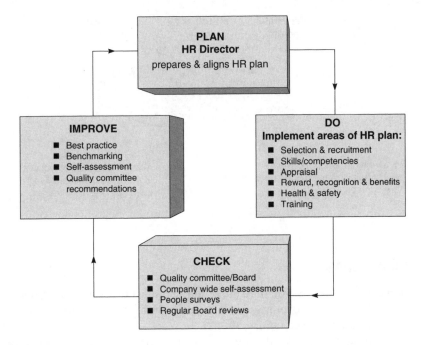

**Figure 14.2**
Human resource process

4. Train all managers and supervisors in interviewing and other selection techniques.
5. Align job descriptions and competencies so that people with the appropriate skills and attributes for the job are identified.
6. Compare the organization's employment terms and conditions (on a regular basis) with published data on best practice and documents to ensure the highest standards are being met.
7. Review HR policies regularly to ensure that they fully reflect legislative and regulatory changes together with known best practice.
8. In the recruitment of new graduates, start early, at least six months before graduation as the best students make selections early to pick the best available jobs.

## Skills/competencies

Since the publication of *The Competent Manager*,[3] the terms competence and competency have been widely used and underpin the work of the specific bodies in any country associated with vocational qualifications and occupational standards. In line with this, good organizations have skills/competence-based human resource management policies underpinning selection and recruitment, training and development, promotion and appraisal.

Although numerous lists of generic management competencies have been published, in essence they are all very similar and are closely allied to the core management competencies underpinning HR policies: leadership, motivation, people management skills, team working skills, comprehensive job knowledge, planning and organizational skills, customer focus, commercial and business awareness, effective communication skills – oral and written and change management skills, coupled with a drive for continuous improvement.

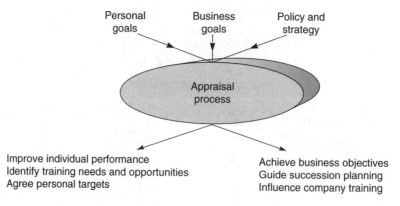

**Figure 14.3**
The appraisal process

## Appraisal process

As with other HR policies, the main thrust of the appraisal process is alignment – alignment of personal, team and corporate goals coupled with appraisals to help individuals achieve their full potential (see Figure 14.3).

Without exception, the appraisal systems described in world-class organizations were based on objectives. Agreed objectives were also time based so that completion dates provided the opportunity for automatic review processes.

Typically, employees are appraised annually and the managers conducting appraisals attend training in appraisal skills. Before each appraisal, the appraisee and appraiser each complete preparation forms thus making the interview a two-way discussion on performance against objectives during, say, the previous 12 months. Training and development work to achieve the objectives is agreed and, if necessary, additional help is available in the form of advice and counselling.

## Employee reward, recognition and benefits

### Reward and recognition

Although an in-depth study of the policies and practices relating to financial reward and recognition was beyond the scope of the research, it is possible to highlight the following activities that were common among the organizations:

- Rewards are based on consistent, quality-based performance.
- Awards are given to employees but also to customers, suppliers, universities, colleges, students, etc.
- Financial incentives are offered for company-wide suggestions and new idea schemes.
- Internal promotion, for example, from non-supervisory roles to divisional managing directors encourage a highly motivated workforce and enhance job security.
- Commendations include ad hoc recognition for length of service, outstanding contributions, etc.
- Recognition is given through performance feedback mechanisms, development opportunities, pay progressions and bonuses.
- Recognition systems operate at all levels of the organization but with particular emphasis on informal recognition ranging from a personal 'thank you' to recognition at team meetings and events.

With regard to employee benefits, it is well documented that benefits are seen as a tangible expression of the psychological bond between employers and employees. However, to maximize effectiveness, benefits packages should be able to be selected on the basis of what is good for the employee as well as the employer. Moreover, when employees can design their own benefits package both they and the company benefit.

Leading edge organizations favour a 'cafeteria' approach to employee benefits and in recent years there has been increasing interest in this idea of cafeteria benefits to maximize flexibility and choice, particularly in the area of fringe benefits, which can make up a high proportion of the total remuneration package. Under this scheme, the company provides a core package of benefits to all employees (including salary) and a 'menu' of other costed benefits (e.g. personal medical care, dental care, company car, health insurance, etc.) from which the employee can select their personal package.

Some of the ideas underpinning cafeteria benefits sit well with the literature on motivation – to emphasize that different individuals have different needs and expectations from work. Moreover, through communicating the benefits package and providing employees with benefit flexibility, the positive impact is further increased; not only are employees more likely to get what benefits they want, but also communication makes them more aware of the benefits they are gaining thus informing and increasing morale.

# Effective communication

Effective communication emerges from the research as an essential facet of people management – be it communication of the organization's goals, vision, strategy and policies or the communication of facts, information and data. For business success, regular, two-way communication, particularly face to face with employees, is an important factor in establishing trust and a feeling of being valued. Two-way communication is regarded as both a core management competency and as a key management responsibility. For example, a typical list of management responsibilities for effective communication is to:

- Regularly meet all their people.
- Ensure people are briefed on key issues in language free of technical jargon.
- Communicate honestly and as fully as possible on all issues which affect their people.
- Encourage team members to discuss company issues and give upward feedback.
- Ensure issues from team members are fed back to senior managers and timely replies given.

Regular two-way communication also involves customers, shareholders, financial communities and the general public.

## Communications process

Successful organizations follow a systematic process for ensuring effective communications as shown in Figure 14.4.

### Plan

Typically, the HR function, e.g. the HR director, is responsible for the communication process. He/she assesses the communication needs of the organization and liaises with

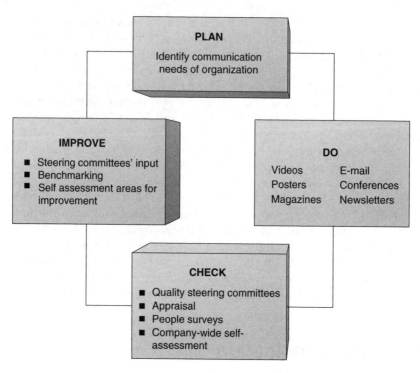

**Figure 14.4**
Best practice communications process

divisional directors, managers or local management teams to ensure that the communication plans are in alignment with overall policy and strategy. A communication programme accompanies any major changes on organization policy or objectives.

## *Do*

A comprehensive mix of diverse media are used to support effective communication throughout their organization. These include:

| | | |
|---|---|---|
| Videos | Posters | Open-door policies |
| Surveys | Campaigns | E-mail |
| Magazines | Briefings | Notice boards |
| Newsletters | Conferences | Internet/Intranet |
| Appraisals | Meetings | Focus groups |

It is evident that the introduction of electronic systems has brought about radical changes in communications. Typically employees are able to access databases, spreadsheets, word processing, e-mail and diary facilities. Information on business performance, market intelligence and quality issues can also be easily and quickly cascaded. Further, video conferencing is used to facilitate internal face-to-face communications with major customers across the world, resulting in substantial savings on travel and associated costs. Furthermore, provision is made for depots, units, regions, divisions, departments, etc., to hold 'virtual' meetings and conferences. Feedback questionnaires then check that events are valuable and help the planning of future events.

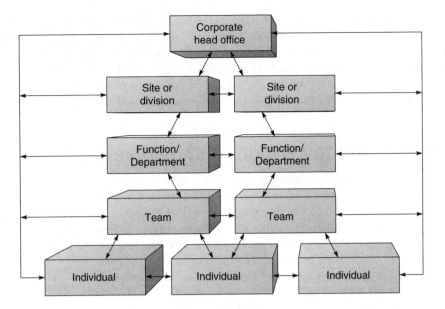

**Figure 14.5**
Multi-directional communications structure

## Check

Quality steering or review committees, people surveys, appraisal and company-wide self-assessment are used to review the effectiveness of the communications process. Appraisal and staff survey data are analysed to ensure that the communications process is continuing to deliver effective upward, downward and lateral communications. Reports are then made quarterly, six-monthly and/or annually to the chief executive and/or the most senior team on the effectiveness and relevance of the communications process. The people survey data are also used to ascertain employee perceptions and to keep in touch with current opinion.

## Improve

The results of the various review processes highlight areas for improvement and results are verified by benchmarking against, for instance, a national survey. Quality steering committees then put forward recommendations for future planning and continuous improvements.

## Communications structure

Successful organizations place great emphasis on communication channels that enable people at all levels in the organization to feel able to talk to each other. Consequently, managers are not only trained but 'are committed to being open-minded, honest, more visible and approachable'. Many formal and informal communication mechanisms exist, all designed to foster an environment of open dialogue, shared knowledge, information and trust in an effective upward, downward, lateral and cross-functional structure such as the one illustrated in Figure 14.5.

TNT, for example, have a regional structure that provides a link between local and central management which ensures that the chain of communication is complete so that information can cascade down and rise up. Briefings of senior managers by executives are followed by briefings of middle managers and so on all the way down the line. The same chain also works in reverse to facilitate bottom-up communications.

(See also Chapter 15 for more detail on the communication process.)

# Employee empowerment and involvement

To encourage employee commitment and involvement, successful organizations place great importance on empowering their employees. The positive effects of employee empowerment are well documented but the notion has been challenged with some writers claiming that it is not possible to empower people – rather, it is possible only to create a climate and a structure in which people will take responsibility. Nonetheless, it is clear that the organizations studied in the research considered empowerment to be a key issue *and* made efforts to create a working environment that was conducive to the employees taking responsibility.

The Dana Commercial Credit Corporation (DCC), for example, subscribe to the importance of employee empowerment for their impressive customer satisfaction scores that are two points higher than the industry average (on a five-point scale). DCC empower their people by encouraging employees to:

- Set their own goals.
- Judge their own performance.
- Take ownership of their actions.
- Identify with DCC (e.g. to become stock shareholders).

Similarly, TNT report that 'all employees are empowered to respond to normal and extraordinary situations without further recourse' and that they have 'worked hard to create a no blame culture where our people are empowered to take decisions to achieve their objectives.'

People empowerment is also a key issue at Texas Instruments. Here empowerment is built into the TI operational approach, organized around processes, by stimulating creativity and encouraging quality teams. Along the same lines, Hewlett-Packard, advocate teamwork and high levels of empowerment combined with a strong setting of objectives and freedom for employees to achieve them. Similarly, the Eastman Chemical Company describe how they focus on employee empowerment and have learnt that a company cannot empower employees who do not care, do not have authority and do not have the appropriate skills.

To address the above issues, management map out processes to provide employees with the necessary authority and skills. In addressing the issue of not caring, employee surveys reveal that appraisal systems can be a major roadblock and the appraisal process may need to be revised.

## Common initiatives

There are three common initiatives in successful organizations which place great store by:

1. *Corporate employee suggestion schemes* – these provide a formalized mechanism for promoting participative management, empowerment and employee involvement.

2. *Company-wide culture change programmes* – in the form of workshops, ceremonies and events used to raise awareness and to empower individuals and teams to practise continuous improvement.
3. *Measurement of key performance indicators* (KPIs) – whereby the effectiveness of staff involvement and empowerment is measured by improvements in human resource KPIs, such as labour turnover, accident rate, absenteeism, and lost time through accidents. Typically, KPI measurements, coupled with appraisal feedback and survey results, are regularly reviewed by the HR director who uses the information as the basis for reports and suggestions for improvements to the board.

On a more general level, successful organizations increase commitment by empowering and involving more and more of their employees in formulating plans that shape the business vision. As more people understand the business and where it is planned to go, the more they become involved in and committed to developing the organization's goals and objectives.

> The Graniterock case study (p. 375) demonstrates this kind of employee involvement in process improvement. It is reported that with an employee base of 700, up to 100 quality improvement teams are at work each year improving processes and solving problems.

# Motivation through recognition of excellence

Publicly recognizing excellent contributions by individuals and teams is an essential part of a HR programme – it is key to motivating people and engendering an ongoing commitment to continuous improvement. This, however, requires that processes be put in place to identify and recognize excellence and that the forums and media for recognizing outstanding achievements be planned rather than ad hoc.

Line managers have the responsibility for mentoring those immediately under them: setting training and development goals and reviewing performance. This also places them in the ideal position to identify outstanding achievements. However, guidance has to be provided by senior HR management and the process of identification has to be regular and ongoing.

Case study company Graniterock, for example, has a Recognition Day, every year in every branch, on which senior executives attend and participate in face-to-face award ceremonies. On this day, people's contribution to the company and their life achievements are publicly recognized. Naturally, recognition is also afforded through the company's regular newsletter; this type of recognition could be done on a company intranet. The most important aspects of a Recognition Day are that senior management be involved and that the process be planned and held regularly.

# Training and development

The training and development of people at work has increasingly come to be recognized as an important part of human resource management. Through the 1980s, major changes in many organizations resulted in increasing workloads, the introduction of

new technology and wider ranges of tasks, all of which required training provision. During the 1990s, initiatives such as ISO 9000, TQM, Investors in People (IiP), benchmarking and self-assessment against frameworks such as the EFQM Excellence Model have further highlighted the need for properly trained employees.

---

Among the case study companies, Babtie (p. 493) is accredited to the IiP Standard. To create strong professional grounding, Babtie requires all new staff joining the company straight out of university to undertake a structured three-year training programme. Graniterock (p. 375) has adopted a strategy implementing individual training programmes for employees.

---

It is widely acknowledged that many writers and practising managers sing the praises of training, saying it is a 'symbol of the employers' commitment to staff', or that it shows an organization's strategy is based on 'adding value rather than lowering costs'. However, others claim that a lack of effective training predominates in many organizations today and that serious doubts remain as to whether or not management actually *does* invest in the training of their human resources.

It is perhaps not surprising then that research on successful and award-winning organizations revealed an ongoing commitment to investing in the provision of planned, relevant and appropriate training. Training was found to be carefully planned through training needs analysis processes that linked the training needs with those of the organization, groups, departments, divisions, and individuals. To maintain training relevancy and currency, databases of training courses are widely available and, to encourage diversification, employees are able to realize their full potential by training in quality, job skills, general education, health and safety and so on, through exams, qualifications, assessor training, etc. Typically, training strategies in these organizations require managers to:

- Play an active role in training delivery (cascade training) and support (including quality tools and techniques).
- Receive training and development based on personal development plans.
- Fund training and improvement activities to allow autonomy at 'local' levels for short payback investments.
- Co-ordinate discussions and peer assessments to develop tailored training plans for individuals.

TI Europe, for instance, has a development and performance management (DPM) process that uses discussion and peer assessment to help create individually tailored training plans with business objectives through policy deployment. Similarly, Trident Precision Manufacturing manages and motivates its employees through training. Training is provided in quality, skills related to the job, general education and safety and since 1989, they have invested an average of *c.* 5 percent of the payroll on training – a figure that represents almost three times the average for all US industries. In addition, Trident encourages employees to diversify their abilities and, typically, 80 percent of their employees are trained in at least two job functions. As a result of their investments these companies boost business benefits such as:

- increases in sales volumes;
- not losing customers to competitors;
- low employee turnover.

Total Quality in the Construction Supply Chain

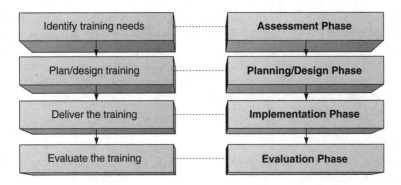

**Figure 14.6**
A systematic model of training

What is particularly noteworthy about the training activities identified is that they are almost identical to those processes and activities commonly found in the management literature on the theory of training. Many writers have developed models of the training process which can be summarized into the four phases shown in Figure 14.6.

The *Assessment Phase* identifies what is needed (the content of the training) at the organizational, group and individual levels. This may involve some overlap, e.g. an individual's poor sales performance may be a symptom of production problems at the group level, as well as a need for more product innovation at the organizational level. The Assessment Phase thus involves identifying training needs by assessing the gaps between future requirements of a job and the current skills, knowledge or attitudes of the person in the job. So the organization looks at what is presently happening and what should or could be happening. Any differences between the two will give some indications of training needs.

The *Planning/Design Phase* identifies where and when the training will take place and involves such questions as:

■ Who needs to be trained?
■ What competencies are required?
■ How long will training take?
■ What are the expected benefits of training?
■ Who/How many will undertake the training?
■ What resources are needed, e.g. money, equipment, accommodation, etc.?

Typically, the organizations involved in the studies planned their training programmes (e.g. annually) according to the needs of the business and circulated lists of available courses well in advance of the training dates. The lists ensure all managers are aware of what is provided so that they are able to schedule attendance by staff. The strategic training plan is supported by an annual budgeting and planning system with quarterly meetings to monitor and review performance. The budgeting and planning process with its integral HR element is cascaded throughout the organization to teams at all locations.

The *Implementation Phase* involves the actual delivery of the training. This might be on site or away from the premises and will include training techniques such as: simulators; business games; case studies; coaching and mentoring; planned experience; computer-assisted instruction. Demonstration or 'sitting next to Nellie' is another commonly used training technique.

Induction and devolved training form an integral part of the training implementation phase. New employees attend induction courses and are issued with personal development documents giving details of what training and assessment will take place in the first few months of employment, as well as copies of the vision and mission statements. At regular intervals throughout the induction period (e.g. every three or four weeks) new employees are then reviewed to identify training and development needs for the remainder of the year.

In addition, much of the training is devolved to line managers through facilitation and facilitator packs. This requires the training and development of all levels of managers and supervisors in facilitation skills. Line managers then identify team members to be trained as facilitators. Adopting this approach is said to create an environment in which everyone is aware of training and development issues for themselves and their colleagues.

The *Evaluation Phase* is widely acknowledged as one of the most critical steps in the training process and can take many forms such as observation, questionnaires, interviews, etc. For example, in this phase the overall effectiveness of training is evaluated and this provides feedback for the trainers, for future improvements to the programme, for senior managers and the trainees themselves. Providing trainees with a set of training objectives will help them know what they need to learn and give them feedback on their progress. This will then influence their attitudes towards future training and even the company itself.

In sum, it seems that successful organizations approach training and development in a planned and systematic way involving training needs analysis, assessment of training content, carefully planned implementation and continuous evaluation and review – a convincing argument for the value of theory when it is put into practice.

(See also Chapter 15 for more detail on the training and development process.)

# Teams and teamwork

It is clear that leading edge organizations place great emphasis on the value of people working together in teams. This is hardly surprising as a great deal of theory and research indicates that people are motivated and work better when they are part of a team. Teams can also achieve more through integrated efforts and problem solving.

Teams are a management tool and are most effective when team activity is clearly linked to organizational strategy. For this, the strategy must be communicated to influence team direction, which then links to the production of team mission statements and the use of team agendas and scorecards. Importantly though, many people emphasize the value of cross-functional teams which proved to be a common feature in many of the organizations studied. Here, teams which have originally evolved out of the old functional departments or units within an organization gain experience and benefit from team building and become cross-functional. For example, in the Eastman company: 'Over the past decade, the Eastman Corporation has developed a quality focus which has expanded from individuals to the concept of interlocking teams.' Each team is required to identify its customers, the customer requirements and what measures need to be used to ensure that those requirements are being satisfied.

Cross-functional teams are also an important feature in TI Europe where almost every employee belongs to at least one team, ranging from managers on quality steering teams, to operators on quality improvement teams, to fully empowered, self-directed work teams. Cross-functional teams are used to address the entire process.

(See also Chapter 15 for more detail on teams and teamwork.)

# The fostering of social cohesion

We live at a time when we hear of high employee mobility and low levels of corporate loyalty. It is said that modern employees are increasingly self-interested and that they develop their competencies and CVs so that they can attract the highest remuneration in the market. While there is evidence to support that many employees do have such a world-view, there is a counter-cyclical trend among employees in world-class companies. These people identify with their work organizations and enjoy being with their colleagues. They do not seek to move at the first opportunity because they enjoy where they are; they receive recognition and they are given challenges.

All the factors in this section combine to form a highly satisfying work environment; however, social cohesion is of equal importance to those already discussed. Therefore, world-class companies take the same planned and structured approach to social activity as they do to everything else they do.

> Graniterock, one of our case study companies (p. 375), assesses the success of its social programme by the level of employee participation. Graniterock, as does CDL in Singapore (p. 389), sees its organization as having a social role that extends beyond the company. Therefore, both companies directly support their employees' efforts in the community.

# Review, continuous improvement and conclusions

In all the organizations involved in the research, processes for reviewing performance and continuous improvement exist at the individual, team, departmental/divisional and organizational levels. These include such processes as:

- Annual staff surveys and subsequent actions, which are viewed as the cornerstones of continuous improvement. The people surveys are also critically reviewed against data from other world-class organizations and benchmarks to determine best practice and feed into the continuous improvement processes.
- Quality committees, the HR department and cross-functional teams drawn from depots, regions and divisions review feedback from surveys as well as the format of the surveys.
- Ongoing performance feedback and development through on the job coaching plus regular one-to-one individual and team reviews.

This chapter so far has highlighted the main people management activities that are currently being used in some world-class organizations. A general conclusion from the research supporting this is that successful organizations pay much more than lip service to the popular idiom 'people are our most important asset'. Indeed successful organizations value and invest in their people in a never-ending quest for effective management and development of their employees. This involves rigorous planning of processes, skilful implementation, regular review of processes, and continuous improvement practices.

From a theoretical viewpoint, these findings about people management activities in successful organizations are hardly surprising, since the management literature is strewn with examples of the benefits of systematic planning, followed by strategic implementation, regular review and continuous improvement. Nonetheless, from a practical viewpoint, the real value of the findings is that they flesh out in some detail those people management activities that are being used to good effect in some world-class organizations which are reaping the benefits of putting theory into practice.

# Organizing people for quality

In many organizations management systems are still viewed in terms of the internal dynamics between marketing, design, sales, production/operations, distribution, accounting, etc. A change is required from this to a larger process-based system that encompasses and integrates the business interests of customers and suppliers. Management needs to develop an in-depth understanding of these relationships and how they may be used to cement the partnership concept. A quality function can be the organization's focal point in this respect, and should be equipped to gauge internal and external customers' expectations and degree of satisfaction. It should also identify deficiencies in all business functions, and promote improvements. The role of the quality function is to make quality an inseparable aspect of every employee's performance and responsibility. The transition in many companies from quality departments with line functions will require careful planning, direction, and monitoring. Quality professionals have developed numerous techniques and skills, focused on product or service quality. In many cases there is a need to adapt these to broader, process applications. The first objectives for many 'quality managers' will be to gradually disengage themselves from line activities, which will then need to be dispersed throughout the appropriate operating departments. This should allow quality to be understood as a 'process' at a senior level, and to be concerned with the following throughout the organization:

- Encouraging and facilitating improvement.
- Monitoring and evaluating the progress of improvement.
- Promoting the 'partnership' in relationships with customers and suppliers.
- Planning, managing, auditing, and reviewing quality management systems.
- Planning and providing training and counselling or consultancy.
- Giving advice to management on:
  - establishment of process management and control;
  - relevant statutory/legislation requirements with respect to quality;
  - quality and process improvement programmes;
  - inclusion of quality elements in all processes, job instructions and procedures.

Quality directors and managers may have an initial task, however, to help those who control the means to implement this concept – the leaders of industry and commerce – to really believe that quality must become an integral part of all the organization's operations.

The authors have a vision of quality as a strategic business management function that will help organizations to change their cultures. To make this vision a reality, quality professionals must expand the application of quality concepts and techniques to all business processes and functions, and develop new forms of providing assurance of quality at every supplier/customer interface. They will need to know the entire cycle of products or services, from concept to the *ultimate* end user. An example of this was observed in the case of a company manufacturing pharmaceutical seals, whose customer expressed concern about excess aluminium projecting below and round a particular type of seal. This was considered a cosmetic defect by the immediate customer, the Health Service, but a safety hazard by a blind patient – the *customer's customer*. The prevention of this 'curling' of excess metal meant changing practices at the mill that rolled the aluminium – at the *supplier's supplier*. Clearly, the quality professional dealing with this problem needed to understand the supplier's processes and the ultimate customer's needs, in order to judge whether the product was indeed capable of meeting the requirements.

The shift in 'philosophy' will require considerable staff education in many organizations. Not only must people in other functions acquire quality management skills, but quality personnel must change old attitudes and acquire new skills – replacing the inspection, calibration, specification-writing mentality with knowledge of defect prevention, wide ranging quality management systems design and audit. Clearly, the challenge for many quality professionals is not so much making changes in their organization, as recognizing the changes required in themselves. It is more than an overnight job to change the attitudes of an inspection police force into those of a consultative, team-oriented improvement resource. This emphasis on prevention and improvement-based systems elevates the role of quality professionals from a technical one to that of general management. A narrow departmental view of quality is totally out of place in an organization aspiring to TQM, and many quality directors and managers will need to widen their perspective and increase their knowledge to encompass all facets of the organization.

To introduce the concepts of process management required for TQM will require not only a determination to implement change but sensitivity and skills in interpersonal relations. This will depend very much of course on the climate within the organization. Those whose management is truly concerned with co-operation and concern for the people will engage strong employee support for the quality manager or director in his catalytic role in the improvement process. Those with aggressive, confrontational management will create for the quality professional impossible difficulties in obtaining support from the 'rank and file'.

# Quality appointments

Many organizations have realized the importance of the contribution a senior, qualified director of quality can make to the prevention strategy. Smaller organizations may well feel that the cost of employing a full-time quality manager is not justified, other than in certain very high-risk areas. In these cases a member of the management team may be appointed to operate on a part-time basis, performing the quality management function in addition to his/her other duties. To obtain the best results from a quality director/manager, he/she should be given sufficient authority to take necessary action to secure the implementation of the organization's quality policy, and must have the personality to be able to communicate the message to all employees, including staff, management and directors. Occasionally the quality director/manager may require some guidance and help on specific technical quality matters, and one of the major attributes required is the knowledge and wherewithal to acquire the necessary information and assistance.

In large organizations, then, it may be necessary to make several specific appointments or to assign details to certain managers. The following actions may be deemed to be necessary.

### *Assign a quality director, manager or co-ordinator*

This person will be responsible for the planning and implementation of TQM. He or she will be chosen first for process, project and people management abilities rather than detailed knowledge of quality assurance matters. Depending on the size and complexity of the organization, and its previous activities in quality management, the position may be either full or part time, but it must report directly to the chief executive.

### Appoint a quality manager adviser

A professional expert on quality management may be required to advise on the 'technical' aspects of planning and implementing TQM. This is a consultancy role, and may be provided from within or without the organization, full or part time. This person needs to be a persuader, philosopher, teacher, adviser, facilitator, reporter and motivator. He or she must clearly understand the organization, its processes and interfaces, be conversant with the key functional languages used in the business, and be comfortable operating at many organizational levels. On a more general level this person must fully understand and be an effective advocate and teacher of TQM, be flexible and become an efficient agent of change.

### Steering committees and teams

Devising and implementing total quality management in an organization takes considerable time and ability. It must be given the status of a senior executive project. The creation of cost-effective performance improvement is difficult, because of the need for full integration with the organization's strategy, operating philosophy and management systems. It may require an extensive review and substantial revision of existing systems of management and ways of operating. Fundamental questions may have to be asked, such as 'do the managers have the necessary authority, capability, and time to carry this through?'

Any review of existing management and operating systems will inevitably 'open many cans of worms' and uncover problems that have been successfully buried and smoothed over – perhaps for years. Authority must be given to those charged with following TQM through with actions that they consider necessary to achieve the goals. The commitment will be continually questioned and will be weakened, perhaps destroyed, by failure to delegate authoritatively.

The following steps are suggested in general terms. Clearly, different types of organization will have need to make adjustments to the detail, but the component parts are the basic requirements.

A disciplined and systematic approach to continuous improvement may be established in a quality or business excellence 'steering committee or council' (Figure 14.7). The committee/council should meet at least monthly to review strategy, implementation progress, and improvement. It should be chaired by the chief executive, who must attend every meeting – only death or serious illness should prevent him/her being there. Clearly, postponement may be necessary occasionally, but the council should not carry on meeting without the chief executive present. The council members should include the top management team and the chairmen of any 'site' steering committees or process management teams, depending on the size of the organization. The objectives of the council are to:

- Provide strategic direction on quality for the organization.
- Establish plans for quality on each 'site'.
- Set up and review the process teams that will own the key or critical business processes.
- Review and revise quality plans for implementation.

The process management teams and any site steering committees should also meet monthly, shortly before the senior steering committee/council meetings. Every senior manager should be a member of at least one such team. This system provides the

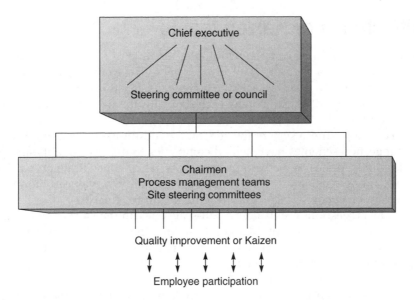

**Figure 14.7**
Employee participation through the team structure

'top-down' support for employee participation in process management and development. It also ensures that the commitment to quality at the top is communicated effectively through the organization.

The three-tier approach of steering committee, process management teams and quality improvement teams allows the first to concentrate on quality strategy, rather than become a senior problem-solving group. Progress is assured if the team chairmen are required to present a status report at each meeting.

The process management teams or steering committees control all the quality improvement teams and have responsibility for:

- The selection of projects for the teams.
- Providing an outline and scope for each project to give to the teams.
- The appointment of team members and leaders.
- Monitoring and reviewing the progress and results from each team project.

As the focus of this work will be the selection of projects, some attention will need to be given to the sources of nominations. Projects may be suggested by:

- Steering committee/council members representing their own departments, process management teams, their suppliers or their customers, internal and external.
- Quality improvement teams.
- Kaizen teams or quality circles (if in existence).
- Suppliers.
- Customers.

The process team members must be given the responsibility and authority to represent their part of the organization in the process. The members must also feel that they represent the team to the rest of the organization. In this way the team will gain knowledge and respect and be seen to have the authority to act in the best interests of the organization, with respect to their process.

# Quality circles or Kaizen teams

No book on TQM would be complete without a mention of Kaizen and quality circles. Kaizen is a philosophy of continuous improvement of all the employees in an organization so that they perform their tasks a little better each day. It is a never-ending journey centred on the concept of starting anew each day with the principle that methods can always be improved.

*Kaizen teian* is a Japanese system for generating and implementing employee ideas. Japanese suggestion schemes have helped companies to improve quality and productivity, and reduced prices to increase market share. They concentrate on participation and the rates of implementation, rather than on the 'quality' or value of the suggestion. The emphasis is on encouraging everyone to make improvements.

Kaizen teian suggestions are usually small-scale ones, in the worker's own area, and are easy and cheap to implement. Key points are that the rewards given are small, and implementation is rapid, which results in many small improvements that accumulate to massive total savings and improvements.

One of the most publicized aspects of the Japanese approach to quality has been these quality circles or Kaizen teams. The quality circle may be defined then as a group of workers doing similar work who meet:

- voluntarily;
- regularly;
- in normal working time;
- under the leadership of their 'supervisor';
- to identify, analyse, and solve work-related problems;
- to recommend solutions to management.

Where possible quality circle members should implement the solutions themselves.

The quality circle concept first originated in Japan in the early 1960s, following a post-war reconstruction period during which the Japanese placed a great deal of emphasis on improving and perfecting their quality control techniques. As a direct result of work carried out to train foremen during that period, the first quality circles were conceived, and the first three circles registered with the Japanese Union of Scientists and Engineers (JUSE) in 1962. Since that time the growth rate has been phenomenal. The concept has spread to Taiwan, the USA and Europe, and circles in many countries have been successful. Many others have failed.

It is very easy to regard quality circles as the magic ointment to be rubbed on the affected spot, and unfortunately many managers in the West first saw them as a panacea for all ills. There are no panaceas, and to place this concept into perspective, Juran, who has been an important influence in Japan's improvement in quality, has stated that quality circles represent only 5–10 percent of the canvas of the Japanese success. The rest is concerned with understanding quality, its related costs and the organization, systems and techniques necessary for achieving customer satisfaction.

Given the right sort of commitment by top management, introduction, and environment in which to operate, quality circles can produce the 'shop floor' motivation to achieve quality performance at that level. Circles should develop out of an understanding and knowledge of quality on the part of senior management. They must not be introduced as a desperate attempt to do something about poor quality. The term 'quality circle' may be replaced with a number of acronyms but the basic concepts and operational aspects may be found in many organizations.

## The structure of a quality circle organization

The unique feature about quality circles or Kaizen teams is that people are asked to join and not told to do so. Consequently, it is difficult to be specific about the structure of such a concept. It is, however, possible to identify four elements in a circle organization:

- members;
- leaders;
- facilitators or co-ordinators;
- management.

*Members* form the prime element of the concept. They will have been taught the basic problem-solving and process control technique and, hence, possess the ability to identify and solve work-related problems.

*Leaders* are usually the immediate supervisors or foremen of the members. They will have been trained to lead a circle and bear the responsibility of its success. A good leader, one who develops the abilities of the circle members, will benefit directly by receiving valuable assistance in tackling nagging problems.

*Facilitators* are the managers of the quality circle programmes. They, more than anyone else, will be responsible for the success of the concept, particularly within an organization. The facilitators must co-ordinate the meetings, the training and energies of the leaders and members, and form the link between the circles and the rest of the organization. Ideally the facilitator will be an innovative industrial teacher, capable of communicating with all levels and with all departments within the organization.

*Management* support and commitment are necessary to quality circles or, like any other concept, they will not succeed. Management must retain its prerogatives, particularly regarding acceptance or non-acceptance of recommendations from circles, but the quickest way to kill a programme is to ignore a proposal arising from it. One of the most difficult facts for management to accept, and yet one forming the cornerstone of the quality circle philosophy, is that the real 'experts' on performing a task are those who do it day after day.

## Training quality circles

The training of circle/Kaizen leaders and members is the foundation of all successful programmes. The whole basis of the training operation is that the ideas must be easy to take in and be put across in a way that facilitates understanding. Simplicity must be the key word, with emphasis being given to the basic techniques. Essentially there are eight segments of training:

1. Introduction to quality circles.
2. Brainstorming.
3. Data gathering and histograms.
4. Cause and effect analysis.
5. Pareto analysis.
6. Sampling and stratification.
7. Control charts.
8. Presentation techniques.

Managers should also be exposed to some training in the part they are required to play in the quality circle philosophy. A quality circle programme can only be effective if management believes in it and is supportive and, since changes in management style may be necessary, managers' training is essential.

## Operation of quality circles/Kaizen teams

There are no formal rules governing the size of a quality circle/Kaizen team. Membership usually varies from three to 15 people, with an average of seven to eight. It is worth remembering that, as the circle becomes larger, it becomes increasingly difficult for all members of the circle to participate.

Meetings should be held away from the work area, so that members are free from interruptions, and are mentally and physically at ease. The room should be arranged in a manner conducive to open discussion, and any situation that physically emphasizes the leader's position should be avoided.

Meeting length and frequency are variable, but new circles meet for approximately one hour once per week. Thereafter, when training is complete, many circles continue to meet weekly; others extend the interval to two or three weeks. To a large extent the nature of the problems selected will determine the interval between meetings, but this should never extend to more than one month, otherwise members will lose interest and the circle will cease to function.

Great care is needed to ensure that every meeting is productive, no matter how long it lasts or how frequently is it held. Any of the following activities may take place during a circle meeting:

- Training – initial or refresher.
- Problem identification.
- Problem analysis.
- Preparation and recommendation for problem solution.
- Management presentations.
- Quality circle administration.

A quality circle usually selects a project to work on through discussion within the circle. The leader then advises management of this choice and, assuming that no objections are raised, the circle proceeds with the work. Other suggestions for projects come from management, quality assurance staff, the maintenance department, various staff personnel, and other circles.

It is sometimes necessary for quality circles to contact experts in a particular field, e.g. engineers, quality experts, safety officers, maintenance personnel. This communication should be strongly encouraged, and the normal company channels should be used to invite specialists to attend meetings and offer advice. The experts may be considered to be 'consultants', the quality circle retaining responsibility for improving a process or solving the particular problem. The overriding purpose of quality circles or Kaizen teams is to provide the powerful motivation of allowing people to take some part in deciding their own actions and futures.

# Acknowledgement

The authors are grateful to Dr Susan Oakland for the contribution she made to this chapter.

# References

1. Maguire, S. (1995) 'Learning to Change', *European Quality*, Vol. 2, No. 6, p. 8.
2. Marchington, M. and Wilkinson, A. (1997) *Core Personnel and Development*, London, Institute of Personnel and Development.
3. Boyatsis, R. (1982) *The Competent Manager: A Model for Effective Performance*, New York, Wiley.

Total Quality in the Construction Supply Chain

## Chapter highlights

### Introduction

■ In recent years the way people are managed has been recognized as a key to improving performance. Recent research (EC*for*BE) on world-class, award-winning organizations has identified the main people management activities used in leading edge organizations.

■ World-class organizations value and invest in people through: strategic alignment of HRM policies, effective communications, employee empowerment and involvement, training and development, teams and teamwork and review and continuous improvement.

### Strategic alignment of HRM policies

■ Leading edge organizations adopt a common approach to aligning HR policies with business strategy. Key elements of policy such as skills, recruitment and selection, training health and safety, appraisal, employee benefits and remuneration are first identified. The HR plan is then devised as part of the strategic planning process, following a Plan, Do, Check, Improve (PDCI) cycle.

### Effective communication

■ Regular two-way communication, particularly face to face, is essential for success.

■ Again the PDCI cycle provides a systematic process for ensuring effective communications, which uses benchmarking and self-assessment as part of the improvement effort.

### Employee empowerment and involvement

■ To encourage employee commitment and involvement, successful organizations place great importance on empowering employees. This can include people setting own goals, judging own performance, taking ownership of actions, and identifying with the organization itself (perhaps as shareholders).

■ Common initiatives include: employee suggestion schemes, culture change programmes and measurement of KPIs. Generally commitment is increased by involving more employees in planning and shaping the vision.

### Motivation

■ Publicly recognising excellent contributions by individuals and teams is an essential part of a HR program.

### Training and development

■ Training and development has been highlighted by many initiatives as a critical success factor, although lack of effective training still predominates in many organizations.

■ In successful organizations, training is planned through needs analysis, use of databases, training delivery at local levels, and peer assessments for evaluation.

### Teams and teamwork

■ Leading edge organizations place great value in people working in teams, because this motivates and causes them to work better.

■ Teams are most effective when their activities are clearly linked to the strategy, which in turn is communicated to influence direction. Cross-functional teams are particularly important to address end-to-end processes.

### The fostering of social cohesion

■ World-class companies take the same planned and structured approach to social activity as they do to everything else they do.

### Review, continuous improvement and conclusions

■ Effective organizations use processes for reviewing performance and continuous improvement at the individual, team, divisional/departmental and organizational levels. These include surveys of staff, committees/teams, and ongoing performance feedback.

### Organizing people for quality

■ The quality function should be the organization's focal point of the integration of the business interests of customers and suppliers into the internal dynamics of the organization.

■ Its role is to encourage and facilitate quality and process improvement; monitor and evaluate progress; promote the quality chains; plan, manage, audit and review systems; plan and provide quality training, counselling and consultancy; and give advice to management.

■ In larger organizations a quality director will contribute to the prevention strategy. Smaller organizations may appoint a member of the management team to this task on a part-time basis. An external TQM adviser is usually required.

■ In devising and implementing TQM for an organization, it may be useful to ask first if the managers have the authority, capability and time to carry it through.

■ A disciplined and systematic approach to continuous improvement may be established in a steering committee/council, whose members are the senior management team.

■ Reporting to the steering committee are the process management teams or any site steering committees, which in turn control the quality improvement or Kaizen teams and quality circles.

### Quality circles or Kaizen teams

■ Kaizen is a philosophy of small step continuous improvement, by all employees. In Kaizen teams the suggestions and rewards are small but the implementation is rapid.

■ A quality circle or Kaizen team is a group of people who do similar work meeting voluntarily, regularly, in normal working time, to identify, analyse and solve work-related problems, under the leadership of their supervisor. They make recommendations to management. Alternative names may be given to the teams, other than 'quality circles'.

# 15

# Culture change through teamwork

# The need for teamwork

The complexity of most of the processes that are operated in industry, commerce and the services places them beyond the control of any one individual. Furthermore, as any supply chain becomes more fragmented, the interface between organizations gives rise to many of the more significant problems and opportunities. This is typical of building and construction projects, and developing effective solutions requires the involvement of all the stakeholders in the problem area. The only really efficient way to tackle process management and improvement is through the use of some form of teamwork which has many advantages over allowing individuals to work separately:

- A greater variety of complex processes and problems may be tackled – those beyond the capability of any one individual or even one department or organization – by the pooling of expertise and resources.
- Processes and problems are exposed to a greater diversity of knowledge, skill, experience, and are solved more efficiently.
- The approach is more satisfying to team members, and boosts morale and ownership through participation in process management, problem solving and decision-making.
- Processes and problems that cross departmental or functional boundaries can be dealt with more easily, and the potential/actual conflicts are more likely to be identified and solved.
- The recommendations are more likely to be implemented than individual suggestions, as the quality of decision-making in *good teams* is high.
- On construction projects processes and problems that cross organizational boundaries can only be effectively dealt with by interorganizational teams.

Most of these factors rely on the premise that people are willing to support any effort in which they have taken part or helped to develop.

When properly managed and developed, teams improve the process of problem solving, producing results quickly and economically. Teamwork throughout any organization is an essential component of the implementation of TQM and process management, for it builds trust, improves communications and develops interdependence. Much of what has been taught previously in management has led to a culture in the West of independence, with little sharing of ideas and information. Knowledge is very much like organic manure – if it is spread around it will fertilize and encourage growth, if it is kept closed in, it will eventually fester and rot.

Good teamwork changes the independence to interdependence through improved communications, trust and the free exchange of ideas, knowledge, data and information (Figure 15.1). The use of the face-to-face interaction method of communication, with a common goal, develops over time the sense of dependence on each other. This forms a key part of any quality improvement process, and provides a methodology for employee recognition and participation, through active encouragement of group activities.

Teamwork provides an environment in which people can grow and use all the resources effectively and efficiently to make continuous improvements. As individuals grow, the organization grows. It is worth pointing out, however, that employees will not be motivated towards continual improvement in the absence of:

- Commitment from top management.
- The right organizational 'climate'.
- A mechanism for enabling individual contributions to be effective.

All these are focused essentially at enabling people to feel, accept, and discharge responsibility. More than one organization has made this part of their strategy – to 'empower

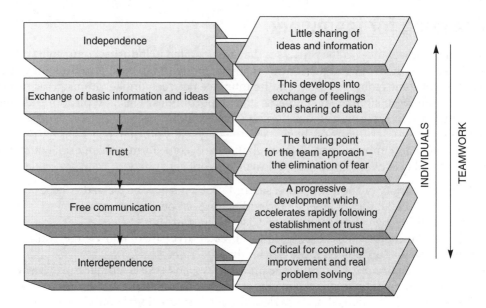

**Figure 15.1**
Independence to interdependence through teamwork

people to act'. If one hears from employees comments such as 'We know this is not the best way to do this job, but if that is the way management want us to do it, that is the way we will do it', then it is clear that the expertise existing at the point of operation has not been harnessed and the people do not feel responsible for the outcome of their actions. Responsibility and accountability foster pride, job satisfaction, and better work.

Empowerment to act is very easy to express conceptually, but it requires real effort and commitment on the part of all managers and supervisors to put into practice. Recognition that only partially successful but good ideas or attempts are to be applauded and not criticized is a good way to start. Encouragement of ideas and suggestions from the workforce, particularly through their part in team or group activities, requires investment. The rewards are total commitment, both inside the organization and outside through the supplier and customer chains.

Teamwork to support process management and improvement has several components. It is driven by a strategy, needs a structure, and must be implemented thoughtfully and effectively. The strategy that drives the improvement comprises the:

- vision and mission of the organization;
- critical success factors;
- core process framework.

These components have been dealt with in other chapters. The structural and implementation aspects of teamwork are the subject of the remainder of this chapter.

# Running process management and improvement teams

Process management and improvement teams are groups of people with the appropriate knowledge, skills, and experience who are brought together specifically by management to improve processes and/or tackle and solve particular problems, usually on a project basis. They are cross-functional and often multi-disciplinary.

The 'task force' has long been a part of the culture of many organizations at the 'technology' and management levels. But process teams go a step further; they expand the traditional definition of 'process' to cover the entire end-to-end operating system. This includes technology, paperwork, communication and other units, operating procedures, and the process equipment itself. By taking this broader view, the teams can address new problems. The actual running of process teams calls several factors into play:

- team selection and leadership;
- team objectives;
- team meetings;
- team assignments;
- team dynamics;
- team results and reviews.

## Team selection and leadership

The most important element of a process team is its members. People with knowledge and experience relevant to the process of solving the problem are clearly required. However, there should be a limit of five to ten members to keep the team small enough to be manageable but allow a good exchange of ideas. Membership should include appropriate people from groups outside the operational and technical areas directly 'responsible' for the process, if their presence is relevant or essential. In the selection of team members it is often useful to start with just one or two people who are clearly concerned directly with the process. If they try to draw maps or flowcharts (see Chapter 10) of the relevant processes, the requirement to include other people, in order to understand the process and complete the charts, will aid the team selection. This method will also ensure that all those who can make a significant contribution to the process and its improvement are represented. In construction settings, it will often be important to have representation from other companies in the supply chain – selecting the right individual is crucial.

The process owner has a primary responsibility for team leadership, management and maintenance, and his/her selection and training is crucial to success. The leader need not be the highest ranking person in the team, but must be concerned about accomplishing the team objectives (this is sometimes described as 'task concern') and the needs of the members (often termed 'people concern'). Weakness in either of these areas will lessen the effectiveness of the team in solving problems or making breakthroughs. The need for team leadership training is often overlooked; never assume that just because people have been elevated to supervisory or even project management roles, they necessarily know how to lead a team – many companies do not train in these basic skills. Skill development may be needed in areas such as facilitation, meeting management, and motivation. Needs should be identified and training directed at correcting deficiencies in these crucial aspects.

## Team objectives

At the beginning of any process improvement project it is important that the objective should be clearly defined and agreed. This may be in problem or performance improvement terms and it may take some time to define – but agreement is important. Also at the start of every meeting the objectives should be stated as clearly as possible by the leader. This can take a simple form: 'This meeting is to continue the discussion from last Tuesday on the development of our design manual and its trial and adoption throughout the company. Last week we agreed on the overall structure of the manual and today we will look in detail at the structure of the first section.' Project and/or meeting objectives enable the team members to focus thoughts and efforts on the aims, which may need to be restated if the team becomes distracted by other issues.

## Team meetings

Meetings need to be seen as part of a process working towards a longer-term goal – and hence, planning for each meeting and maintaining the continuity between meetings is important. An agenda should be prepared by the leader and distributed to each team member before every meeting. It should include the following information:

- Meeting place, time and how long it will be.
- A list of members (and co-opted members) expected to attend.
- Any preparatory assignments for individual members or groups.
- Any supporting material to be discussed at the meeting.

Early in a project the leader should orient the team members in terms of the approach, methods, and techniques they will use to solve the problem. This may require a review of the:

- Systematic approach (Chapter 13).
- Procedures and rules for using some of the basic tools, e.g. brainstorming – no judgement of initial ideas.
- Role of the team in the continuous improvement process.
- Authority of the team.

To make sure that the meeting process is used to maximum advantage it is important that the team leader manages the meeting process; there are several important aspects to this. First of all bear in mind the overall meeting plan (including an approximate timeframe), then for each topic that is addressed:

- maintain the participation of everyone;
- maintain focus on the topic being considered;
- maintain momentum, keep the process moving forward;
- achieve closure, before moving on; capture where the group is up to and where and how it will proceed.

A team secretary should be appointed to take the minutes of meeting and distribute them to members as soon as possible after each meeting. The minutes should not be formal, but reflect decisions and carry a clear statement of the action plans, together with assignments of tasks. They may be hand-written initially, copied and given to team members at the end of the meeting, to be followed later by a more formal document that will be seen by any member of staff interested in knowing the outcome of the meeting. In this way the minutes form an important part of the communication system, supplying information to other teams or people needing to know what is going on.

# Team assignments

It is never possible to solve problems by meetings alone. What must come out of those meetings is a series of action plans that assigns specific tasks to team members. This is the responsibility of the team leader. Agreement must be reached regarding the responsibilities for individual assignments, together with the time scale, and this must be made clear in the minutes. Task assignments must be decided while the team is together and not by separate individuals in after-meeting discussions. Make sure that task assignments are realistic to the timeframe and resources available. This may need the allocation of additional resources and the team leader may need to negotiate this with senior management.

# Team dynamics

In any team activity the interactions between the members are vital to success. If solutions to problems are to be found, the meetings and ensuing assignments should assist and harness the creative thinking process. This is easier said than done, because many people have either not learned or been encouraged to be innovative. The team leader clearly has a role here to:

- Create a 'climate' for creativity.
- Encourage all team members to speak out and contribute their own ideas or build on others.
- Allow differing points of view and ideas to emerge.
- Remove barriers to idea generation, e.g. incorrect preconceptions that are usually destroyed by asking 'Why?'
- Support all team members in their attempts to become creative.

In addition to the team leader's responsibilities, the members should:

- Prepare themselves well for meetings, by collecting appropriate data or information (*facts*) pertaining to a particular problem.
- Share ideas and opinions.
- Encourage other points of view.
- Listen 'openly' for alternative approaches to a problem or issue.
- Help the team determine the best solutions.
- Reserve judgement until all the arguments have been heard *and* fully understood.
- Accept individual responsibility for assignments and group responsibility for the efforts of the team.

# Team results and reviews

A process approach to improvement and problem solving is most effective when the results of the work are communicated and acted upon. Regular feedback to the teams, via their leaders, will assist them to focus on objectives, and review progress.

Reviews also help to deal with certain problems that may arise in teamwork. For example, certain members may be concerned more with their own personal objectives than those of the team. This may result in some manipulation of the problem-solving process to achieve different goals, resulting in the team splitting apart through self-interest. If recognized, the review can correct this effect and demand greater openness and honesty.

A different type of problem is the failure of certain members to contribute and take their share of individual and group responsibility. Allowing other people to do their work results in an uneven distribution of effort, and leads to bitterness. The review should make sure that all members have assigned and specific tasks, and perhaps lead to the documentation of duties in the minutes. A team roster may even help. If some members of a team are not contributing and cannot be induced to do so, consideration should be given to their replacement. However, this can become a more complex issue, if the team leader does not manage team processes well and people believe they are wasting their time. There may be a high level of frustration that could lead to some members withdrawing support.

A third area of difficulty, which may be improved by reviewing progress, is the ready-fire-aim syndrome of action before analysis. This often results from team leaders being too anxious to deal with a problem. A review should allow the problem to be redefined adequately and expose the real cause(s). This will release the trap the team may be in of doing something before they really know what should be done. The review will provide the opportunity to rehearse the steps in the systematic approach.

# Teamwork and action-centred leadership

Over the years there has been much academic work on the psychology of teams and on the leadership of teams. Three points on which all authors are in agreement are that teams develop a personality and culture of their own, respond to leadership, and are motivated according to criteria usually applied to individuals.

Key figures in the field of human relations, like Douglas McGregor (Theories X & Y), Abraham Maslow (Hierarchy of Needs) and Fred Hertzberg (Motivators and Hygiene Factors), all changed their opinions on group dynamics over time as they came to realize that groups are not the democratic entity that everyone would like them to be, but respond to individual, strong, well-directed leadership, both from without and within the group, just like individuals (see also *Total Quality Management*, 2nd edition).

## Action-centred leadership

During the 1960s John Adair, senior lecturer in Military History and the Leadership Training Adviser at the Military Academy, Sandhurst, and later assistant director of the Industrial Society, developed what he called the action-centred leadership model, based on his experiences at Sandhurst, where he had the responsibility to ensure that results in the cadet training did not fall below a certain standard. He had observed that some instructors frequently achieved well above average results, owing to their own natural ability with groups and their enthusiasm. He developed this further into a team model, which is the basis of the approach of the authors and their colleagues to this subject.

In developing his model for teamwork and leadership, Adair brought out clearly that for any group or team, big or small, to respond to leadership, they need a clearly defined *task*, and the response and achievement of that task are interrelated to the needs of the *team* and the separate needs of the *individual members* of the team (Figure 15.2).

The value of the overlapping circles is that it emphasizes the unity of leadership and the interdependence and multifunctional reaction to single decisions affecting any of the three areas.

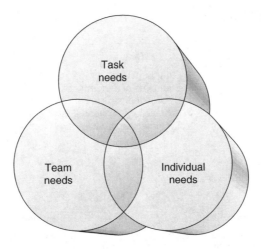

**Figure 15.2**
Adair's model

## *Leadership tasks*

Drawing upon the discipline of social psychology, Adair developed and applied to training the functional view of leadership. The essence of this he distilled into the three interrelated but distinctive requirements of a leader. These are to define and achieve the job or task, to build up and co-ordinate a team to do this, and to develop and satisfy the individuals within the team (Figure 15.3).

1. *Task needs.* The difference between a team and a random crowd is that a team has some common purpose, goal or objective, e.g. a football team. If a work team does not achieve the required results or meaningful results, it will become frustrated. Organizations have to make a profit, to provide a service, or even to survive. So anyone who manages others has to achieve results; in production, marketing, selling or whatever. Achieving objectives is a major criterion of success.
2. *Team needs.* To achieve these objectives, the group needs to be held together. People need to be working in a co-ordinated fashion in the same direction. Teamwork will ensure that the team's contribution is greater than the sum of its parts. Conflict within the team must be used effectively; arguments can lead to ideas or to tension and lack of co-operation.
3. *Individual needs.* Within working groups, individuals also have their own set of needs. They need to know what their responsibilities are, how they will be needed, how well they are performing. They need an opportunity to show their potential, take on responsibility and receive recognition for good work.

The task, team and individual functions for the leader are as follows:

(a) *Task functions*    Defining the task
Making a plan
Allocating work and resources
Controlling quality and tempo of work
Checking performance against the plan
Adjusting the plan

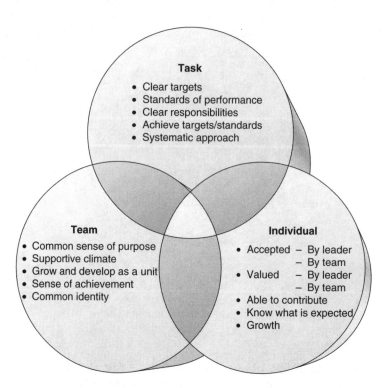

**Figure 15.3**
The leadership needs

(b) *Team functions*　　Setting standards
　　　　　　　　　　Maintaining discipline
　　　　　　　　　　Building team spirit
　　　　　　　　　　Encouraging, motivating, giving a sense of purpose
　　　　　　　　　　Appointing subleaders
　　　　　　　　　　Ensuring communication within the group
　　　　　　　　　　Training the group
(c) *Individual*　　　Attending to personal problems
　　　*functions*　　Praising individuals
　　　　　　　　　　Giving status
　　　　　　　　　　Recognizing and using individual abilities
　　　　　　　　　　Training the individual

The team leader's or facilitator's task is to concentrate on the small central area where all three circles overlap. In a business that is introducing TQM this is the 'action to change' area, where the leaders are attempting to manage the change from *business as usual*, through total quality management, to *TQM equals business as usual*, using the cross-functional quality improvement teams at the strategic interface.

In the action area the facilitator's or leader's task is similar to the task outlined by John Adair. It is to try to satisfy all three areas of need by achieving the task, building the team, and satisfying individual needs. If a leader concentrates on the task, e.g. in going all out for production schedules, while neglecting the training, encouragement and motivation of the team and individuals, (s)he may do very well in the short term. Eventually, however, the team members will give less effort than they are capable of.

Similarly, a leader who concentrates only on creating team spirit, while neglecting the task and the individuals, will not receive maximum contribution from the people. They may enjoy working in the team but they will lack the real sense of achievement that comes from accomplishing a task to the utmost of the collective ability.

So the leader/facilitator must try to achieve a balance by acting in all three areas of overlapping need. It is always wise to work out a list of required functions within the context of any given situation, based on a general agreement on the essentials. Here is Adair's original Sandhurst list, on which one's own adaptation may be based:

- *Planning*, e.g. seeking all available information.
  Defining group task, purpose or goal.
  Making a workable plan (in right decision-making framework).
- *Initiating*, e.g. briefing group on the aims and the plan.
  Explaining why aim or plan is necessary.
  Allocating tasks to group members.
- *Controlling*, e.g. maintaining group standard.
  Influencing tempo.
  Ensuring all actions are taken towards objectives.
  Keeping discussions relevant.
  Prodding group to action/decision.
- *Supporting*, e.g. expressing acceptance of persons and their contribution.
  Encouraging group/individuals.
  Disciplining group/individuals.
  Creating team spirit.
  Relieving tension with humour.
  Reconciling disagreements or getting others to explore them.
- *Informing*, e.g. clarifying task and plan.
  Giving new information to the group, i.e. keeping them 'in the picture'.
  Receiving information from the group.
  Summarizing suggestions and ideas coherently.
- *Evaluating*, e.g. checking feasibility of an idea.
  Testing the consequences of a proposed solution.
  Evaluating group performance.
  Helping the group to evaluate its own performance against standards.

## Situational leadership

In dealing with the task, the team, and with any individual in the team, a style of leadership appropriate to the situation must be adopted. The teams and the individuals within them will, to some extent, start 'cold', but they will develop and grow in both strength and experience. The interface with the leader must also change with the change in the team, according to the Tannenbaum and Schmidt model (Figure 15.4).[1]

Initially a very directive approach may be appropriate, giving clear instructions to meet agreed goals. Gradually, as the teams become more experienced and have some success, the facilitating team leader will move through coaching and support to less directing and eventually a less supporting and less directive approach – as the more interdependent style permeates the whole organization.

This equates to the modified Blanchard model[2] in Figure 15.5, where directive behaviour moves from high to low as people develop and are more easily empowered. When this is coupled with the appropriate level of supportive behaviour, a directing style of leadership can move through coaching and supporting to a delegating style. It must be

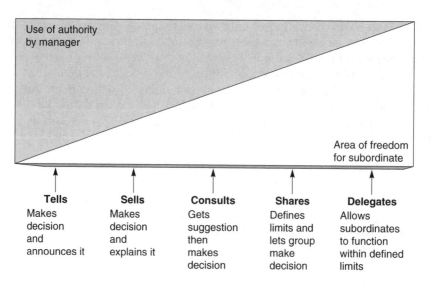

**Figure 15.4**
Continuum leadership behaviour

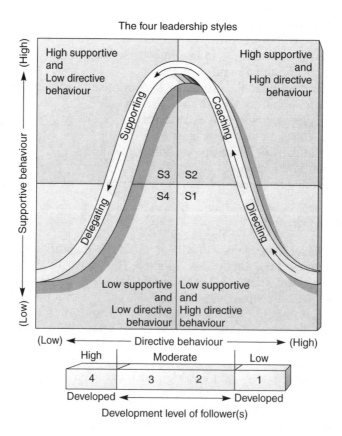

**Figure 15.5**
Situational leadership – progressive empowerment through TQM

stressed, however, that effective delegation is only possible with developed 'followers', who can be fully empowered.

One of the great mistakes in recent years has been the expectation by management that teams can be put together with virtually no training or development (S1 in Figure 15.5) and that they will perform as a mature team (S4). The Blanchard model emphasizes that there is no quick and easy 'tunnel' from S1 to S4. The only route is the laborious climb through S2 and S3.

# Stages of team development

Original work by Tuckman[3] suggested that when teams are put together, there are four main stages of team development, the so-called forming (awareness), storming (conflict), norming (co-operation), and performing (productivity). The characteristics of each stage and some key aspects to look out for in the early stages are given below.

## Forming – awareness

Characteristics:

- Feelings, weaknesses and mistakes are covered up.
- People conform to established lines.
- Little care is shown for others' values and views.
- There is no shared understanding of what needs to be done.

Watch out for:

- Increasing bureaucracy and paperwork.
- People confining themselves to defined jobs.
- The 'boss' is ruling with a firm hand.

## Storming – conflict

Characteristics:

- More risky, personal issues are opened up.
- The team becomes more inward-looking.
- There is more concern for the values, views and problems of others in the team.

Watch out for:

- The team becomes more open, but lacks the capacity to act in a unified, economic, and effective way.

## Norming – co-operation

Characteristics:

- Confidence and trust to look at how the team is operating.
- A more systematic and open approach, leading to a clearer and more methodical way of working.

- Greater valuing of people for their differences.
- Clarification of purpose and establishing of objectives.
- Systematic collection of information.
- Considering all options.
- Preparing detailed plans.
- Reviewing progress to make improvements.

## Performing – productivity

Characteristics:

- Flexibility.
- Leadership decided by situations, not protocols.
- Everyone's energies utilized.
- Basic principles and social aspects of the organization's decisions considered.

The team stages, the task outcomes and the relationship outcomes are shown together in Figure 15.6. This model which has been modified from Kormanski[4] may be used as a framework for the assessment of team performance. The issues to look for are:

1. How is leadership exercised in the team?
2. How is decision-making accomplished?
3. How are team resources utilized?
4. How are new members integrated into the team?

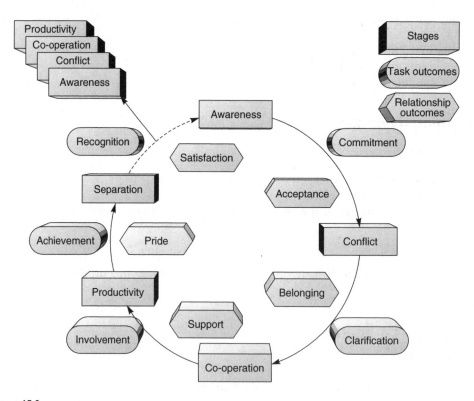

**Figure 15.6**
Team stages and outcomes. (Derived from Kormanski, 1987[4])

Teams, which go through these stages successfully, should become effective teams and display the following attributes.

# Attributes of successful teams

## Clear objectives and agreed goals

No group of people can be effective unless they know what they want to achieve, but it is more than knowing what the objectives are. People are only likely to be committed to them if they can identify with and have ownership of them – in other words, objectives and goals are agreed by team members.

Often this agreement is difficult to achieve but experience shows that it is an essential prerequisite for the effective group.

## Openness and confrontation

If a team is to be effective, then the members of it need to be able to state their views, their differences of opinion, interests and problems, without fear of ridicule or retaliation. No teams work effectively if there is a cut-throat atmosphere, where members become less willing or able to express themselves openly; then much energy, effort and creativity are lost.

## Support and trust

Support naturally implies trust among team members. Where individual group members do not feel they have to protect their territory or job, and feel able to talk straight to other members, about both 'nice' and 'nasty' things, then there is an opportunity for trust to be shown. Based on this trust, people can talk freely about their fears and problems and receive from others the help they need to be more effective.

## Co-operation and conflict

When there is an atmosphere of trust, members are more ready to participate and are committed. Information is shared rather than hidden. Individuals listen to the ideas of others and build on them. People find ways of being more helpful to each other and the group generally. Co-operation causes high morale – individuals accept each other's strengths and weaknesses and contribute from their pool of knowledge of skill. All abilities, knowledge and experience are fully utilized by the group; individuals have no inhibitions about using other people's abilities to help solve their problems, which are shared.

Allied to this, conflicts are seen as a necessary and useful part of the organizational life. The effective team works through issues of conflict and uses the results to help objectives. Conflict prevents teams from becoming complacent and lazy, and often generates new ideas.

## Good decision-making

As mentioned earlier, objectives need to be clearly and completely understood by all members before good decision-making can begin. In making decisions effective, teams develop the ability to collect information quickly then discuss the alternatives openly. They become committed to their decisions and ensure quick action.

### Appropriate leadership

Effective teams have a leader whose responsibility it is to achieve results through the efforts of a number of people. Power and authority can be applied in many ways, and team members often differ on the style of leadership they prefer. Collectively, teams may come to different views of leadership but, whatever their view, the effective team usually sorts through the alternatives in an open and honest way.

### Review of the team processes

Effective teams understand not only the group's character and its role in the organization, but how it makes decisions, deals with conflicts, etc. The team process allows the team to learn from experience and consciously to improve teamwork. There are numerous ways of looking at team processes – use of an observer, by a team member giving feedback, or by the whole group discussing members' performance.

### Sound intergroup relationships

No human being or group is an island; they need the help of others. An organization will not achieve maximum benefit from a collection of quality improvement teams that are effective within themselves but fight among each other.

### Individual development opportunities

Effective teams seek to pool the skills of individuals, and it necessarily follows that they pay attention to development of individual skills and try to provide opportunities for individuals to grow and learn, and of course have FUN.

Once again, these ideas are not new but are very applicable and useful in the management of teams for quality improvements, just as Newton's theories on gravity still apply!

# Personality types and the MBTI

No one person has a monopoly of 'good characteristics'. Attempts to list the qualities of the ideal manager, for example, demonstrate why that paragon cannot exist. This is because many of the qualities are mutually exclusive, for example:

| | | |
|---|---|---|
| Highly intelligent | v. | Not too clever |
| Forceful and driving | v. | Sensitive to people's feelings |
| Dynamic | v. | Patient |
| Fluent communicator | v. | Good listener |
| Decisive | v. | Reflective |

Although no individual can possess all these and more desirable qualities, a team often does.

A powerful aid to team development is the use of the Myers-Briggs Type Indicator (MBTI).[5] This is based on an individual's preferences on four scales for:

- giving and receiving 'energy';
- gathering information;

- making decisions;
- handling the outer world.

Its aim is to help individuals understand and value themselves and others, in terms of their differences as well as their similarities. It is well researched and non-threatening when used appropriately.

The four MBTI preference scales, which are based on Jung's theories of psychological types, represent two opposite preferences:

- *Extroversion–Introversion*   –  how we prefer to give/receive energy or focus our attention.
- *Sensing–iNtuition*         –  how we prefer to gather information.
- *Thinking–Feeling*          –  how we prefer to make decisions.
- *Judgement–Perception*    –  how we prefer to handle the outer world.

To understand what is meant by preferences, the analogy of left- and right-handedness is useful. Most people have a preference to write with either their left or their right hand. When using the preferred hand, they tend not to think about it, it is done naturally. When writing with the other hand, however, it takes longer, needs careful concentration, seems more difficult, but with practice would no doubt become easier. Most people *can* write with and use both hands, but tend to prefer one over the other. This is similar to the MBTI psychological preferences: most people are able to use both preferences at different times, but will indicate a preference on each of the scales.

In all, there are eight possible preferences – E or I, S or N, T or F, J or P, i.e. two opposites for each of the four scales. An individual's *type* is the combination and interaction of the four preferences. It can be assessed initially by completion of a simple questionnaire. Hence, if each preference is represented by its letter, a person's type may be shown by a four letter code – there are 16 in all. For example, ESTJ represents an *extrovert* (E) who prefers to gather information with *sensing* (S), prefers to make decisions by *thinking* (T) and has a *judging* (J) attitude towards the world, i.e. prefers to make decisions rather than continue to collect information. The person with opposite preferences on all four scales would be an INFP, an introvert who prefers intuition for perceiving, feelings or values for making decisions, and likes to maintain a perceiving attitude towards the outer world.

The questionnaire, its analysis and feedback must be administered by a qualified MBTI practitioner, who may also act as external facilitator to the team in its forming and storming stages.

## Type and teamwork

With regard to teamwork, the preference types and their interpretation are extremely powerful. The *extrovert* prefers action and the outer world, while the *introvert* prefers ideas and the inner world.

*Sensing–thinking* types are interested in facts, analyse facts impersonally, and use a step-by-step process from cause to effect, premise to conclusion. The *sensing–feeling* combinations, however, are interested in facts, analyse facts personally, and are concerned about how things matter to themselves and others.

*Intuition–thinking* types are interested in possibilities, analyse possibilities impersonally, and have theoretical, technical, or executive abilities. On the other hand, the *intuition–feeling* combinations are interested in possibilities, analyse possibilities personally, and prefer new projects, new truths, things not yet apparent.

| ISTJ | ISFJ | INFJ | INTJ |
| --- | --- | --- | --- |
| ISTP | ISFP | INFP | INTP |
| ESTP | ESFP | ENFP | ENTP |
| ESTJ | ESFJ | ENFJ | ENTJ |

**Figure 15.7**
MBTI type table form. (Source: Isabel Myers-Briggs, Introduction to type[5])

*Judging* types are decisive and planful, they live in orderly fashion, and like to regulate and control. *Perceivers,* on the other hand, are flexible, live spontaneously, and understand and adapt readily.

As we have seen, an individual's type is the combination of four preferences on each of the scales. There are 16 combinations of the preference scales and these may be displayed on a *type table* (Figure 15.7). If the individuals within a team are prepared to share with each other their MBTI preferences, this can dramatically increase understanding and frequently is of great assistance in team development and good team working. The similarities and differences in behaviour and personality can be identified. The assistance of a qualified MBTI practitioner is absolutely essential in the initial stages of this work.

# Interpersonal relations – FIRO-B and the Elements

The FIRO-B (Fundamental Interpersonal Relations Orientation – Behaviour) is a powerful psychological instrument which can be used to give valuable insights into the needs individuals bring to their relationships with other people. The instrument assesses needs for *inclusion, control* and *openness* and therefore offers a framework for understanding the dynamics of interpersonal relationships.

Use of the FIRO instrument helps individuals to be more aware of how they relate to others and to become more flexible in this behaviour. Consequently it enables people to build more productive teams through better working relationships.

Since its creation by William Schutz in the 1950s, to predict how military personnel would work together in groups, the FIRO-B instrument has been used throughout the world by managers and professionals to look at management and decision-making

| Table 15.1 | The FIRO-B interpersonal dimensions and aspects | | |
|---|---|---|---|
| | **Inclusion** | **Control** | **Openness** |
| **Expressed behaviour** | Expressed inclusion | Expressed control | Expressed openness |
| **Wanted behaviour** | Wanted inclusion | Wanted control | Wanted openness |

Modified from: W. Schutz (1978) *FIRO Awareness Scales Manual*, Palo Alto, CA, Consulting Psychologists Press.

styles. Through its ability to predict areas of probable tension and compatibility between individuals, the FIRO-B is a highly effective team-building tool which can aid in the creation of the positive environment in which people thrive and achieve improvements in performance.

The theory underlying the FIRO-B incorporates ideas from the work of Adomo, Fromm and Bion and it was first fully described in Schutz's book, *FIRO: A Three Dimensional Theory of Personal Behaviour* (1958). In his more recent book *The Human Element,* Schutz developed the instrument into a series of 'elements', B, F, S, etc., and offers strategies for heightening our awareness of ourselves and others.

The FIRO-B takes the form of a simple-to-complete questionnaire the analysis of which provides scores that estimate the levels of behaviour with which the individual is comfortable, with regard to his/her needs for inclusion, control and openness. Schutz described these three dimensions in the form of the decision we make in our relationships regarding whether we want to be:

- 'in' or 'out' – inclusion;
- 'up' or 'down' – control;
- 'close' or 'distant' – openness.

The FIRO-B estimates our unique level of needs for each of these dimensions of interpersonal interaction.

The instrument further divides each of these dimensions into:

- the behaviour we feel most comfortable *exhibiting towards* other people – *expressed* behaviours; and
- the behaviour we *want from* others – *wanted* behaviours.

Hence, the FIRO-B 'measures', on a scale of 0–9, each of the three interpersonal dimensions in two aspects (Table 15.1).

The *expressed* aspect of each dimension indicates the level of behaviour the individual is most comfortable with towards others, so high scores for the expressed dimensions would be associated with:

## High scored expressed behaviours

**Inclusion**  Makes efforts to include other people in his/her activities – tries to belong to or join groups and to be with people as much as possible.

**Control**  Tries to exert control and influence over people and tell them what to do.

**Openness**  Makes efforts to become close to people – expresses friendly open feelings, tries to be personal and even intimate.

Low scores would be associated with the opposite expressed behaviour.

The *wanted* aspect of each dimension indicates the behaviour the individual prefers others to adopt towards him/her, so high scores for the wanted dimensions would be associated with:

## High scored wanted behaviours

**Inclusion**  Wants other people to include him/her in their activities – to be invited to belong to or join groups (even if no effort is made by the individual to be included).

**Control**  Wants others to control and influence him/her and be told what to do.

**Openness**  Wants others to become close to him/her and express friendly, open, even affectionate feelings.

Low scores would be associated with the opposite wanted behaviours.

It is interesting to look at typical manager FIRO-B profiles, based on their scores for the six dimensions/aspects in Table 15.1. Figure 15.8 shows the average of a sample of 700 middle and senior managers in the UK with boundaries at one sigma, plotted on expressed/wanted scales for the three dimensions.

On average the managers show a higher level of expressed inclusion – including people in his/her activities – than wanted inclusion. Similarly, and not surprisingly perhaps, expressed control – trying to exert influence and control over others – is higher in managers than wanted control. When it comes to openness, the managers tend to want others to be open, rather than be open themselves.

It is even more interesting to contemplate these results when one considers the demands of some of the recent popular management programmes, such as total quality

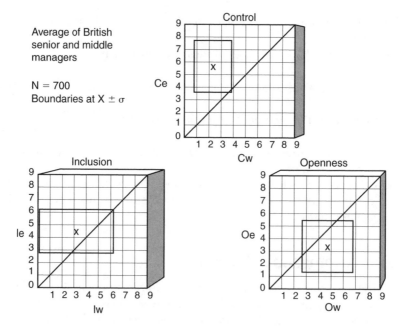

**Figure 15.8**
Typical manager profiles (FIRO-B)

management, employment involvement, and self-directed teams. These tend to require from managers certain behaviours, for example lower levels of expressed control and higher levels of wanted control, so that the people feel empowered. Similarly, managers are encouraged to be more open. These, however, are *opposite* to the apparent behaviours of the sample of managers shown graphically in Figure 15.8. It is not surprising then that TQM has failed in some organizations where managers were being asked to empower employees and be more open – and who can argue against that – yet their basic underlying needs caused them to behave in the opposite way.

Understanding what drives these behaviours is outside the scope of this book but other FIRO and Element instruments can help individuals to further develop understanding of themselves and others. FIRO and Schutz's Elements instruments for measuring *feelings* (F) and *self-concept* (S) can deepen the awareness of what lies behind our behaviours with respect to inclusion, control and openness. The reader is advised to undertake further reading and seek guidance from a trained administrator of these instruments, but the overall relationship between the B and F instruments is given below:

**Behaviours related to:**

Inclusion

Control

Openness

**Feelings related to:**

Significance

Competence

Likeability

Issues around control behaviour then may arise because of underlying feelings about competence. Similarly, underlying feelings concerning significance may lead to certain inclusion behaviours.

## FIRO-B in the workplace

The inclusion, control and openness (I–C–O) dimensions form a cycle (Figure 15.9), which can help groups of people to understand how their individual and joint behaviour develops as teams are formed. Given in Table 15.2 are the considerations, questions and outcomes under each dimension. If inclusion issues are resolved first it is possible to progress to dealing with the control issues, which in turn must be resolved if the openness issues are to be dealt with successfully. As a team develops, it travels around the inclusion, control and openness cycle time and time again. If the issues are not resolved in each dimension, further progress in the next dimension will be hindered – it is difficult to deal with issues of control if unresolved inclusion issues are still around and people do not know whether they are 'in' or 'out' of the group. Similarly it is difficult to be open if it is not clear where the power base is in the group.

This I–C–O cycle has led to the development by John Oakland and his colleagues of an 'openness model' which is in three parts. Part 1 based on the premise that to participate productively in a team individuals must first be involved and then committed. Figure 15.10 shows some of the questions which need to be answered and the outcomes from this stage. Part 2 deals with the control aspects of empowerment and management and Figure 15.11 summarizes the questions and outcomes. Finally Part 3, summarized in Figure 15.12, ensures openness through acknowledgement and trust. The full openness cycle (Figure 15.13) operates in a clockwise direction so that trust leads to more involvement, further commitment, increased empowerment, etc. Of course, if progress is not made round the cycle and trust is replaced by fear, it is possible to send the whole process into reverse – a negative cycle of suspicion, fault-finding, abdication and confusion

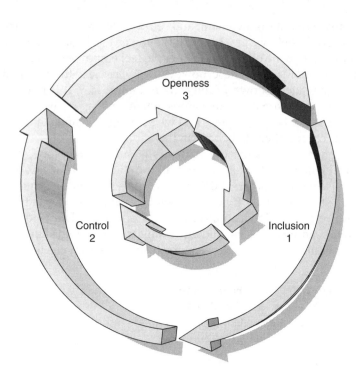

**Figure 15.9**
The inclusion, control and openness cycle

(Figure 15.14). Unfortunately this will be recognized as the culture in some organizations where the focus of enquiry is 'what has gone wrong' leading to 'whose fault was it?'

Fortunately, organizations and individuals seem keen to learn ways to change these negative communications that sour relationships, not dampen personal satisfaction and reduce productivity. The inclusion, control, openness cycle is a useful framework for helping teams to pass successfully through the forming and storming stages of team development. As teams are disbanded for whatever reason, the process reverses and the first thing that goes is the openness.

## The five 'A' stages for teamwork

The awareness provided by the use of the MBTI and FIRO-B instruments helps people to appreciate their own uniqueness and the uniqueness of others – the foundation of mutual respect and for building positive, productive and high performing teams.

For any of these models or theories to benefit a team, however, the individuals within it need to become *aware* of the theory, e.g. the MBTI or FIRO-B. They then need to *accept* the principles as valid, *adopt* them for themselves in order to *adapt* their behaviour accordingly. This will lead to individual and team *action* (Figure 15.15).

In the early stages of team development particularly, the assistance of a skilled facilitator to aid progress through these stages is necessary. This is often neglected, causing failure in so many team initiatives. In such cases the net output turns out to be lots of nice warm feelings about 'how good that team workshop was a year ago', but the nagging reality that no action came out and nothing has really changed.

Total Quality in the Construction Supply Chain

**Table 15.2** Considerations, questions and outcomes for the FIRO-B dimensions

| Dimension | Considerations | Some typical questions | If resolved we get | If not resolved we get |
|---|---|---|---|---|
| Inclusion | Involvement – how much you want to include other people in your life and how much attention and recognition you want | Do I care about this? Do I want to be involved? Does this fit with my values? Do I matter to this group? Can I be committed? … leading to … Am I 'in' or 'out'? | A feeling of belonging A sense of being recognized and valued Willingness to become committed | A feeling of alienation A sense of personal insignificance No desire for commitment or involvement |
| Control | Authority, responsibility, decision-making, influence | Who is in charge here? Do I have power to make decisions? What is the plan? When do we start? What support do I have? What resources do I have? … leading to … Am I 'up' or 'down'? | Confidence in self and others Comfort with level of responsibility Willingness to belong | Lack of confidence in leadership Discomfort with level of responsibility – fear of too much – frustration with too little 'Griping' between individuals |
| Openness | How much are we prepared to express our true thoughts and feelings with other individuals | Does she like me? Will my work be recognized? Is he being honest with me? How should I show appreciation? Do I appear aloof? … leading to … Am I 'open' or 'closed'? | Lively and relaxed atmosphere Good-humoured interactions Open and trusting relationships | Tense and suspicious atmosphere Flippant or malicious humour Individuals isolated |

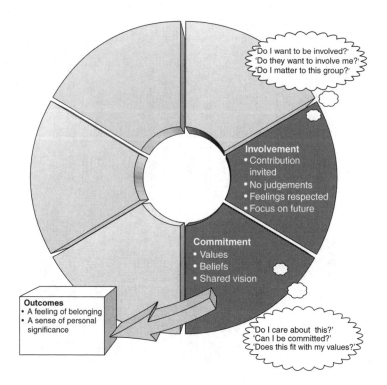

**Figure 15.10**
The openness model, Part 1 Inclusion: involvement, inviting contribution, responding

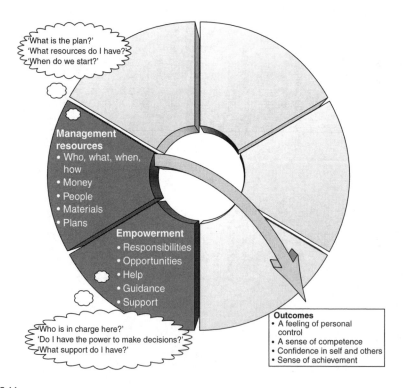

**Figure 15.11**
The openness model, Part 2 Control: choice, influence, power

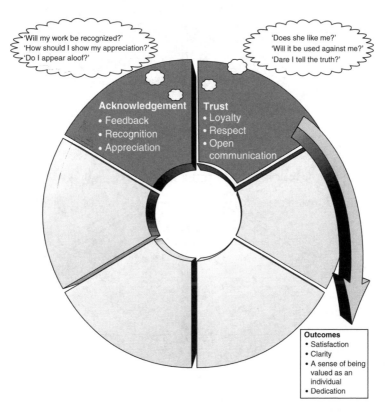

**Figure 15.12**
The openness model, Part 3 Openness: expression of true thoughts and feelings with respect for self and others

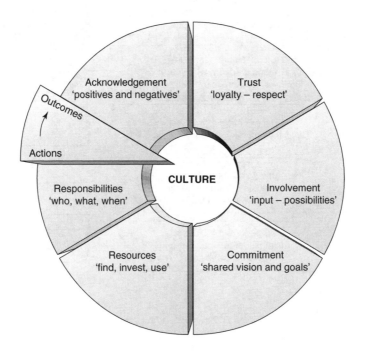

**Figure 15.13**
The full openness model

**Figure 15.14**
The negative cycle

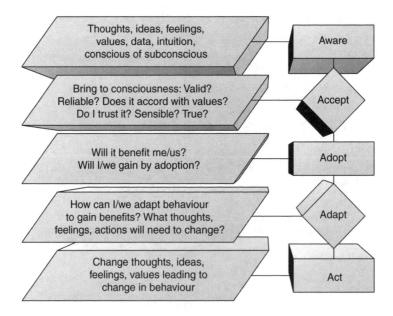

**Figure 15.15**
The five 'A' stages for teamwork

# References

1. Tannenbaum, R. and Schmidt, W.H. (1973) 'How to choose a leadership pattern', *Harvard Business Review*, May-June.
2. Blanchard, K. and Herrsey, P. (1982) *Management of Organizational Behaviour: Utilizing Human Resources* (4th edn), Prentice-Hall, Englewood Cliffs, NJ.
3. Tuckman, B.W. and Jensen, M.A. (1977) 'States of small group development revisited', *Group and Organizational Studies*, 2(4), pp. 419–427, New York.
4. Kormanski, C. (1985) 'A situational leadership approach to groups using the Tuckman Model of Group Development', *The 1985 Annual: Developing Human Resources*, University Associates, San Diego, CA.
5. *Myers-Briggs Type Indicator*. (1987) Consulting Psychologists Press, Palo Alto.

## *Chapter highlights*

### The need for teamwork

■ The only efficient way to tackle process improvement or complex problems is through teamwork. The team approach allows individuals and organizations to grow.

■ Within fragmented supply chains there is often a need for effective teams that cross-organizational boundaries.

■ Employees will not engage continual improvement without commitment from the top, a quality 'climate', and an effective mechanism for capturing individual contributions.

■ Teamwork for quality improvement is driven by a strategy, needs a structure, and must be implemented thoughtfully and effectively.

### Running process management and improvement teams

■ Process management and improvement teams are groups brought together by management to improve a process or tackle a particular problem on a project basis. The running of these teams involves several factors: selection and leadership, objectives, meetings, assignments, dynamics, results and reviews.

■ The need for training in the basic skills of team leadership should not be underestimated if successful outcomes are sought.

### Teamwork and action-centred leadership

■ Early work in the field of human relations by McGregor, Maslow, and Hertzberg was useful to John Adair in the development of his model for teamwork and action-centred leadership.

■ Adair's model addresses the needs of the task, the team, and the individuals in the team, in the form of three overlapping circles. There are specific task, team and individual functions for the leader, but (s)he must concentrate on the small central overlap area of the three circles.

■ The team process has inputs and outputs. Good teams have three main attributes: high task fulfilment, high team maintenance, and low self-orientation.

■ In dealing with the task, the team and its individuals, a situational style of leadership must be adopted. This may follow the Tannenbaum and Schmidt, and Blanchard models through directing, coaching, and supporting to delegating.

### Stages of team development

■ When teams are put together, they pass through Tuckman's forming (awareness), storming (conflict), norming (co-operation), and performing (productivity) stages of development.

- Teams that go through these stages successfully become effective and display clear objectives and agreed goals, openness and confrontation, support and trust, co-operation and conflict, good decision-making, appropriate leadership, review of the team processes, sound relationships, and individual development opportunities.

### Personality types and the MBTI
- A powerful aid to team development is provided by the Myers-Briggs Type Indicator (MBTI).
- The MBTI is based on individuals' preferences on four scales for giving and receiving 'energy' (extroversion – E or introversion – I), gathering information (sensing – S or intuition – N), making decisions (thinking – T or feeling – F) and handling the outer world (judging – J or perceiving – P).
- An individual's type is the combination and interaction of the four scales and can be assessed initially by completion of a simple questionnaire. There are 16 types in all, which may be displayed for a team on a type table.

### Interpersonal relations – FIRO-B and the Elements
- The FIRO-B (Fundamental Interpersonal Relations Orientation – Behaviour) instrument gives insights into the needs individuals bring to their relationships with other people.
- The FIRO-B questionnaire assesses needs for inclusion, control and openness, in terms of expressed and wanted behaviour.
- Typical manager FIRO-B profiles conflict with some of the demands of TQM and can, therefore, indicate where particular attention is needed to achieve successful TQM implementation.
- The inclusion, control, and openness dimensions form an 'openness' cycle which can help groups to understand how to develop their individual and joint behaviours as the team is formed. An alternative negative cycle may develop if the understanding of some of these behaviours is absent.
- The five As: for any of the teamwork models and theories, the individuals must become aware, need to accept, adopt and adapt, in order to act. A skilled facilitator is always necessary.

# 16

# Communications, innovation and learning

# Communicating the quality strategy

People's attitudes and behaviour clearly can be influenced by communications; one has to look only at the media or advertising to understand this. The essence of changing attitudes is to gain acceptance for the need to change, and for this to happen it is essential to provide relevant information, convey good practices, and generate interest, ideas and awareness through excellent communication processes. This is possibly the most neglected part of many organizations' operations, yet failure to communicate effectively creates unnecessary problems, resulting in confusion, loss of interest and eventually in declining quality through apparent lack of guidance and stimulus.

In the construction industry, projects are often delivered by highly fragmented supply chains, making the need for communication even greater. Managers and workers from myriad organizations work together to achieve a shared project outcome. The achievement of the vision in terms of product and process goals and the quality of those outcomes depend largely on the effectiveness of communications at the project level.

Total quality management will significantly change the way many organizations operate and 'do business'. This change will require direct and clear communication from the top management to all staff and employees, to explain the need to focus on processes. Everyone will need to know their roles in understanding processes and improving their performance. This applies as much to every subcontract designer or contract employee as it does to every person working directly for the principal contracting organization.

Whether a strategy is developed by top management for the direction of the business/organization as a whole, or specifically for the introduction of TQM, that is only half the battle. An early implementation step must be the clear widespread communication of the strategy.

An excellent way to accomplish this first step is to issue a total quality message that clearly states top management's commitment to quality and outlines the role everyone must play. This can be in the form of a quality policy (see Chapters 3 and 4) or a specific statement about the organization's intention to integrate quality into the business operations. Such a statement might read:

> The board of directors (or appropriate title) believe that the successful implementation of total quality management is critical to achieving and maintaining our business goals of leadership in quality, delivery and price competitiveness.
>
> We wish to convey to everyone our enthusiasm and personal commitment to the total quality approach, and how much we need your support in our mission of business improvement. We hope that you will become as convinced as we are that business and process improvement is critical for our survival and continued success.
>
> We can become a total quality organization only with your commitment and dedication to improving the processes in which you work. We will help you by putting in place a programme of education, training, and teamwork development, based on business and process improvement, to ensure that we move forward together to achieve our business goals.

The quality director or TQM co-ordinator should then assist the senior management team to prepare a directive. This must be signed by all business unit, division, or process leaders, and distributed to everyone in the organization. The directive should include the following:

- need for improvement;
- concept for total quality;

- importance of understanding business processes;
- approach that will be taken and people's roles;
- individual and process group responsibilities;
- principles of process measurement.

The systems for disseminating the message should include all the conventional communication methods of seminars, departmental meetings, posters, newsletters, intranet, etc. First line supervision will need to review the directive with all the staff, and a set of questions and answers may be suitably pre-prepared in support.

Once people understand the strategy, the management must establish the infrastructure (see Chapter 14). The required level of individual commitment is likely to be achieved, however, only if everyone understands the aims and benefits of TQM, the role they must play, and how they can implement process improvements. For this understanding a constant flow of information is necessary, including:

1. When and how individuals will be involved.
2. What the process requires.
3. The successes and benefits achieved.

The most effective means of developing the personnel commitment required is to ensure people know what is going on. Otherwise they will feel left out and begin to believe that TQM is not for them, which will lead to resentment and undermining of the whole process. The first line of supervision again has an important part to play in ensuring key messages are communicated and in building teams by demonstrating everyone's participation and commitment.

Naturally the extent to which such policy is formally set out and formally communicated is influenced by the size of the organization and the stage of organizational development at which TQM values are first introduced. For example, an interesting case study is the one of Mirvac, the Australian property group that started as a developer and builder of residential medium density housing in 1972. From its inception the company CEO and his co-founder saw the opportunity for their business as creating housing of high quality, not Rolls-Royce quality but a dependable, well-designed defect-free product. Their motto from day one has been: 'If you wouldn't live in it yourself, you should not be building it.' This has influenced what land they buy, how many units they build in a development, how they orient them and the quality of workmanship. Till the present, the simple but unwavering focus of senior management on quality in every decision has created a company whose brand image for quality housing is next to none. It is acknowledged that their product holds its value better than any other in the market. They have achieved this through clear sighted management focus, no formally documented systems, selecting suppliers and employees carefully, and quickly learning from mistakes. It is only recently, since the company has grown to 220 employees in the construction division, that the company is beginning to document its practices and processes, the need being driven by growth in size. This case study demonstrates very well the need for clear focus and clear unambiguous messages from senior management.

In the Larkins' excellent book *Communicating Change*[1], the authors refer to three 'facts' regarding the best ways to communicate change to employees:

1. Communicate directly to supervisors (first line).
2. Use face-to-face communication.
3. Communicate relative performance of the local work area.

338

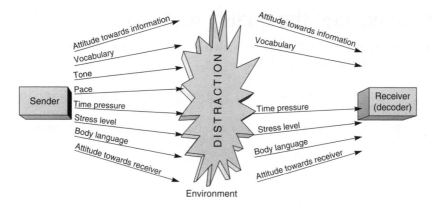

**Figure 16.1**
Communication model

The language used at the 'coal face' will need attention in many organizations. Reducing the complexity and jargon in written and spoken communications will facilitate comprehension. When written business communications cannot be read or understood easily, they receive only cursory glances, rather than the detailed study they require. *Simplify and shorten* must be the guiding principles. The communication model illustrated in Figure 16.1 indicates the potential for problems through environmental distractions, mismatches between sender and receiver (or more correctly, decoder) in terms of attitudes – towards the information and each other – vocabulary, time pressures, etc.

All levels of management should introduce and stress 'open' methods of communication, by maintaining open offices, being accessible to staff/employees and taking part in day-to-day interactions and the detailed processes. This will lay the foundation for improved interactions *between* staff and employees, which is essential for information flow and process improvement. Opening these lines of communication may lead to confrontation with many barriers and much resistance. Training and the behaviour of supervisors/managements should be geared to helping people accept responsibility for their own behaviour, which often creates the barriers, and for breaking the barriers down by concentrating on the process rather than 'departmental' needs.

Resistance to change will always occur and is to be expected. Again first-line management should be trained to help people deal with it. This requires an understanding of the dynamics of change and the support necessary – not an obsession with forcing people to change. Opening up lines of communication through a previously closed system, and publicizing people's efforts to change and their results, will aid the process. Change can be – even should be – exciting if employees start to share their development, growth, suggestions, and questions. Management needs to encourage and participate in this by creating the most appropriate communication systems.

In construction this resistance to change is likely to be significant, especially across the supply chain. The cost-driven focus and exploitative management practices in this industry have created a great deal of cynicism. Subcontractors are well used to hearing lip service to quality, but seeing a shortsighted management focus that accepts high rework costs and expensive legal costs in litigation as normal business practice. The implication of this situation for management is that as much effort has to be invested in bringing the supply chain partners along on the quality journey as on training internal employees. A strategy that might be considered is the joint training of some people from supply chain partners; after all, they work next to each other on a daily basis.

# Communicating the quality message

The people in most organizations fall into one of four 'audience' groups, each with particular general attitudes towards TQM:

- *Senior managers,* who should see TQM as an opportunity, both for the organization and themselves.
- *Middle managers,* who may see TQM as another burden without any benefits, and may perceive a vested interest in the status quo.
- *Supervisors* (first-line or junior managers), who may see TQM as another 'flavour of the period' or campaign, and who may respond by trying to keep heads down so that it will pass over.
- *Other employees,* who may not care, so long as they still have jobs and get paid, though these people must be the custodians of the delivery of quality to the customer and own that responsibility.

Senior management needs to ensure that each group sees TQM as being beneficial to them. Total quality training material and support (whether internal from a quality director and team or from external consultants) will be of real value only if the employees are motivated to respond positively to them. The implementation strategy must then be based on two mutually supporting aspects:

1. 'Marketing' any TQM initiatives.
2. A positive, logical process of communication designed to motivate.

There are of course a wide variety of approaches to, and methods of, TQM and business improvement. Any individual organization's quality strategy must be designed to meet the needs of its own structure and business, and the state of commitment to continuous improvement activities. These days very few organizations are starting from a greenfield site. The key is that groups of people must feel able to 'join' the quality process at the most appropriate point for them. For middle managers to be convinced that they must participate, TQM must be presented as the key to help them turn the people who work for them into 'total quality employees'. At the senior management level, an early and critical task is to win the support of management from supply chain partners to a joint initiative. Without co-operation the task of implementing a total quality initiative is much greater.

The noisy, showy, hype-type activity is not appropriate to any aspect of TQM. TQM 'events' should of course be fun, because this is often the best way to persuade and motivate, but the value of any event should be judged by its ability to contribute to understanding and the change to TQM. Key words in successful exercises include 'discovery', 'affirmation', 'participation', and 'team-based learning'. In the difficult area of dealing with middle and junior managers, who can and will prevent change with ease and invisibility, the recognition that progress must change from being a threat to a promise will help. In any workshops designed for them managers and supervisors should be made to feel recognized, not victimized and the programmes should be delivered by specially trained people. The environment and conduct of the workshops must also demonstrate the organization's concern for quality.

The key medium for motivating the employees and gaining their commitment to quality is face-to-face communication and *visible* management commitment. Much is written and spoken about leadership, but it is mainly about communication. If people are good leaders, they are invariably good communicators. Leadership is a human interaction depending on the communications between the leaders and the followers. It calls for many skills that can be *learned* from education and training, but must be *acquired* through practice.

# Communication, learning, education and training

It may be useful to consider why people learn. They do so for several reasons, some of which include:

- Self-betterment.
- Self-preservation.
- Need for responsibility.
- Saving time or effort.
- Sense of achievement.
- Pride of work.
- Curiosity.

So communication and training can be a powerful stimulus to personal development at the workplace, as well as achieving improvements for the organization. This may be useful in the selection of the appropriate method(s) of communication, the principal ones being:

- *Verbal communication* either between individuals or groups, using direct or indirect methods, such as public address and other broadcasting systems and recordings.
- *Written communication* in the form of notices, bulletins, information sheets, reports, e-mail and recommendations.
- *Visual communication* such as posters, films, video, internet/intranet, exhibitions, demonstrations, displays and other promotional features. Some of these also call for verbal and written communication.
- *Example*, through the way people conduct themselves and adhere to established working codes and procedures, through their effectiveness as communicators and ability to 'sell' good practices.

The characteristics of each of these methods should be carefully examined before they are used in helping people to learn.

It is the authors' belief that education and training are the single most important factors in actually improving quality and business performance, once there has been commitment to do so. For education and training to be the effective, however, they must be planned in a systematic and objective manner to provide the right sort of learning experience. Education and training must be continuous to meet not only changes in technology but also changes in the environment in which an organization operates, its structure, and perhaps most important of all the people who work there.

An interesting paradox is that while industry in general tends to underinvest in training, world-class companies invest far above the industry average. Cost-focused organizations are concerned about the payback on training and see it as overhead; managers tend to want to minimize investment. Value-focused companies see training as the key to maximizing the potential of their workforce; hence, they are active in identifying priorities and ensuring that training is effective. All the case study companies invest far above the industry average in training – how far above varies – but the Graniterock case study (p. 375) is instructive and underlines the importance of training. As its business is in quarrying and concrete construction, the company benchmarks itself against

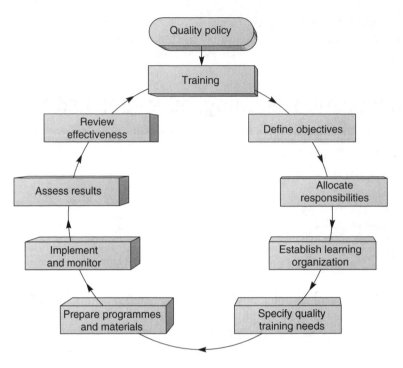

**Figure 16.2**
The quality training circle

construction and mining. Graniterock employee training is at five times the US national average for mining and 20 times the construction industry average.

## Education and training cycle of improvement

Education and training activities can be considered in the form of a cycle of improvement (Figure 16.2), the elements of which are the following.

### *Ensure education and training are part of the policy*

Every organization should define its policy in relation to education and training. The policy should contain principles and goals to provide a framework within which learning experiences may be planned and operated. This policy should be communicated to all levels.

### *Establish objectives and responsibilities for education and training*

When attempting to set education and training objectives three essential requirements must be met:

1. Senior management must ensure that learning outcomes are clarified and priorities set.
2. The defined education and training objectives must be realizable and attainable.

3. The main objectives should be 'translated' for all functional areas in the organization. Large organizations may find it necessary to promote a phased plan to identify these.

The following questions are useful first steps when identifying education and training objectives:

- How are the customer requirements transmitted through the organization?
- Which areas need improved performance?
- What changes are planned for the future?
- What are the implications for the process framework?

Education and training must be the responsibility of line management, but there are also important roles for the individuals concerned.

## Establish the platform for a learning organization

The overall responsibility for seeing that education and training are properly organized must be assumed by one or more designated senior executives. All managers have a responsibility for ensuring that personnel reporting to them are properly trained and competent in their jobs. This responsibility should be written into every manager's job description. The question of whether line management requires specialized help should be answered when objectives have been identified. It is often necessary to use specialists, who may be internal or external to the organization.

Construction organizations have a particular problem: site staff tend to be completely focused on the delivery process. Therefore, unless training is seen as a site responsibility led by senior project management, it will tend to be seen as a head office requirement and a distraction, creating a tension between production and personnel development. To overcome this, certain training should be work based and problem based, using the challenges on the project as the setting for shared learning within production teams.

## Specify education and training needs

The next step in the cycle is to assess and clarify specific education and training needs. The following questions need to be answered:

- Who needs to be educated/trained?
- What competencies are required?
- How long will the education/training take?
- What are the expected benefits?
- Is the training need urgent?
- How many people are to be educated/trained?
- Who will undertake the actual education/training?
- What resources are needed, e.g. money, people, equipment, accommodation, outside resources?

## Prepare education/training programmes and materials

Senior management should participate in the creation of overall programmes, although line managers should retain the final responsibility for what is implemented, and they will often need to create the training programmes themselves.

Training programmes should include:

- The training objectives expressed in terms of the desired behaviour.
- The actual training content.
- The methods to be adopted.
- Who is responsible for the various sections of the programme?

### Implement and monitor education and training

The effective implementation of education and training programmes demands considerable commitment and adjustment by the trainers and trainees alike. Training is a progressive process, which must take into account any learning problems of the trainees. The authors believe that in construction, given the fragmentation of the supply chain and the consequent need to develop deep relationships within the supply chain, some joint training could be an important part of an effective organizational development strategy.

### Assess the results

In order to determine whether further education or training is required, line management should themselves review performance when training is completed. However good the training may be, if it is not valued and built upon by managers and supervisors, its effect can be severely reduced.

### Review effectiveness of education and training

Senior management will require a system whereby decisions are taken at regular fixed intervals on:

- the policy;
- the education and training objectives;
- the education/training organization;
- the progress towards a learning organization.

Even if the policy remains constant, there is a continuing need to ensure that new education and training objectives are set either to promote work changes or to raise the standards already achieved.

The purpose of management system audits and reviews is to assess the effectiveness of the management effort. Clearly, adequate and refresher training in these methods is essential if such checks are to be realistic and effective. Audits and reviews can provide useful information for the identification of changing quality-training needs.

The education/training organization should similarly be reviewed in the light of the new objectives, and here again it is essential to aim at continuous improvement. Training must never be allowed to become static, and the effectiveness of the organization's education and training programmes and methods must be assessed systematically.

# A systematic approach to education and training for quality

Education and training for quality should have, as its first objective, an appreciation of the personal responsibility for meeting the 'customer' requirements by everyone from the most senior executive to the newest and most junior employee. Responsibility for the training of employees in quality rests with management at all levels and, in particular, the person nominated for the co-ordination of the organization's quality effort. Education and training will not be fully effective, however, unless responsibility for the deployment of the policy rests clearly with the chief executive. One objective of this policy should be to develop a *climate* in which everyone is quality conscious and acts with the needs of the customer in mind at all times.

The main elements of effective and systematic quality training may be considered under four broad headings:

- Error/defect/problem prevention.
- Error/defect/problem reporting and analysis.
- Error/defect/problem investigation.
- Review.

The emphasis should obviously be on error, defect, or problem prevention, and hopefully what is said under the other headings maintains this objective.

## Error/defect/problem prevention

The following contribute to effective and systematic training for prevention of problems in the organization:

1. An issued quality policy.
2. A written management system.
3. Job specifications that include quality requirements.
4. Effective steering committees, including representatives of both management and employees.
5. Efficient housekeeping standards.
6. Preparation and display of maps, flow diagrams and charts for all processes.
7. Simple transparent performance measures in all critical outcome areas.

## Error/defect/problem reporting and analysis

It will be necessary for management to arrange the necessary reporting procedures, and ensure that those concerned are adequately trained in these procedures. All errors, rejects, defects, defectives, problems, waste, etc., should be recorded and analysed in a way that is meaningful for each organization, bearing in mind the corrective action programmes that should be initiated at appropriate times.

This area is a critical challenge on construction projects simply because of the high degree of supply chain fragmentation and the independent way in which subcontractors build on each other's work. This means that effective error identification and reporting has to be done by the employees of subcontractors, who are driven by piece rate payment for their output and hence are generally not motivated to report a problem as it may delay their work. The biggest challenge is to develop motivation and procedures across the supply chain. One of the authors is currently researching this very question on construction projects in Sydney: where with a group of 18 contractors, general contractors and subcontractors, they are experimenting with the areas of motivation, performance measurement and improved information capture and information flow to identify error detection and improve management response.

# Error/defect/problem investigation

The investigation of errors, defects, and problems can provide valuable information that can be used in their prevention. Participating in investigations offers an opportunity for training. The following information is useful for the investigation:

- Nature of problem.
- Date, time and place.
- Product/service with problem.
- Description of problem.
- Causes and reasons behind causes.
- Action advised.
- Action taken to prevent recurrence.

Once again this area is one of the construction industry's great challenges. Most project managers rely on error detection and rectification to get the job done and never take their project organizations into this critical phase of continuous improvement. Through their experience, the authors are aware of countless examples of quality and safety errors that have been allowed to happen again and again on projects (often detected and corrected), but at times ultimately leading to very significant injury or cost impacts. Fragmentation and time pressure are the two main challenges to overcome.

# Review of quality training

Review of the effectiveness of quality training programmes should be a continuous process. However, the measurement of effectiveness is a complex problem. One way of reviewing the content and assimilation of a training course or programme is to monitor behaviour during quality audits. This review can be taken a stage further by comparing employees' behaviour with the objectives of the quality-training programme. Other measures of the training processes should be found to establish the benefits derived.

Once again the challenge on construction projects is to ensure that a similar approach to training is implemented across the entire project workforce. Most organizations consider only the training of their own people; however, when more than three quarters of the workforce on a project are employed by subcontractors, limiting the training strategy to any one organization, limits the potential benefits of the training programme.

# Education and training records

All organizations should establish and maintain procedures for the identification of education and training needs and the provision of the actual training itself. These procedures should be designed (and documented) to include all personnel. In many situations it is necessary to employ professionally qualified people to carry out specific tasks, e.g. accountants, lawyers, engineers, chemists, etc., but it must be recognized that all other employees, including managers, must have or receive from the company the appropriate education, training and/or experience to perform their jobs. This leads to the establishment of education and training records.

Once an organization has identified the special skills required for each task, and developed suitable education and training programmes to provide competence for the tasks to be undertaken, it should prescribe how the competence is to be demonstrated. This can be by some form of examination, test or certification, which may be carried out in-house or by a recognized external body. In every case, records of personnel qualifications, education, training, and experience should be developed and maintained. National vocational qualifications may have an important role to play here.

At the simplest level this may be a record of tasks and a date placed against each employee's name as he/she acquires the appropriate skill through education and training. Details of attendance on external short courses, in-house induction or training schemes complete such records. What must be clear and easily retrievable is the status of training and development of any single individual, related to the tasks that he/she is likely to encounter. For example, in a factory producing contact lenses that has developed a series of well-defined tasks for each stage of the manufacturing process, it would be possible, by turning up the appropriate records, to decide whether a certain operator is competent to carry out a lathe-turning process. Clearly, as the complexity of jobs increases and managerial activity replaces direct manual skill, it becomes more difficult to make decisions on the basis of such records alone. Nevertheless, they should document the basic competency requirements and assist the selection procedure.

On a construction project, training of subcontract staff is as critical to project success as the training of the general contractor's staff. This includes critical areas such as safety and quality as well as areas of technical competency. However, a prerequisite to ensuring that all key subcontractors do maintain training records is a close and durable relationship between the general contractor and the subcontract supply chain.

# Starting where and for whom?

Education and training needs occur at four levels of an organization:

- *Very senior management* (strategic decisions-makers).
- *Middle management* (tactical decision-makers or implementers of policy).
- *First level supervision and quality team leaders* (on-the-spot decision-makers).
- *All other employees* (the doers).

Neglect of education/training in any of these areas will, at best, delay the implementation of TQM and the improvements in performance. It is also critical that organizations within the supply chain have the same commitment at all levels of the organization. The provision of training for each group will be considered in turn, but it is important to realize that an integrated programme is required, one that includes follow-up learning-based activities and encourages exchange of ideas and experience.

## Very senior management

The chief executive and his team of strategic policy makers are of primary importance, and the role of education and training here is to provide awareness and instil commitment to quality. The importance of developing real commitment must be established and often this can only be done by a free and frank exchange of views between trainers and trainees. This has implications for the choice of the trainers themselves, and the fresh-faced graduate, sent by the 'package consultancy' operator into the lion's den of a boardroom, will not make much impression with the theoretical approach that he or she is obliged to bring to bear. The authors recall thumping many a boardroom table, and using all their experience and whatever presentation skills they could muster, to convince senior managers that without a TQM-based approach they would fail. It is a sobering fact that the pressure from competition and customers has a much greater record of success than enlightenment, although dragging a team of senior managers down to the shop floor to show them the results of poor management was successful on more than one occasion.

Executives responsible for marketing, sales, finance, design, operations, purchasing, personnel, distribution, etc., all need to understand quality. They must be shown how to define the policy and objectives, how to establish the appropriate organization, how to clarify authority, and generally how to create the atmosphere in which total quality will thrive. This is the only group of people in the organization that can ensure that adequate resources are provided and directed at:

1. Meeting customer requirements – internally and externally.
2. Setting standards to be achieved – zero failure.
3. Monitoring of quality performance – quality costs.
4. Introducing a good quality management system – prevention.
5. Implementing process control methods – SPC.
6. Spreading the idea of quality throughout the whole workforce – TQM.

The senior management of any principal contracting organization has the task of finding like-minded suppliers: suppliers who, because of their commitment to the same quality ideals, will deliver services and products that support the overall customer requirements.

## Middle management

The basic objectives of management quality training should be to make managers conscious and anxious to secure the benefits of the total quality effort. One particular 'staff' manager will require special training – the quality manager, who will carry the responsibility for management of the quality management system, including its design, operation, and review.

The middle managers should be provided with the technical skills required to design, implement, review, and change the parts of the quality management system that will be under their direct operational control. It will be useful throughout the training programmes to ensure that the responsibilities for the various activities in each of the functional areas are clarified. The presence of a highly qualified and experienced quality manager should not allow abdication of these responsibilities, for the internal 'consultant' can easily create not-invented-here feelings by writing out procedures without adequate consultation of those charged with implementation.

Middle management should receive comprehensive training on the philosophy and concepts of teamwork, and the techniques and applications of statistical process control (SPC). Without the teams and tools, the quality management system will lie dormant and lifeless. It will relapse into a paper generating system, fulfilling the needs of only those who thrive on bureaucracy. They need to learn how to put this lot together in a planning–process–people–performance value chain that is sustainable for the future.

A suggestion for organizations with fragmented supply chains is to develop some joint training, in quality management, with their suppliers; not only to ensure that values and goals are aligned, but also to build the supply team.

## First-level supervision

There is a layer of personnel in many organizations which plays a vital role in their inadequate performance – foremen and supervisors – the forgotten men and women of industry and commerce. Frequently promoted from the 'shop floor' (or recruited as graduates in a flush of conscience and wealth!), these people occupy one of the most crucial managerial roles, often with no idea of what they are supposed to be doing, without an identity, and without training. If this behaviour pattern is familiar and is continued, then TQM is doomed.

The first level of supervision is where the implementation of total quality is actually 'managed'. Supervisors' training should include an explanation of the principles of TQM, a convincing exposition on the commitment to quality of the senior management, and an explanation of what the quality policy means for them. The remainder of their training needs to be devoted to explaining their role in the operation of the quality management system, teamwork, SPC, etc., and to gaining *their* commitment to the concepts and techniques of total quality.

It is often desirable to involve the middle managers in the training of first-line supervision in order to:

- Ensure that the message they wish to convey through their tactical manoeuvres is not distorted.
- Indicate to the first-line supervision that the organization's whole management structure is serious about quality, and intends that everyone is suitably trained and concerned about it too. One display of arrogance towards the training of supervisors and the workforce can destroy such careful planning, and will certainly undermine the educational effort.

## All other employees

Awareness and commitment at the point of production or service delivery is just as vital as at the very senior level. If it is absent from the latter, the TQM programme will not begin; if it is absent from the shop floor, total quality will not be implemented. The training here should include the basics of quality and particular care should be given to using easy reference points for the explanation of the terms and concepts. Most people can relate to quality and how it should be managed, if they can think about the applications in their own lives and at home. Quality is really such common sense that, with sensitivity and regard to various levels of intellect and experience, little resistance should be experienced.

All employees should receive detailed training on the processes and procedures relevant to their own work. Obviously they must have appropriate technical or 'job' training, but they must also understand the requirements of their customers. This is frequently a difficult concept to introduce, particularly in the non-manufacturing areas, and time and follow-up assistance need to be given if TQM is to take hold. It is always bad management to ask people to follow instructions without understanding why and where they fit into their own scheme of things.

This is where the idea of internal and external suppliers within the supply chain is important in helping workers to see their role within a supply continuum. One problem on construction projects is that while subcontractors may be contractually bound to the general contractor or client, in their supply relationships they build on each other's work and, hence, they need to understand and work towards the needs of their immediate customer – the following trade.

# Turning education and training into learning

For successful learning training must be followed up. This can take many forms, but the managers need to provide the lead through the design of improvement projects and 'surgery' workshops.

In introducing statistical methods of process control, for example, the most satisfactory strategy is to start small and build up a bank of knowledge and experience. Sometimes it is necessary to introduce SPC techniques alongside existing methods of control (if they exist), thus allowing comparisons to be made between the new and old methods. When confidence has been established from these comparisons, the SPC methods will almost take over the control of the processes themselves. Improvements in one or two areas of the organization's operations by means of this approach will quickly establish the techniques as reliable methods of controlling quality, and people will learn how to use them effectively.

The authors and their colleagues have found that a successful formula is the in-company training course plus follow-up workshops. Usually a workshop or seminar is followed within a few weeks by a 'surgery' workshop at which participants on the initial training course present the results of their efforts to improve processes, and use the various methods. The presentations and specific implementation problems are discussed. A series of such workshops will add continually to the follow-up, and can be used to initiate process or quality improvement teams. Wider organizational presence and activities are then encouraged by the follow-up activities.

## Information and knowledge

Information and knowledge are two words used very frequently in organizations, often together in the context of 'knowledge management' and 'information technology', but how well are they managed and what is their role in supporting TQM?

Recent researchers and writers on knowledge management (e.g. Dawson 2000[1]) have drawn attention to the distinction between explicit knowledge – that which we can express to others – and tacit knowledge – the rest of our knowledge, which we cannot easily communicate in words or symbols.

If much of our knowledge is tacit, perhaps we do not fully know what we know and it can be very difficult to explain or communicate what we know. Explicit knowledge can be put into a form that we can communicate to others – the words, figures and models in this book are an example of that. In many organizations, however, especially the service sector, much of people's valuable and useful knowledge is tacit rather than explicit.

The creation and expression of knowledge takes place through social interaction between tacit and explicit knowledge and the matrix in Figure 16.3 shows this as four modes of knowledge conversion.

*Socialization* allows the conversion of tacit knowledge in one individual into tacit knowledge in other people, primarily through sharing experiences. The conversion of tacit knowledge to explicit knowledge is *externalization*, which is the process of making it readily communicable. *Internalization* converts explicit knowledge to tacit knowledge by translating it into personal knowledge – this could be called learning. The conversion of forms of explicit knowledge, such as creating frameworks, is *combination*.

## Explicit knowledge as information

When knowledge is made explicit by putting it into words, diagrams, or other representations, it can then be typed, copied, stored, and communicated electronically – it becomes *information*. Perhaps then a useful definition for information is something that is or can be made explicit. Information, which represents captured knowledge has value as an input to human decision-making and capabilities. Tacit knowledge remains intrinsic to individuals and only they have the capacity to act effectively in its use.

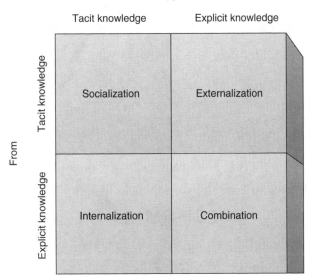

To

Tacit knowledge | Explicit knowledge

From — Tacit knowledge

| Socialization | Externalization |

From — Explicit knowledge

| Internalization | Combination |

**Figure 16.3**
Modes of knowledge conversion. (From *The Knowledge-Creating Company: How Japanese Companies Create the Dynamics of Innovation*, by Ikujiro Nonaka and Hirotaka Takeuchi © 1995, Oxford University Press, Inc. (Source: R. Dawson, Reference 2))

These ideas about information and knowledge enable us to substitute information for explicit knowledge, and simply *knowledge* – in the business sense of capacity to act effectively – as tacit knowledge. This clarification of the distinction between information and knowledge makes the knowledge conversion framework more directly applicable to interorganizational interaction.

In the same way, externalization is capturing people's knowledge – their capacity to act in their business roles – by making it explicit and turning it into information, as in the form of written documentation or structured business processes. This remains information until other people internalize it to become part of their own knowledge – or capacity to act effectively. Having a document on a server or bookshelf does not make individuals knowledgeable, nor does reading it. Knowledge comes from understanding the document by integrating the ideas into existing experience and knowledge, and thus providing the capacity to act usefully in new ways. In the case of written documents, language and diagrams are the media by which the knowledge is transferred. The information presented must be actively interpreted and internalized, however, before it becomes new knowledge to the reader.

The process of internalization is essentially that of knowledge acquisition, which is central to the whole idea of learning, knowledge management and knowledge transfer. Understanding the nature of this process is extremely valuable in implementing effective business improvements and in adding greater value to customers.

Socialization refers to the transfer of one person's knowledge to another person, without an intermediary of captured information in documents. It is a most powerful form of knowledge transfer. As we know from childhood people learn from other people far more effectively than they learn from books and documents, in both obvious and subtle ways. Despite technological advances that allow people to telecommute and work in different locations, organizations function effectively mainly because people who work closely together have the opportunity for rich interaction and learning on an

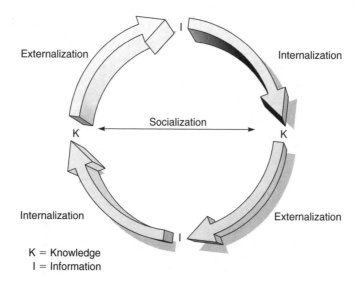

Externalization

Internalization

Socialization

K             K

Internalization           Externalization

K = Knowledge
I = Information

**Figure 16.4**
The knowledge management cycle. (Source: R. Dawson, Reference 2)

ongoing and often informal basis. This presents challenges, of course, in today's 'virtual organization'.

## The learning – knowledge management cycle

One way of thinking about learning and knowledge management is as a dynamic cycle from tacit knowledge to explicit knowledge and back to tacit knowledge. In other words, people's knowledge is externalized into information, which to be useful must then be internalized by others (learned) to become part of their knowledge, as illustrated in Figure 16.4. This flow from knowledge to information and back to knowledge constitutes the heart of organizational learning and knowledge management. Direct sharing of knowledge through socialization is also vital.

In large organizations, however, capturing whatever is possible in the form of documents and other digitized representations means that information can be stored, duplicated, shared, and made available to people on whatever scale desired. The construction process is extremely information intensive. This presents a challenge in that the relationship between much of the information in the process and knowledge is very tenuous. The storing and analysing of all the information may create more problems than it solves: important 'needles' of knowledge may be lost in the 'haystack' of information generated.

The fields of learning and knowledge management encompass all the human issues of effective externalization, internalization, and socialization of knowledge. As subsets of those fields, information management and document management address the middle part of the cycle, in which information is stored, disseminated, and made easily available on demand. It is a misnomer to refer to information sharing technology, however advanced, as knowledge management. Effective implementation of those systems must address how people interact with technology in an organizational context, which only then is beginning to address the real issues of knowledge.[1]

# The practicalities of sharing knowledge and learning

In world-class organizations there is clear evidence that knowledge is shared to maximize performance, with learning, innovation and improvement encouraged. In such establishments information (explicit knowledge) is collected, structured and managed in alignment with and in support of the organization's policies and strategies. These days this is often achieved in large organizations through 'intranet' and/or common network mounted file servers. These can provide common access to on-line reports, training material and performance figures versus targets.

The key is to provide appropriate access for both internal and external users to relevant information, with assurance required of its validity, integrity and security.

Another important element of knowledge sharing in large organizations is the effective exchange of tacit information among people – creating networks within the organization. This is probably best supported by the 'intranet'; however, the goal, rather than capturing the knowledge, is to identify knowledge sources within the organization and put people in touch with others within the organization who may be able to assist in solving a problem.

Cultivating, developing and protecting intellectual property can be the key in many sectors to maximizing customer value, for example in professional services, such as legal and consultancy. Firms in this sector can survive only if they are constantly seeking to acquire, increase, use and transfer knowledge effectively. This in turn has strong links with learning and innovation. The proper use and management of relevant information and knowledge resources leads to the generation of creative thinking and innovative solutions. The aim must be to make information available as widely as possible on the most appropriate basis to improve general understanding and increase efficiency. If this is to be linked to questions in employee surveys, internal valuation that knowledge is being captured and used effectively can take place.

In the EFQM Excellence Model there are now clear feedback loops of 'innovation and learning', both from the results – on people, customers, society and key performance – to the enablers and within the enablers themselves. For example, key performance outcomes such as profit market share and cash flow can and should be used to identify problems, areas for improvements or even strengths in aspects of the organization's leadership, its policies and strategies or its processes. Equally the way people are managed and partnerships established can be improved, innovatively based on perception measures from customers. The results from the society in which we operate can lead to innovations in resource management.

In some companies and industries, for example the retail or mail order businesses, information on each customer is vital to identify specific needs and to aid communication, planning and monitoring progress. The whole area of customer's relationship management (CRM) relies on some sort of personal records for each customer being kept. This presents modern methods of knowledge management and information technology with the opportunity to demonstrate their powers.

There may also be learning and innovation loops and opportunities with the enabler's criteria. For example, key performance indicators related to the management of processes, such as cycle time or amount of businesses won, can help to generate innovations in strategies, particularly if there is a good alignment between the two.

Information on staff, on customers and other sources of information can be readily kept on computer databases, and their effective and efficient use can mean the difference between success and failure in many industries and sectors.

Communications, innovation and learning

# References

1. Larkin, T.J. and Larkin, S. (1994) *Communicating Change – winning employee support for new business goals.* McGraw-Hill, New York.
2. Dawson, R. (2000) *Developing Knowledge-based Client Relationships – The Future of Professional Services,* Butterworth-Heinemann, Oxford. (The authors have drawn heavily on the material in this excellent book for the section of this chapter on turning education and training into learning.)

## *Chapter highlights*

### Communicating the quality strategy

- People's attitudes and behaviour can be influenced by communication, and the essence of changing attitudes is to gain acceptance through excellent communication processes.
- The strategy and changes to be brought about through TQM should be clearly and directly communicated from top management to all staff/employees both within the organization and across the supply chain. The first step is to issue a 'total quality message'. This should be followed by a signed TQM directive.
- People must know when and how they will be brought into the TQM process, what the process is, and the successes and benefits achieved. First-line supervision has an important role in communicating the key messages and overcoming resistance to change both within the organization and among the employees of supply chain partners.
- The complexity and jargon in the language used between functional groups need to be reduced in many organizations. Simplify and shorten are the guiding principles.
- 'Open' methods of communication and participation should be used at all levels. Barriers may need to be broken down by concentrating on process rather than 'departmental' issues.
- There are four audience groups in most organizations – senior managers, middle managers, supervisors, and employees – each with different general attitudes towards TQM. The senior management must ensure that each group sees TQM as being beneficial.
- Good leadership is mostly about good communications, the skills of which can be learned through training but must be acquired through practice.

### Communication, learning, education and training

- There are four principal types of communication: verbal (direct and indirect), written, visual, and by example. Each has its own requirements, strengths, and weaknesses.
- Education and training are the single most important factors in improving quality and performance, once commitment is present. They must be objectively, systematically, and continuously performed.
- All education and training should occur in an improvement cycle of ensuring it is part of policy, establishing objectives and responsibilities, establishing a platform for a learning organization, specifying needs, preparing programmes and materials, implementing and monitoring, assessing results, and reviewing effectiveness.

### A systematic approach to education and training for quality

- Responsibility for education and training of employees rests with management at all levels. The main elements should include error/defect/problem prevention, reporting and analysis, investigation, and review.

- Education and training procedures and records should be established to show how job competence is demonstrated.


**Starting where and for whom?**
- Education and training needs occur at four levels of the organization: very senior management, middle management, first level supervision and quality team leaders, and all other employees.


**Turning education and training into learning**
- For successful learning all quality training should be followed up with improvement projects and 'surgery' workshops.
- It is useful to draw the distinction between explicit knowledge (that which we can express to others) and tacit knowledge (the rest of our knowledge which cannot be communicated in words or symbols).
- The creation and expression of knowledge takes place through social interaction between tacit and explicit knowledge, which takes the form of socialization, externalization, internalization and combination.
- When knowledge is made explicit it becomes 'information', which in turn has value as an input to human decision-making and capability. Tacit knowledge (simply 'knowledge') remains intrinsic to individuals who have the capacity to act effectively in its use.
- One way of thinking about learning and knowledge management is as a dynamic cycle from tacit knowledge to explicit knowledge (information) and back to tacit knowledge.


**The practicalities of sharing knowledge and learning**
- In world-class organizations there is clear evidence that knowledge is shared to maximize performance, with learning, innovation and improvement encouraged. This is often achieved through an 'intranet' or common network-mounted file servers, providing common on-line access to information.
- Managing intellectual property is key to success in many sectors and this has strong links with learning and innovation. Where information must be made available as widely as possible, internal performance of this aspect can be valuable.
- The clear feedback loops of 'innovation and learning' in the EFQM Excellence Model drive increased understanding of the linkages between the results and the enablers, and between the results and the enablers, and between the enabler criteria themselves.

# 17

# Implementing TQM

# TQM and the management of change

The author recalls the managing director of a large support services group who decided that a major change was required in the way the company operated if serious competitive challenges were to be met. The board of directors went away for a weekend and developed a new vision for the company and its 'culture'. A human resources director was recruited and given the task of managing the change in the people and their 'attitudes'. After several 'programmes' aimed at achieving the required change, including a new structure for the organization, a staff appraisal system linked to pay, and seminars to change attitudes, very little change in actual organizational behaviour had occurred.

Clearly something had gone wrong somewhere. But what, who, where? Everything was wrong, including what needed changing, who should lead the changes and, in particular, how the change should be brought about. This type of problem is very common in organizations which desire to change the way they operate to deal with increased competition, a changing marketplace, new technology, and different business rules. In this situation many organizations recognize the need to move away from an autocratic management style, with formal rules and hierarchical procedures, and narrow work demarcations. Some have tried to create teams, to delegate (perhaps for the first time), and to improve communications.

Some of the senior managers in such organizations recognize the need for change to deal with the new realities of competitiveness, but they lack an understanding of how the change should be implemented. They often believe that changing the formal organizational structure, having 'culture change' programmes and new payment systems, will, by themselves, make the transformations. In much research work carried out by the European Centre for Business Excellence, the research and education division of Oakland Consulting plc, it has been shown that there is almost an inverse relationship between successful change and having formal organization-wide 'change programmes'. This is particularly true if one functional group, such as HR, 'owns' the programme.

In several large organizations in which TQM has been used successfully to effect change, the senior management did not focus on formal structures and systems, but set up *process-management* teams to solve real business or organization problems. The key to success in this area is to align the employees of the business, their roles and responsibilities with the organization and its *processes*. When an organization focuses on its key processes, that is the activities and tasks themselves, rather than on abstract issues such as 'culture' and 'participation', then the change process can begin in earnest.

---

This applies equally to the ongoing implementation of improvement strategies. Look, for example, at the case studies of Graniterock (p. 375) and Boldt (p. 479). Each describes the very detailed process through which the company creates its plans for the forthcoming year, defining detailed targets, roles and responsibilities, and then leading to specific activities and tasks that are designed to achieve them.

---

An approach to change based on process alignment, and starting with the vision and mission statements, analysing the critical success factors, *and* moving on to the core processes, is the most effective way to engage the staff in an enduring change process (see Chapter 4).

Many change programmes do not work because they begin by trying to change the knowledge, attitudes and beliefs of individuals. This is based on the theory that changes in these areas will lead to changes in behaviour throughout the organization. It relies on a form of religion spreading through the people in the business.

What is often required, however, is virtually the opposite process, based on the recognition that people's behaviour is determined largely by the roles they have to take up. If we create for them new responsibilities, team roles, and a process-driven environment, a new situation will develop, one that will force their attention and work on the processes. This will change the culture. *Teamwork* is an especially important part of the TQM model in terms of bringing about change. If changes are to be made in quality, costs, market, product or service development, close co-ordination among the marketing, design, production/operations and distribution groups is essential. This can be brought about effectively by multifunctional teams working on the processes and understanding their interrelationships. *Commitment* is a key element of support for the high levels of co-operation, initiative, and effort that will be required to understand and work on the labyrinth of processes existing in most organizations. In addition to the knowledge of the business as a whole, which will be brought about by an understanding of the mission, CSF, process breakdown links, certain *tools*, *techniques*, and *interpersonal skills* will be required for good *communication* around the processes. These are essential if people are to identify and solve problems as teams.

If any of these elements are missing the total quality underpinned change process will collapse. The difficulties experienced by many organizations' formal change processes are that they tackle only one or two of these necessities. Many organizations trying to create a new philosophy based on teamwork fail to recognize that the employees do not know which teams need to be formed round their process – which they begin to understand together, perhaps for the first time – and further recognition that they then need to be helped as individuals through the forming–storming–norming–performing sequence, will generate the interpersonal skills and attitude changes necessary to make the new 'structure' work.

Organizations will avoid the problems of 'change programmes' by concentrating on 'process alignment' – recognizing that people's roles and responsibilities must be related to the processes in which they work. Senior managers may begin the task of process alignment by developing a self-reinforcing cycle of *commitment*, *communication*, and *culture* change. In the introduction of total quality for managing change, timing can be critical and an appropriate starting point can be a broad review of the organization and the changes required by the top management team. By gaining this shared diagnosis of what changes are required, what the 'business' problems are, and/or what must be improved, the most senior executive mobilizes the initial commitment that is vital to begin the change process. An important element here is to get the top team working as a team, and techniques such as MBTI and/or FIRO-B will play an important part (see Chapter 15).

# Planning the implementation of TQM

The task of implementing TQM can be daunting and the chief executive faced with this may draw little comfort from the 'quality gurus'. The first decision is where to begin and this can be so difficult that many organizations never get started. This has been called TQP – total quality paralysis!

The chapters of this book have been arranged in an order which should help senior management bring total quality into existence. The preliminary stages of understanding and commitment are vital first steps which also form the foundation of the whole TQM structure. Too many organizations skip these phases, believing that they have the right attitude and awareness, when in fact there are some fundamental gaps in their 'quality credibility'. These will soon lead to insurmountable difficulties and collapse of the process.

While an intellectual understanding of quality provides a basis for TQM, it is clearly only the planting of the seed. The understanding must be translated into commitment, policies, plans and actions for TQM to germinate. Making this happen requires not only commitment, but a competence in leadership and in making changes. Without a strategy to implement TQM through process management, capability, and control, the expended effort will lead to frustration. Poor quality management can become like poor gardening – a few weed leaves are pulled off only for others to appear in their place days later, plus additional weeds elsewhere. Problem solving is very much like weeding; tackling the root causes, often by digging deep, is essential for better control.

Individuals working on their own, even with a plan, will never generate optimum results. The individual effort is required in improvement but it must be co-ordinated and become involved with the efforts of others to be truly effective. The implementation begins with the drawing up of a quality policy statement, and the establishment of the appropriate organizational structure, both for managing and encouraging involvement in quality through teamwork. Collecting information on how the organization operates, including the cost of quality, helps to identify the prime areas in which improvements will have the largest impact on performance. Planning improvement involves all managers but a crucial early stage involves putting quality management systems in place to drive the improvement process and make sure that problems remain solved forever, using structured corrective action procedures.

Once the plans and systems have been put into place, the need for continued education, training, and communication becomes paramount. Organizations which try to change the culture, operate systems, procedures, or control methods without effective, honest two-way communication, will experience the frustration of being a 'cloned' type of organization which can function but inspires no confidence in being able to survive the changing environment in which it lives.

An organization may, of course, have already taken several steps on the road to TQM. If good understanding of quality and how it should be managed already exists, there is top management commitment, a written quality policy, and a satisfactory organizational structure, then the planning stage may begin straight away. When implementation is contemplated, priorities among the various projects must be identified. For example, a quality system which conforms to the requirements of ISO 9000 may already exist and the systems step will not be a major task, but introducing a quality costing system may well be. It is important to remember, however, that a review of the current performance in all the areas, even when well established, should be part of normal operations to ensure continuous improvement.

These major steps may be used as an overall planning aid for the introduction of TQM, and they should appear on a planning or Gantt chart. Major projects should be time-phased to suit individual organizations' requirement, but this may be influenced by outside factors, such as the need to operate a quality system which meets the requirements of a standard. The main projects may need to be split into smaller subprojects, and this is certainly true of management system work, the introduction of SPC, and improvement teams.

The education and training part will be continuous and draw together the requirements of all the steps into a cohesive programme of introduction. The timing of the training inputs, follow-us sessions, and advisory work should be co-ordinated and reviewed, in terms of their effectiveness, on a regular basis. It may be useful at various stages of the implementation to develop checks to establish the true progress. For example, before moving from understanding to trying to obtain top management commitment, objective evidence should be obtained to show that the next stage is justified.

Following commitment being demonstrated by the publication of a signed quality policy, there may be the formation of a council and/or steering committee(s). Delay here will prevent real progress being made towards TQM through teamwork activities.

The launch of process improvement requires a balanced approach and the three major components must be 'fired' in the right order to lift the campaign off the ground. If teams are started before the establishment of a good system of management, there will be nothing to which they can adhere. Equally, if management strategies are initiated without matching performance measures being in place, the effectiveness of strategies will not be assessable and hence support for the initiatives will weaken with time. A quality management system on its own will give only a weak thrust which must have the boost of improvement teams and performance measurement to make it come 'alive'.

An effective co-ordination of these three components will result in quality improvement through increased capability. This should in turn lead to consistently satisfied customers and, where appropriate, increase or preservation of market share.

TQM may be integrated into the strategy of any organization through an understanding of the core business processes and involvement of the people. This leads through process analysis, self-assessment and benchmarking, to identifying the improvement opportunities for the organization, including people development.

The identified processes should be prioritized into those that require continuous improvement, those that require re-engineering or redesign, and those that lead to a complete rethink or visioning of the business.

Performance-based measurement of all processes and people development activities is necessary to determine progress so that the vision, goals, mission, and critical success factors may be examined and reconstituted if necessary to meet new requirements for the organization and its customers, internal and external. This forms the basis of a new implementation framework for TQM (Figure 17.1).

This all starts with the vision, goals, strategies and mission which should be fully thought through, agreed, and shared in the business. What follows determines whether

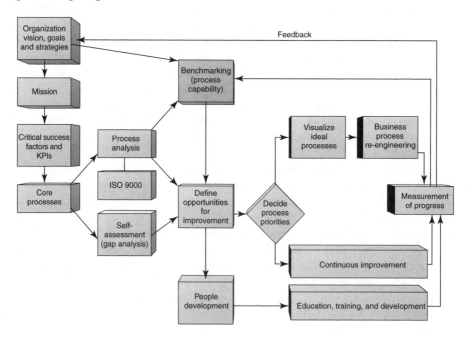

**Figure 17.1**
The framework for implementation of TQM

these are achieved. The factors which are critical to success, the CSFs – the building blocks of the mission – are then identified. The key performance indicators (KPIs), the measures associated with the CSFs, tell us whether we are moving towards or away from the mission or just standing still.

Having identified the CSFs and KPIs, the organization should know what are its *core processes*. This is an area of potential bottleneck for many organizations because, if the core processes are not understood, the rest of the framework is difficult to implement. If the processes are known, we can carry out process analysis, sometimes called mapping and flowcharting, to fully understand our business and identify opportunities for improvement. By the way, ISO 9000 standard-based systems should drop out at this stage, rather than needing a separate and huge effort and expense.

Self-assessment to the European (EFQM) Excellence Model or Baldrige quality model, and benchmarking, will identify further improvement opportunities. This will create a very long list of things to attend to, many of which require people development, training and education. What is clearly needed next is prioritization to identify those processes which are run pretty well – they may be advertising/promoting the business or recruitment/selection processes – and subject them to a continuous improvement regime. For those processes that we identify as being poorly carried out, perhaps forecasting, training, or even financial management, we may subject them to a complete revisioning and redesign activity. That is where BPR comes in. What must happen to all processes, of course, is performance measurement, the results of which feed back to our benchmarking and strategic planning activities.

World-class organizations, of which there need to be more in most countries, are doing all of these things. They have implemented their version of the framework and are achieving world-class performance and results. What this requires first, of course, is world-class leadership and commitment.

In many successful companies TQM is not the very narrow set of tools and techniques often associated with failed 'programmes' in organizations in various parts of the world. It is part of a broad-based approach used by world-class companies, such as Hewlett Packard, Milliken, TNT, and Yellow Pages, to achieve organizational excellence, based on customer results, the highest weighted category of all the quality and excellence awards. TQM embraces all of these areas. If used properly, and fully integrated into the business, TQM will help any organization deliver its goals, targets and strategy, including those in the public sector. This is because it is about people and their identifying, understanding, managing and improving processes – the things any organization has to do particularly well to achieve its objectives.

# Sustained improvement

Never-ending or continuous improvement is probably the most powerful concept to guide management. It is a term not well understood in many organizations, although that must begin to change if those organizations are to survive. To maintain a wave of sustained improvement, it is necessary to develop generations of managers who not only understand but are dedicated to the pursuit of never-ending improvement in meeting external and internal customer needs.

The concept requires a systematic approach to quality management that has the following components:

- *Planning* the processes and their inputs.
- *Providing* the inputs.

- *Operating* the processes.
- *Evaluating* the outputs.
- *Examining* the performance of the processes.
- *Modifying* the processes and their inputs.

This system must be firmly tied to a continuous assessment of customer needs, and depends on a flow of ideas on how to make improvements, reduce variation, and generate greater customer satisfaction. It also requires a high level of commitment, and a sense of personal responsibility in those operating the processes.

The never-ending improvement cycle ensures that the organization learns from results, standardizes what it does well in a documented quality management system, and improves operations and outputs from what it learns. But the emphasis must be that this is done in a planned, systematic, and conscientious way to create a climate – a way of life – that permeates the whole organization.

There are four basic principles of sustained improvement:

- Focusing on the *customer*.
- Understanding the *process*.
- All *employees* committed to quality.
- All *suppliers* committed to quality.

## Focusing on the customer

An organization must recognize, throughout its ranks, that the purpose of all work and all efforts to make improvements is to serve the customers better. This means that it must always know how well its outputs are performing, in the eyes of the customer, through measurement and feedback. The most important customers are the external ones, but the quality chains can break down at any point in the flows of work. Internal customers therefore must also be well served if the external ones are to be satisfied.

## Understanding the process

In the successful operation of any process it is essential to understand what determines its performance and outputs. This means intense focus on the design and control of the inputs, working closely with suppliers, and understanding process flows to eliminate bottlenecks and reduce waste. If there is one difference between management/supervision in the Far East and the West, it is that in the former management is closer to, and more involved in, the processes. It is not possible to stand aside and manage in never-ending improvement. TQM in an organization means that everyone has the determination to use their detailed knowledge of the processes and make improvements, and use appropriate statistical methods to analyse and create action plans.

## All employees committed to quality

Everyone in the organization, from top to bottom, from offices to technical service, from headquarters to local sites, must play their part. People are the source of ideas and innovation, and their expertise, experience, knowledge, and co-operation have to be harnessed to get those ideas implemented.

When people are treated like machines, work becomes uninteresting and unsatisfying. Under such conditions it is not possible to expect quality services and reliable products. The rates of absenteeism and of staff turnover are measures that can be used in determining the strengths and weaknesses, or management style and people's morale, in any company.

The first step is to convince everyone of their own role in total quality. Employers and managers must of course take the lead, and the most senior executive has a personal responsibility. The degree of management's enthusiasm and drive will determine the ease with which the whole workforce is motivated.

Most of the work in any organization is done away from the immediate view of management and supervision, and often with individual discretion. If the co-operation of some or all of the people is absent, there is no way that managers will be able to cope with the chaos that will result. This principle is extremely important at the points where the processes 'touch' the outside customer. Every phase of these operations must be subject to continuous improvement, and for that everyone's co-operation is required.

Never-ending improvement is the process by which greater customer satisfaction is achieved. Its adoption recognizes that quality is a moving target, but its operation actually results in quality.

## All suppliers committed to quality

Every supplier, both in the planning and design phase as well as the construction and maintenance phase, is a key player in your supply chain, every one of them must share your commitment to quality.

They are a critical part of your team and your ability to provide your end customers with quality services and products depends on your team's commitment to quality at every step. Furthermore your ability to solve problems and improve relies on your team's ability to innovate.

Your risks are often in the hands of your supplier's managers and employees. You must develop close relationships with each of them to ensure that they share your values and that their employees have the skills you need.

Here too you must employ all the strategies for continuous improvement to ensure that the entire team is performing as well as possible.

# A model for TQM

The concept of TQM is basically very simple. Each part of an organization has customers, whether within or without, and the need to identify what the customer requirements are, and then set about meeting them, forms the core of a total quality approach. Good *performance* requires the three hard management necessities: *planning*, including the right policies and strategies, *processes* and supporting management systems and improvement tools, and *people* with the right knowledge, skills and training (Figure 17.2). These are complementary in many ways, and they share the same requirement for an uncompromising top level commitment, the right culture and good communications. This must start with the most senior management and flow down through the organization. Having said that, teamwork, SPC, or a quality management system, may be used as a spearhead to drive TQM through an organization. The attention to many aspects of a company's operations – from purchasing through to distribution, from data recording to control chart plotting – which are required for the successful introduction of a good quality management system, or the implementation of SPC, will have a 'Hawthorne effect', concentrating everyone's attention on the customer/supplier interface, both inside and outside the organization.

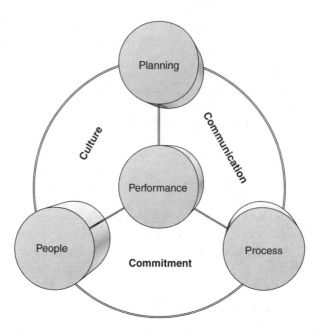

**Figure 17.2**
A model for TQM

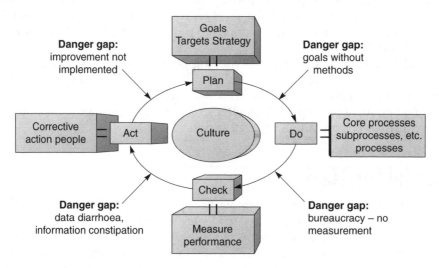

**Figure 17.3**
TQM implementation – all done with the Deming continuous improvement cycle

TQM calls for consideration of processes in all the major areas: marketing, design, procurement, operations, distribution, etc. Clearly, these each require considerable expansion and thought, but if attention is given to all areas, using the concepts of TQM, then very little will be left to chance. Much of industry commerce and the public sector would benefit from the continuous improvement cycle approach represented in Figure 17.3, which also shows the 'danger gaps' to be avoided. This approach will ensure the implementation of the management commitment represented in the quality policy, and provide the environment and information base on which teamwork thrives.

## Chapter highlights

### TQM and the management of change

- Senior managers in some organizations recognize the need for change to deal with increasing competitiveness, but lack an understanding of how to implement the changes.
- Successful change is effected not by focusing on formal structures and systems, but by aligning process management teams. This starts with writing the mission statement, analysis of the critical success factors (CSFs) and understanding the critical or key processes.
- Senior managers may begin the task of process alignment through a self-reinforcing cycle of commitment, communication and culture change.

### Planning the implementation of TQM

- Making quality happen requires not only commitment but competence in the mechanics of TQM. Crucial early stages will comprise establishment of the appropriate organization structure; collecting information, including quality costs; teamwork; quality systems; and training.
- The launch of quality improvement requires a balanced approach, through systems, teams and tools.
- A new implementation framework allows the integration of TQM into the strategy of an organization through an understanding of the core business processes and involvement of people. This leads through process analysis, self-assessment and benchmarking to identifying opportunities for improvement, including people development.
- The process opportunities should be prioritized into continuous improvement and re-engineering/redesign. Performance-based measurement determines progress, and feeds back to the strategic framework.

### Sustained improvement

- Managers need to understand and pursue never-ending improvement. This should cover planning and operating processes, providing inputs, evaluating outputs, examining performance, and modifying processes and their inputs.
- There are four basic principles of continuous improvement: focusing on the customer, understanding the process, seeing that all employees are committed to quality and working with suppliers similarly committed to quality.
- In the model for TQM the customer/supplier chains form the core, which is surrounded by the hard management necessities of planning, processes and people. These are complementary and share the same needs – for top level commitment, the right culture and good communications.

# Case studies

# Reading, using and analysing the cases

The cases in this book provide a description of what occurred in 14 different organizations, regarding various aspects of their quality and performance improvement efforts. They may each be used as a learning vehicle as well as providing information and description which demonstrate the application of the concepts and techniques of TQM and business excellence.

The objective of writing the cases has been to offer a resource through which the student of TQM, including the practicing manger, understands how organizations which adopt the approach operate. It is hoped that the cases provide a useful and distinct contribution to TQM education and training.

The case materials is suitable for practicing managers, students on undergraduate and postgraduate courses, and all teachers of the various aspects of business management and TQM. The cases have been written so that they may be used in three ways.

1. As orthodox cases for student preparation and discussion.
2. As illustrations, which teachers may also use as support for their other methods of training and education.
3. As supporting/background reading on TQM.

If used in the orthodox way, it is recommended that first the case is read to gain an understanding of the issues and to raise questions which may lead a collective and more complete understanding of the organization, TQM and the issues in the particular case. Second, case discussion or presentations in groups will give practice in putting forward thoughts and ideas persuasively.

The greater the effort put into case study preparation, analysis and discussions in groups, the greater will be the individual benefit. There are, of course, no perfect or tidy cases in any subject area. What the directors and managers of an organization actually did is not necessarily the best way forward. One object of the cases is to make the reader think about the situation, the problems and the progress made, and what improvements or developments are possible.

Each case emphasizes particular problems or issues which were apparent for the organization. This may have obscured other more important ones. Imagination, innovation and intuition should be as much a part of the study of a case as observation and analysis of the facts and any data available.

TQM cases, by their nature, will be very complicated and, to render the cases in this book useful for study, some of the complexity has been omitted. This simplification is accompanied by the danger of making the implementation seem clear-cut and obvious, but that is never the case with TQM!

The main objective of each description is to enable the reader to understand the situation and its implications, and to learn from the particular experiences. The cases are not, in the main, offering specific problems to be solved. In using the cases, the reader/student should try to:

- *Recognize or imagine* the situation in the organization.
- *Understand* the context and objectives of the approaches adopted.
- *Analyse* the different parts of the case (including any data) and their inter-relationships.
- *Determine* the overall structure of the situation/problem/case.
- *Consider* the different options facing the organization.
- *Evaluate* the options and the course of action chosen, using any results stated.
- *Consider any recommendations* which should be made to the organization for further work, action, or implementation.

The set of cases has been chosen to provide a good coverage across different types of industry and organization, including those in the service, manufacturing and public sectors. The value of illustrative cases in an area such as TQM is that they inject reality into the conceptual frameworks developed by authors on the subject. The cases are all based on real situations and are designed to illustrate certain aspects of managing change in organizations, rather than to reflect good or poor management practice. The cases may be used for analysis, discussion, evaluation, and even decision-making within groups without risk to the individuals, groups, or organization(s) involved. In this way, students of TQM and business excellence may become 'involved' in many different organizations and their approaches to TQM implementation, over a short period and in an efficient and effective way.

The organizations described here have faced translating TQM theory into practice, and the description of their experiences should provide an opportunity for the reader of TQM literature to test his/her preconceptions and understanding of this topic. All the cases describe real TQM processes in real organizations and we are grateful to the people involved for their contribution to this book.

*John Oakland*
*Marton Marosszeky*

# Further reading

Easton, G., *Learning from Case Studies* (2nd edition), Prentice-Hall, UK, 1992.

# Case study order

Each of the case studies in this section may be used as illustrations and the basis of discussions on a number of parts of the text. The cases are listed in the order given below – this also indicates which parts of the text are illustrated by each case.

*1. Continuous improvement and growth at Graniterock, pages 375–88*
Internal and external customer focus (Chapter 1)
Excellent leadership process for setting goals (Chapter 3)
Clear articulation of core values (Chapter 4)
Public service and sustainability focus (Chapter 4)
Continuous improvement demonstrated (Chapters 7 and 9)
Development of people, empowerment and recognition (Chapter 14)

*2. Growth of a commitment to total quality at CDL, pages 389–98*
End customer satisfaction (Chapters 1 and 7)
Client showing leadership in construction safety (Chapter 3)
Public service and sustainability focus (Chapter 4)
Good use of performance measurement in a RE company (Chapters 7 and 9)
Construction quality (Chapters 12 and 13)
Design briefing and design management (Chapter 6)

*3. Future growth at Gammon underpinned by a commitment to excellence, pages 399–408*
Client focus (Chapter 1)
Sustainability focus (Chapter 4)
Good use of performance measurement based on TBL (Chapters 7 and 9)
Safety focus, performance measurement (Chapters 7 and 14)
Product quality focus (Chapters 7 and 12)
Transparent reporting (Chapter 16)
Excellence in recruitment and training of new graduates (Chapter 14)
Innovation in construction technology – building systems (Chapter 11)

*4. World-class project performance from a partnership between IDC and TNC, pages 409–14*
Acceleration to shorten time to market (Chapter 1)
Supply chain innovation (Chapters 11 and 15)
Interorganizational team building (Chapter 15)
Lean construction (Chapter 11)
Quality focus (Chapter 12)

*5. Commitment to total quality at Mirvac, pages 415–20*
Leadership (Chapter 3)
Design management (Chapter 6)
Managing design related construction risk (Chapter 6)
TQM commitment with informal systems (Chapter 12)

*6. Triple bottom line reporting as a catalyst for organizational culture change at Landcom, pages 421–36*
Leadership in community and industry education and reform (Chapter 3)
Sustainability performance (Chapter 4)
TBL based performance measurement and company development (Chapter 7)
Cultural change (Chapter 15)
Transparent reporting (Chapter 16)

*7. Best value in Harrogate Borough Council, pages 437–48*
Best value framework (Chapter 4)
Social and environmental sustainability focus (Chapter 4)
Performance measurement (Chapter 7)

*8. Takenaka – 400 years old and a commitment to innovation and excellence ensures the future, pages 449–58*
Long-term strategic vision (Chapter 4)
Community and sustainability focus (Chapters 4, 7 and 9)
Whole of life performance commitment (Chapters 4 and 6)
Overview of maturation of TQM implementation over 20 years (Chapter 17)
Training at all levels (Chapters 14 and 16)

*9. Business improvement strategies in the Highways Agency, pages 459–67*
Strategy for business improvement in the public sector (Chapter 4)
Self assessment to EQFM (Chapter 8)

*10. Sekisui Heim, pages 469–77*
Long-term customer quality commitment (Chapters 4 and 12)
Community and sustainability focus (Chapters 4, 7 and 9)
Whole of life performance commitment (Chapter 6)
Innovation in product development and production (Chapters 10, 11 and 12)

*11. Flower and Samios* (included in Chapter 11), *pages 211–12*
IT enabled stakeholder engagement (Chapters 4 and 16)
Early technology adoption (Chapters 6 and 11)
Supply chain re-engineering (Chapters 6 and 11)
High performance supply chain (Chapter 11)

*12. Boldt leverages lean thinking to drive efficiency and growth, pages 479–85*
3D CAD based integration, virtual prototyping (Chapter 6)
Value engineering (Chapter 6)
Lean process implementation (Chapter 11)
Modularization and JIT (Chapter 11)

*13. Re-engineering timber floors in the Australian housing sector: an example of process innovation, pages 487–92*
Construction process re-engineering (Chapter 11)
Waste elimination (Chapter 11)

*14. Early adoption of TQM values at Babtie Asia pays off, pages 493–8*
Focus on training, special program for new graduates (Chapter 14)
Excellence in personnel management (Chapter 14)
Early TQM adoption (Chapter 17)

*15. Quality management at Contractor Woh Hup, pages 499–510*
Leadership, policy and strategy in implementing change (Chapters 3 and 4)
Performance measurement and benchmarking (Chapters 7 and 9)
Early TQM adoption (Chapter 17)

# Continuous improvement and growth at Graniterock

# Company background

On 14 February 1900, in days of manual labour and horse-drawn wagons, the Granite Rock Company was incorporated as a quarrying business. By 1924, both the quarry and transport system had become increasingly automated and the business had expanded to include the Granite Construction Company and the Central Supply Company. In the 1930s, Central Supply opened California's first asphaltic concrete plant, which then continued to automate and expand through to 1960.

The tremendous development of the Monterey and San Francisco Bay areas during the 1960s and 1970s contributed to the continuing growth of Graniterock. Central Supply merged with Graniterock to form one company for construction materials production and sales while numerous new plants were opened for the sand, concrete, asphaltic concrete and building materials operations.

Prior to 1987, construction material needs were served by a number of medium-sized, family-owned businesses; most of these were locally based and served local demand. However, in 1987, competitive conditions changed dramatically: large multi-national corporations acquired and consolidated smaller suppliers into large businesses. Although Graniterock remained a California business, it now had to compete with the world.

Despite price and market share-driven competitors, Graniterock has continued to grow (today it has over 700 employees) and increase its share of the market. Due to its values, the quality of its people, products and services, and its innovation, Graniterock has become a commercially successful enterprise that is repeatedly listed in the top 100 employers in the USA. Under pressure, Graniterock increased its focus on its people and on its innovation in every area of its business, resulting in increased competitiveness and higher than industry performance levels: Graniterock won the *National Baldrige Award for Quality* in 1992.

# Basis of achievements

These achievements are the result of the collective efforts of individuals – each person having the freedom to do the job in the best way he or she knows; however, few achievements are derived purely by the effort of individuals. For a company to translate such individual achievements into maximum value, teamwork is essential. Therefore, when financial resources, human talent and energy are directed by *core values, core purpose, corporate objectives* and *annual baseline goals*, the co-operative effort and teamwork result in both a greater individual and company achievement.

Corporate objectives delineate the ways by which a company is committed to achieving leadership excellence. Annual baseline goals are set so that company-wide, improvement activities are focused on achieving the corporate objectives. As market conditions and opportunities change, the company's *executive committee* periodically reviews and updates the corporate objectives.

# Core values

Graniterock staff and the company are expected to adhere to the following core values.

## Safety before all else

Core values are placed ahead of achieving any other plan or goal; this commitment extends beyond employees to customers and the public. All work plans are expected to implement safety as the primary goal: they are to improve work-site conditions so that potential safety risks are eliminated. The company combines effective safety training (health and wellness education and support) with finance for plant and equipment safety innovations.

## Dedication to customer service excellence

Customer service excellence requires staff to anticipate and exceed customer needs and expectations. They work hard to build long-term valuable business partnerships with their customers and to live by their 'yes, we will' standard of responding positively and creatively to every ethical request for special products and services.

## People growth and development

Through the application of a 'life-long learning' strategy and a trust in people to do a good job, Graniterock has created an environment of freedom: individuals are encouraged to direct their own work and achieve work improvements. The company facilitates, encourages personal growth and rewards professional achievement: job responsibilities are structured so that they are worthwhile and challenging. This approach does not only promote respect and caring among the members of the Graniterock team but it also extends to their families, neighbours, customers and suppliers.

## Honesty and integrity

The company and staff are exhorted to conduct their activities with uncompromising honesty and integrity. People in every job are expected to adhere to the highest standards of business ethics and fairness in all of their dealings with each other, customers, suppliers, communities, government officials and agencies. The company also expects these same high standards from the people and organizations with which they work.

## Continuous improvement as a way of life

The staff are trained to be achievement oriented and is expected to be unsatisfied with 'things the way they are' when improvements are possible. Rather than preserving or protecting the work of the past, the company encourages and supports both incremental and sweeping change. It recognizes that risk-taking and honest mistakes are unavoidable parts of achieving these aims. In everything they do the company compares its results with those achieved by role model, high performance companies in any industry. They are not satisfied with simply being 'good': Graniterock must be among the 'best' in every important thing they do. The company consistently provide a 'can do' commitment to individual and team-driven improvements.

# Core purpose

The company's core purpose is to nurture and promote exceptional creativity and innovation. To achieve this, Graniterock supports the continuous and ambitious personal and professional growth of its staff. Through these pursuits it is able to produce the

**Figure C1**
Graniterock corporate objectives

highest quality products and provide services that advance global business standards and practices.

# Corporate objectives

The corporate objectives are presented as a part of a circle (Figure C1) rather than a list to accentuate the equality of their importance to the company's long-term success. Each objective is equally weighted in business performance assessments; each receives prime consideration in business investment and in all other decisions. In addition, the circle format illustrates the supportive and synergistic relationship between the objectives.

## Safety

On all Graniterock facilities and job sites safety is the primary goal. Meeting production schedules or customer commitments is secondary. Every person is expected to put safety first – before all else.

## Customer satisfaction and service

The company prides itself on earning the respect of its customers by partnering with them and providing them with products and services in a timely manner that anticipate and exceed their needs. Graniterock provides its customers with product and service advantages, which, in turn, contribute to their customers' business success.

## Financial performance and growth

Graniterock's commitment to knowledge development, coupled with available financial resources, drive the company's business growth strategies and its expanding new product offering. The company plans diligently and implements very long-term mining and resource plans to ensure its supply of construction materials for the next 100 or more years.

## People

The knowledge that they make a difference to the success of the company encourages the staff to carry out their jobs more effectively. Every person in the organization is given the opportunity to attain a sense of satisfaction and accomplishment from work achievements. Individual and team accomplishments are recognized and rewarded according to demonstrated skills and job ownership performance.

## Product quality assurance

The company ensures that it supplies products and services which provide lasting value to its customers and end-point users (e.g. owners and architects): at all times bettering industry standards and practices, aiming towards 'zero defects'.

## Community contribution and responsibility

Graniterock strives to be an exemplary citizen in all of the communities in which it operates by participating in community goals. The environment is important to the company, therefore its actions are always consistent with environmental responsibility.

## Efficiency

The company is committed to producing and delivering its services and products more efficiently than anyone else in the world.

## Management

Initiative, creativity and commitment are three main aims of management, therefore the company expects each individual to desire personal and professional growth. In return, the company provides all staff with the freedom of job ownership – so they can establish work improvement objectives; and the freedom of action – so they can implement individual work improvement methods.

## Profit

Graniterock aims to earn a fair profit so that it can fund other company objectives.

# Quality in management

## Leadership

Information gained from its customers is the foundation upon which objectives, goals and quality values are built. While the executive leadership defines and redefines quality values and expectations through the articulation of the company core values and core purpose, the nine corporate objectives are critical to the future of the company as they provide the focus areas for quality improvement.

Annually, the *executive committee* establishes baseline criteria within these corporate objectives, thereby focusing even more on the specific goals through associated and measurable performance criteria. These baseline goals are communicated throughout the organization, ensuring alignment of effort and creativity with top management's ethics, quality values and customer focus. Achievements towards the baseline goals are progressively evaluated every quarter, and reviewed annually. Annual performance achievements are shared with everyone in the organization. Executive remunerations/compensation and promotional opportunities are directly affected by the baseline goal performance.

Senior management are encouraged to attend appropriate courses and to be involved in the detail of company training: assisting to design courses and teaching in some instances. This reinforces the company's strong life-long learning focus. Annual, Recognition Day celebrations are held in every branch, creating the opportunity for senior management to acknowledge, face to face, the successes of individuals and teams throughout the company.

The leadership also makes it clear that Graniterock is a company of impeccable integrity; companies that do not meet Graniterock standards do not qualify as customers or suppliers. Graniterock's quality commitment to the end-point customer is communicated throughout the market and widely acknowledged.

## Strategic quality planning

Quality and business plans start from customer needs. These are established through regular surveys, which are developed and aligned with information gathered from other relevant sources both from the supply side and from the perspective of the market trends. Inputs into the process include feedback from customers, employees, suppliers and *quality teams*. Strategic quality goals with measurable performance outcomes are set and they form the basis of performance reporting.

All processes are evaluated regularly through the process shown in Figure C2. Information is gathered early to help establish goals that are most often set to exceed current industry capability. Benchmarks, supplier capability and customer aspirations are routinely sought to assist in the systematic planning that will ensure superiority of performance.

*Statistical process* control and R&D through the Research Technical Services Department underpin continuous improvement and growth. A commitment to continuous improvement through technological advancement ensures timely service improvements in areas such as product load-out and on-time delivery, billing speed and accuracy, and problem resolution.

Plans are deployed as outlined in the flowchart in Figure C3.

Once customer needs are established the planning and strategy formulation stages begin. After annual improvement targets are set in baseline goals, branches and divisions develop their own implementation plans. Ten *corporate quality teams* oversee and help align improvement efforts across the entire organization and foster co-ordination across divisions. Although senior executives chair committees, members include managers, salaried professional and technical workers, and hourly union employees. Teams carry out quality improvement projects as well as many day-to-day activities and operations. Weekly and monthly meetings of teams and task forces monitor progress to ensure goal achievement.

## Information and analysis

A comprehensive measurement system is in place to enable management to make decisions based on facts. The system identifies the key areas for which information is to be

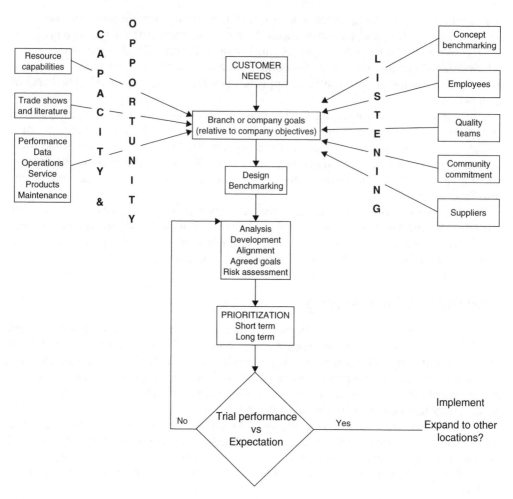

**Figure C2**
Flowchart model of planning/improvement process

gathered and measured, and allows the comparison between results and goals and benchmarks.

Criteria for selecting key data are:

- *Customer needs* – established through *surveys and regular focus groups.*
- *Product quality assurance* – measures of process quality to ensure downstream excellence.
- *Production efficiency* to ensure lowest cost production.
- *Supplier performance* to ensure the highest quality from suppliers.
- *Financial performance* to maintain business efficiency in terms of resource utilization.
- *People development* to ensure the development, health, welfare and safety of team members.

The use of key data in the management system is critical to achieving Graniterock's corporate objectives. They provide the information needed to guide the development of the company and to deploy its resources wisely and effectively. Benchmarking with the

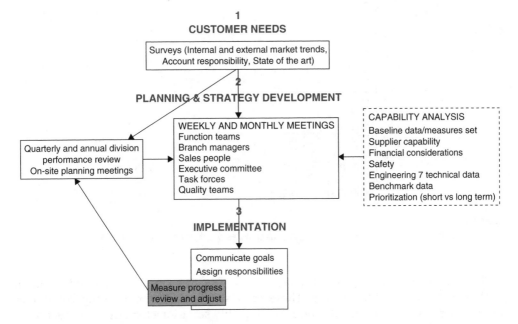

**1**
**CUSTOMER NEEDS**

Surveys (Internal and external market trends, Account responsibility, State of the art)

**2**
**PLANNING & STRATEGY DEVELOPMENT**

WEEKLY AND MONTHLY MEETINGS
Function teams
Branch managers
Sales people
Executive committee
Task forces
Quality teams

Quarterly and annual division
performance review
On-site planning meetings

CAPABILITY ANALYSIS
Baseline data/measures set
Supplier capability
Financial considerations
Safety
Engineering 7 technical data
Benchmark data
Prioritization (short vs long term)

**3**
**IMPLEMENTATION**

Communicate goals
Assign responsibilities
Measure progress review and adjust

**Figure C3**
Model of plan deployment

world's best companies is also undertaken to provide an external yardstick to evaluate progress.

# Quality in the marketplace

## Customer focus and satisfaction

Graniterock's key driver is value creation for the end customer, not just the immediate customer (e.g. the contractor), for the products or services. Focusing on both of these customer groups (each with its own perception of quality) gives the company clear direction for the establishment of marketplace plans which will support current and evolving customer needs.

Customer survey destinations are determined through customer files; trade association membership, project bids and building permit lists as well as supplier input and telephone directories. While the company surveys its principal customer groups at least once a year, it also solicits data from all potential customers, annually. Customer product and service factors are prioritized and customers are asked to rank Graniterock performance against that of market competitors.

Validation of survey accuracy and assessment of customer service levels are accomplished through two mechanisms: a *short pay* system – any customer can omit payment on any product or service that does not meet his or her quality expectations – which is a service and product guarantee to customers, allowing them to be the judge as well as ensuring a rapid response from the company to any complaint; a *product/service discrepancy* (PSD) system – any complaint is root cause analysed, rapidly resolved and a preventive action plan generated to eliminate recurrence – is used for overall product and process improvement as well as for immediate problem solving.

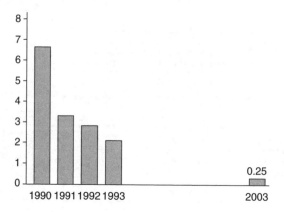

**Figure C4**

Concrete batching accuracy – all ingredients (average of all branches coefficient of variation (percent))

*Statistical process control* (SPC) and *numerical performance measures* are used to monitor performance and product quality. These are also the basis for ensuring continuous improvement in all the targeted product and quality service factors. Internal customer/supplier relationships receive the same attention as external ones do.

Being a leader in new product and process development has also helped to build the company's competitive edge. For example, they developed GraniteXpress 2, an automatic customer card controlled loading system that enables regular customers to drive up to the loading facility 24 hours a day, 7 days a week and accurately fill their truck with the materials required. This reduces waiting time for trucks and increases flexibility for customers.

## Quality and operational results

Quality performance is assessed both in terms of physical product quality and service quality. Both areas being represented by hard measures that can be compared to other products and organizations. In the quarries, SPC is used to control variables such as aggregate gradation and cleanliness. Graniterock has reached quality levels that are not generally attained by industry competitors. Variability in some key measures has been reduced to six sigma levels – a standard that is not generally seen in such a 'low-tech' industry.

One of the best measures of overall quality in concrete is low drying shrinkage. Shrinkage creates cracking and concrete 'curl' problems – both an aesthetic and a durability concern. Crack elimination or at least minimization is in the interests of all customers and users, present and future. To achieve this there has been a focus on improving the quality of the ingredients, the control of the mix proportions and improving mix design. Graniterock traditionally leads the industry with shrinkage values that are less than half of those of their competitors.

Improvements in service quality receive the same attention as improvements in product quality. Graniterock's on-time-delivery performance record illustrates this (see Figure C5). This on-time-delivery performance – getting concrete into mixers on time, delivered to the job site and unloaded quickly – has been listed as the number one concern of its contractor customers: a concern that is even above price and concrete quality. In contrast, owners' and architects' (representing the end-use customers) interests are most concerned with concrete quality.

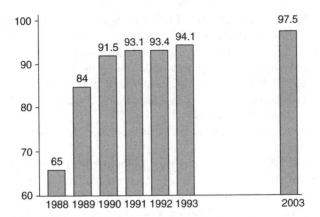

**Figure C5**
Concrete on-time delivery (percent on time)

Similar measures are used in each business area. In the manufacture of asphaltic concrete, SPC has been used to improve batching reliability. Whereas, previously, it was rare to complete an asphaltic concrete project without some rejected material, product quality has now improved to the extent that it is rare to have a single specification discrepancy on any project.

Plant production and operation measures such as variable costs and production efficiency show generally improving trends. Particularly important is the trend in market share, which has been improving steadily since 1986.

A key function of product quality is the quality of suppliers' products and services. Partnering arrangements with key suppliers have been put into place and are working well. Three of Graniterock's outstanding relationships are with its primary cement supplier, trucking company and railroad company. These relationships have created real benefits for all involved.

Graniterock gives a Golden Chain award to its best suppliers in recognition of the important part they play in helping Graniterock deliver quality to the end-use customer. To assure effective vendor contact, the number of suppliers has been reduced. For example, the company only has two cement suppliers and one petroleum product supplier, and ten building materials suppliers provide 80 percent of the products purchased for resale.

# Quality in the workplace

## Human resource development and management

The company has a policy of favouring internal applicants for senior positions. Therefore, already at the entry level, it seeks to hire people who demonstrate the capacity for senior management.

Graniterock encourages all its team members to use *individual professional development plans* (IPDP) to plan and track their own skill and knowledge development. This is a comprehensive process for integrating the company's human resource needs and quality objectives with the individual's aspirations and abilities through education and training. The effectiveness of this process is assured through the presence of the executive committee at the IPDP *round table*. The IPDP is a forward-looking system because individuals can use it in alignment with company needs as a tool to plan their professional development goals. It is a contrast to the *past performance assessment* system, in that it represents a proactive approach to personnel development.

The backbone of the IPDP is a simple four-page plan drafted by individual team members and their immediate supervisors. Short- and long-term goals are selected and a training/development plan is created to help the individual achieve these goals. Finally, observable measures (i.e. new skills incorporated into the job) are chosen to confirm goal achievement.

The thrust of the IPDP is to develop people to their fullest – setting goals beneficial to both the individual and the company. Although percentage of goals achieved is tracked, there is no stigma attached to failure. Currently IPDP goal attainment is at 91 percent – a measure of the success of both planning and implementation. The programme is voluntary and currently enjoys a 92 percent participation level.

Apart from helping to set Graniterock repeatedly in the top 100 employers in the US (*Fortune* magazine), the process has benefited both managers and team members. It has improved the coaching and personnel management skills of managers; it has helped to identify and highlight the talent within the company; and it has helped to avoid the underutilization of people throughout the company.

Other performance measures used to track the company's human resource development practices are:

- Number attending the Graniterock university and other seminars.
- Number of cross-trained people.
- Number requesting special education and training.
- Number attending company picnics and parties.
- Number of absentees.
- Number of turnover rates.
- Analysis of exit interviews and employee surveys.
- Satisfaction with benefit surveys.
- Speed of processing accident and health insurance claims.

Teamwork is widely used within the quality improvement processes of the company. The types of teams include quality teams, function teams and task force teams. There are normally up to 100 teams active within the company at any one time, and a high level of involvement reflects the company's team culture.

Recognition Day, held annually at each branch and department, plays the most significant role in the all-important recognition process. It allows face-to-face presentations to recognize all team member achievements in the past year. Senior management attends these days to send a clear message of support. This process is just one part of the overall communication and support network that makes up the company's employment culture. Other activities include company awards, recognition of achievements in the company's two quarterly publications, a weekly publication and occasional letters from the president of the company to all team members.

## Education and training

Annual expenditure per team member on quality-related education has increased consistently since 1985: then, the company spent slightly above the construction industry average. Now, the expenditure per head is more than 20 times the USA construction industry average and more than five times the mining industry average. By 1993, the average team member spent 40 hours per annum on education and training (not including safety, which is mandatory). All team members receive more than 8 hours of quality training per annum. As a result of this focus on safety and training, the company has a much better than industry average safety record: Graniterock's Safety Experience Modification on its safety insurance in 1993 was 48 percent – a saving of more than 50 percent on its insurance premiums, allowing it to pay less than half the industry average.

### Employee morale

The company arranges many recreational opportunities for its team members, encouraging them to play together as well as work together. Events range from concerts, parties at elegant hotels with famous speakers to parties for the whole family. In this way Graniterock is committed to building a community of excellence among its team members.

## Quality in the community

Graniterock is an exemplary corporate citizen. Company team members provided record financial support for United Way agencies, in 2003, with a total contribution of over $200 000. Graniterock matched this contribution dollar for dollar, resulting in a doubling of overall community financial support.

The team members demonstrate the values of their employer by investing their own time and money in building the communities in which they live and work. This is exhibited by the more than 700 members of the Graniterock team who have volunteered their personal talents and shared their financial support with more than 500 organizations in Monterey, San Benito, Santa Cruz, and Santa Clara counties in 2003. Graniterock matches employee donations to charities dollar for dollar.

Graniterock people respond to a broad spectrum of needs within the community with real-life benefits: they coach little league teams, provide senior care, staff domestic abuse hotlines, and support children's hospitals. They raise funds for community recreational programmes and deliver food, companionship and good cheer to people and families in need. They heighten awareness for safety programmes and volunteer in fire-fighting. They apply their skills to the advancement of women in construction, the protection and cleaning of US coastlines, to emergency flood repairs and relief work and to the recycling effort. Graniterock people apply the total quality management approaches that they have learned in business to their dealings with government, educational, and non-profit organizations.

## Recognition for excellence

Graniterock has received more than 50 awards in recognition of the group's performance. In the category of business awards these have come from Californian and national organizations for:

- Accuracy in company accounts.
- Excellence in business.
- Outstanding business leadership.

- Excellence in engineering management.
- Excellence in personnel management.
- Outstanding customer service.
- Outstanding website.
- Producer of the year.
- Sixteenth best employer in the USA.

Since 1993 the company has won numerous awards for excellence for its products and services, from Californian and national organizations for:

- Construction innovation.
- Extremely high standards of safety.
- High quality products and services.
- High quality plant operations.
- Information executive top 100 list.
- Winner of the US Baldrige award for overall quality.

In the area of community awards the company has gained recognition for a broad range of outstanding performance achievements:

- Environmental responsibility through waste reduction.
- Excellence in community service.
- Outstanding community relations.
- Partners in educational excellence.
- Support for charities.
- Resource conservation for the use of water.

## Review questions

1. Consider how Graniterock uses performance measurement to drive improvement and innovation.
2. In the construction supply chain it is often said that one of the impediments to quality is the fact that everyone focuses on the immediate customer and only the builder is focused on the end customer. Explain and evaluate how Graniterock uses end-customer satisfaction as the main driver for its improvement processes.
3. How does the planning/improvement process compare with 'theory' – identify the strengths and areas for improvement in the Graniterock approach.

## Acknowledgements

The authors gratefully acknowledge the assistance of the Graniterock team led by president and CEO Bruce W. Woolpert. Some of the text and illustrations have been taken from company information. Further information on Graniterock may be obtained from its website: www.graniterock.com.

# Growth of a commitment to total quality at CDL

# Company background

Established in 1963, CDL is an international property and hotel conglomerate involved in real estate development and investment, hotel ownership, and operations and provision of hospitality solutions.

As a leading residential developer with a track record of close to 40 years, CDL has built more than 15 000 quality homes in over 60 projects in Singapore and the region. The group is also one of Singapore's biggest commercial landlords, with approximately 5 million square feet of office, industrial and retail space.

The CDL Group's London-listed subsidiary, Millennium & Copthorne Hotels PLC (M&C), is one of the largest hotel groups in the world. It has over 100 hotels in major cities in 16 countries including London, New York, Paris, Sydney, Singapore, Hong Kong and Taipei.

Operating in 18 countries spanning Asia, Europe, North America and Australasia, the CDL Group has over 200 subsidiaries and associated companies, eight of which are listed on stock exchanges in Singapore, London, Amsterdam, Hong Kong, Auckland and Manila.

The Group has set up a dedicated subsidiary; namely, City e-Solutions Limited (CeS) geared to provide technology solutions for the global hospitality industry.

Apart from its business operations, CDL is committed to sharing the benefits of its success with the less fortunate in the community. Their contribution focuses on charity, environmental protection, the arts, education, sports and youth development. CDL has also formed a staff volunteer club – City Sunshine Club – to extend a helping hand to the sick and needy.

This case study is about CDL as a client for medium- and high-density residential building projects in the region.

# Recognition for excellence

The CDL's core values are expressed as *competent, committed, creative, co-operative* and *caring*. The company has achieved recognition for its excellence in terms of a number of key business, operational and community service parameters, reflecting the company's commitment to the broader framework of total quality. Its vision is to maintain industry leadership in innovation, product quality, service standards, profitability and community work. In detail, this translates into a guarantee for the:

- *customer* of quality and innovative products, unsurpassed service and value for money;
- *investor* of maintained profitability and optimum returns for investment;
- *employee* of maximized staff potential and care for personal well-being and career development;
- *community* of shared benefits of CDL's success with the needy, and of caring for youth development and the environment.

## Customer service

Commitment to customer satisfaction resulted in the company winning the prestigious Building and Construction Authority's (BCA) Best Practice Award in 2003. The CDL Customer Centre launched in June 2001 houses a state-of-the-art call centre supported by SAP software, which provides links between customers and all operating departments within the group, enabling prompt feedback and action.

## Creating value for investors

For three consecutive years, CDL has been voted one of Asia's Best Managed Companies in the Asiamoney poll conducted among fund managers from around the world. The poll was based on the criteria of business strategy, financial management, fairness to minorities and investor relations. In end-1998, CDL was conferred the special award for Highest Growth in Market Capitalization in 25 years by the Stock Exchange of Singapore, reflecting strong investor confidence in the company's track record, financial strength and professional management.

The company is regularly recognized in market evaluations and in regional business surveys as a leader for its high standards of governance, the transparency of its reporting, growth in asset backing, growth in sales and profitability and in its general business performance.

## Commitment to sustainability

As CDL sees environmental protection as a core corporate responsibility, the company has been committed to sustainability values since the late 1990s and was awarded the Green Leaf Award 2000 by the Singapore Ministry of Environment for its outstanding contribution and commitment to environmental conservation and protection. The award was principally awarded because CDL is the first developer in Singapore to install eco-friendly lifts and innovative pneumatic refuse collection systems in selected condominiums, and for setting aside a large percentage of site area for landscaped gardens. Also, CDL homebuyers are encouraged to recycle, reduce and reuse.

CDL is among a small group of companies worldwide that meets FTSE4 Good's Socially Responsible Investments (SRI) standards of corporate social responsibility. CDL is recognized by FTSE4 for working towards environmental sustainability, developing positive relationships with stakeholders and upholding and supporting universal human rights.

The company was awarded the Merit Award for Best Public Relations Work for the Environment at the PR in the Service of Mankind (PRISM) Awards 2002. This biennial award recognized CDL's extensive PR Eco-Campaign over the previous two years, including its seminal Shop & Conserve Programme, Singapore Green Map, Nature Series Corporate Calendars and other eco-friendly initiatives.

In 2003 the company launched the Eco Office website and the Eco Office Rating System, which enables companies to perform a self-audit to qualify for a Green Office Label.

## Occupational health and safety

CDL's has also been committed to lifting environmental and occupational health and safety standards in all its projects. In 2003, it was awarded both the BCA ISO 14000 and OHSAS 18000 Certification for Property Development and Project Management services – the first private property developer in Singapore to be thus honoured. Also in 2003, the company was the first of its kind in Singapore to be conferred the UK's Royal Society for the Prevention of Accidents (RosPA) Silver Award for excellence in managing health and safety issues.

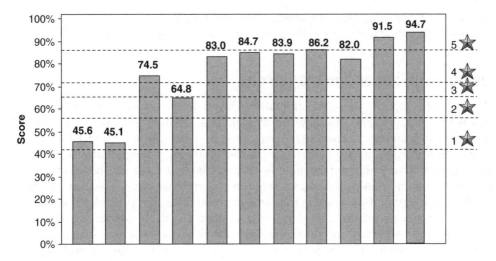

**Figure C6**
CDL 5-star EHS assessment results on individual projects

## Environmental, health and safety benchmarking

CDL started benchmarking environmental, health and safety performance on its projects some three years ago. The graph in Figure C6 shows the company's rapid improvement in its composite environmental health and safety performance on individual projects over a three-year period.

## Project awards

A company project was awarded the highest construction quality assessment score (CONQUAS) in 1997 and it was the first developer to receive the Singapore BCA Quality Mark Award in 2002, which recognizes high workmanship standards for all units in a residential project.

Projects have won numerous awards for construction excellence, buildability, design excellence, fire safety excellence, and for outstanding safety in 2002 for having accomplished at least 100 000 accident-free man-hours in 30 consecutive days through the implementation of sound safety management systems.

## Community awards

The company has won several awards for its commitment to community service in the areas of health, support for the arts, the volunteering efforts of its employees, CDL's support and that of its employees as volunteers for charities and for the support it gives its employees when working for charity.

# Construction-related processes at CDL

CDL believes that, to a great degree, the measure of quality depends on satisfying client expectations and therefore a significant effort is invested in knowing what clients want.

The following description of the construction-oriented processes of CDL is based on an extensive interview with Eddie Wong, the General Manager of Projects at CDL.

## Design briefing and management

At CDL the company believes that quality starts with the client; therefore, even before pen is put to paper, the company conducts market research to identify demands. The market is segmented and analysed to determine the brief for the architect. Every four or five years thorough market research is conducted by an independent market research organization to redefine market preferences and trends. In refining the tenancy mix for a project, both market pull and industry push factors are considered, anchor tenants carefully selected and destination marketing is used wherever possible. The brief is not too prescriptive, as architects need autonomy if their creativity is to have free rein; however, customer expectations and the market mix in terms of space size and customer requirements are clearly specified. In some projects the floor plate size and its primary proportions are approximately defined.

Just as quality starts with the client at CDL, project success relies on identifying and satisfying customer needs. Therefore, CDL is committed to sustainability and having a whole-of-life approach to issues such as maintainability and the operating costs of the future owners; customer requirements drive CDL decision-making throughout the process. Fortunately, since Singapore has a very flexible planning environment, limitations on new projects are few as long as environmental performance limitations are met.

In cases where the long-term occupant is known, such as the design and construction of a corporate head office, CDL, as the initial developer, remains closely involved to ensure that the occupant's needs are fully addressed during the design development phase. In most projects, normal development projects that represent more than 90 percent of CDL operations, the architect has a more open hand in addressing the design brief.

Even so, CDL has developed, and uses, design checklists along with a digital library of problems compiled from previous projects to guide the design through the many pitfalls inherent in the building construction process.

## Construction quality management

Construction quality management at CDL starts with a careful prequalification of tenderers, leaving only a short list from which to choose. Considerations include technical, human and financial resources as well as current workload.

Wherever there is scope for a contractor to develop more efficient methods of construction, the design documentation defines performance levels, allowing the contractor the flexibility to optimize processes. In these areas the tender scope includes both detailed design and construction.

In order to encourage contractors to build to a high quality standard, target levels are defined in terms of the Building Construction Authority CONQUAS measure – bonuses and penalties are payable based on the standard of product quality achieved. In some instances CDL has as much as a 1 percent tender advantage to a builder they believe will build a better quality product than a competitor.

When the Building Construction Authority introduced its CONQUAS score for build quality, in the early 1990s, CDL embraced it as a performance driver to motivate its internal improvement and simultaneously differentiate CDL from its competitors.

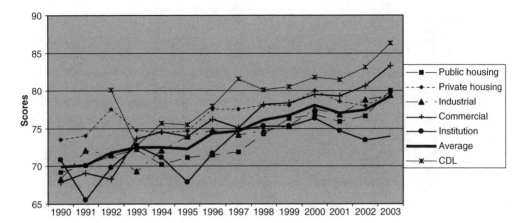

**Figure C7**
CDL's quality relative to the market. (Source: BCA)

The graph Figure C7 shows how CDL has performed relative to its market competitors over a 13-year period.

CDL has a number of other strategies to ensure that a high standard of quality is achieved on site:

■ As client, they always employ a quality inspector to represent their interests on site.
■ In general, the philosophy is adopted that if something is simple to build it is easier to build well; hence, simplicity in process and detail is actively sought.
■ Hidden details are standardized wherever possible and unnecessary differentiation is avoided.
■ Construction is planned in detail and method statements are developed and prototypes built wherever possible, as they are the best way to identify problems before they occur on site and they are an effective way of defining quality standards to be achieved.
■ Wherever possible, prefabrication is encouraged because quality is easier to ensure using prefabricated technologies.

CDL also studies defects and their causes so that it can focus the attention of its designers and contractors on the development of defect-free solutions. The graph in Figure C8 illustrates the most common issues on their projects. Each of these is analysed to determine the root causes and to develop strategies for improvement together with their suppliers, whether they lie in design, construction technique or worker skill. In some instances customer education is needed and this is addressed through the pre-purchase workshops.

## Measures to improve customer satisfaction

A number of initiatives have been introduced in recent years with a view to improving the customer experience in a CDL development:

■ A call centre has been established to provide 24-hour response to problems.
■ Customers are invited to attend seminars before occupancy. Where occupancy and fit-out issues are discussed, advice is available in relation to furniture, fit-out and landscape.

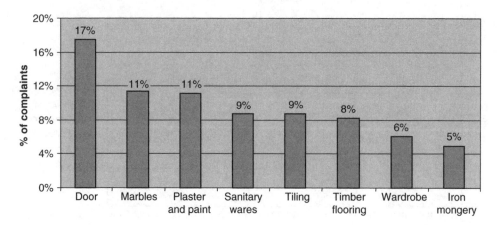

**Figure C8**
Common complaints by customers

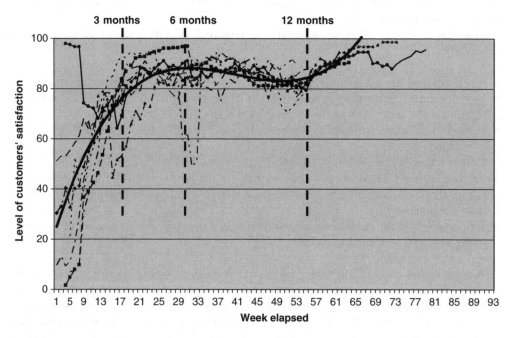

**Figure C9**
CDL measures customer satisfaction

- Contractor performance is assessed against a number of customer-oriented performance measures:
  - average number of complaints per unit;
  - number of weeks from occupancy to full defect rectification;
  - response time to individual defects.

One of the performance measures used to drive customer satisfaction is the time taken to have all customer complaints fully rectified. The graph in Figure C9 shows how this kind of information is plotted for each project. This data is used to assess contractor performance and to drive improvement, the goal being to achieve the resolution of all customer complaints as early as possible.

- Common area gardens have been expanded both on the ground and on roofs to create a cooler and more pleasant environment. This has been done in response to customer feedback and recognition that hard building surfaces build up heat in urban environments.

# The push for ongoing improvement

The drive to differentiate themselves from their competitors is a continuous goal on the journey for improvement at CDL.

The push for change is driven by customers looking to buy the best properties and investors looking for product differentiation. It is a whole-team effort for CDL management, from the managing director to the heads of department.

# Business awards

As an international property and hotel conglomerate, CDL has received numerous awards in recognition of the group's performance.

In the category of business awards, the company has won more than 20 within Singapore and from international organizations for such aspects of its performance as:

- Business performance and profitability.
- Excellence in business management.
- Growth in market capitalization.
- High quality products and services.
- High standards of corporate governance.
- Long-term vision.
- Transparency of reporting.

In recent years the company has won some 30 awards for excellence for its projects: while most of these have been from Singapore authorities and associations some have been from international organizations. They have been for issues such as:

- Buildable designs.
- Fire safety excellence.
- High standards of construction quality.
- Landscape design excellence.
- Outstanding OH&S performance.

In the area of community awards the company has gained recognition for a broad range of outstanding performance achievements, including:

- Environmental consciousness of their project designs.
- Commitment to a family-friendly culture for their employees.
- Delivery of community service projects.
- Health and safety.
- Outstanding staff volunteer contributions to the community.

- Socially responsibility in their investments.
- Support for the arts.
- Support for charities.

## Some specific awards

Voted one of Asia's Best Managed Companies in the Asiamoney poll.

Awarded the highest construction quality assessment score (CONQUAS).

Conferred the Stock Exchange of Singapore's special award for Highest Growth in Market Capitalization in 25 years, reflecting strong investor confidence in the company's track record, financial strength and professional management.

Awarded the Green Leaf Award by the Singapore Ministry of Environment.

Awarded the Merit Award for Best Public Relations Work for the Environment at the PR in the Service of Mankind (PRISM) Awards.

Received the Singapore BCA Quality Mark Award, recognizing high workmanship standards for all units in a residential project. CDL was the first developer to receive this award. Awarded for outstanding safety for having accomplished at least 100 000 accident-free man-hours in 30 consecutive days through the implementation of sound safety management systems. CDL projects have won numerous awards for construction excellence, buildability, design excellence and fire safety excellence.

Winner of the prestigious Building & Construction Authority's Best Practice Award (BCA) in commitment to customer satisfaction.

Awarded both the BCA ISO 14000 and OHSAS 18000 Certification for Property Development and Project Management Services. It was the first private property developer in Singapore to receive this.

Awarded the UK's Royal Society for the Prevention of Accidents (RosPA) Silver Award for excellence in managing health and safety issues. CDL was the first company of its kind in Singapore to be thus honoured.

# Review questions

1. Consider how CDL use performance measurement to drive improvement and innovation across a wide range of issues that impact on their overall performance.
2. CDL has been innovative in its focus on customer satisfaction, developing novel improvement strategies ahead of the rest of the industry. Present an overview of these strategies.
3. Suggest a 'quality framework' that would convey how CDL have travelled their journey and evaluate their performance against it.

# Acknowledgements

The authors gratefully acknowledge the assistance of Eddie Wong, the GM of the Projects Division at City Developments Limited. Some of the text and illustrations are taken from company information.

# Future growth at Gammon underpinned by a commitment to excellence

# Company background

The company originated from a construction business founded in India in 1919 by John C. Gammon. In 1958, after having built a new runway at Kai Tak Airport in Hong Kong, it was decided to establish a permanent presence in Hong Kong as Gammon Construction Limited. By the late 1970s, the company had developed a recognized expertise in piling, foundations, substructures, tunnelling, bridges, buildings, marine works and water storage schemes, and had established a reputation as the leading contractor in Hong Kong. Gammon Construction's further participation in the many major infrastructure projects in HK and expansion into South East Asia and China have consolidated this position. After a number of changes in ownership, today, the business is jointly owned by two private companies: Jardine Matheson and Balfour Beatty. With an annual turnover around US$1 billion, the company employs approximately 2000 full-time staff, including some 650 professional engineers and builders, making it one of the strongest technical teams in Asia.

In the late 1990s senior management realized that the tender-based construction market was becoming increasingly competitive and price driven so they decided that growth and survival depended on the company's ability to differentiate its services from that of its competitors. Gammon has committed itself to serve its clients through innovation and excellence, offering value that is unmatched by competitors. Today the focus is on how to add value to its customers by making the best use of the abilities and resources of the group – working with them to manage risk and develop innovative solutions. The focus on customer satisfaction, quality, safety and environmental management has created a situation where, in 2003, the company was able to win more than 50 percent of its new work from other than the lowest bid position.

## Mission

To develop, build and service Asia's physical infrastructure for living, working and travelling.

## Vision

To be the leading provider of construction services in Asia.

## Strategies

- To focus on customer needs and develop long-term relationships.
- To have a comprehensive range of business streams delivering a diversified range of construction services.
- To have best in class management and technical competence.
- To be technically innovative and flexible in its approach to business.
- To participate throughout the project life cycle from idea through construction, operation and maintenance to decommissioning and renewal.
- To grow organically and through acquisition in existing and new markets.

# Group quality policy

To realize its vision – to be the leading provider of construction services in Asia – the company believes that Gammon must differentiate itself by quality of service, customer focus, innovation and better value. Internally its processes and use of resources must also be effective and efficient. To underpin its whole philosophy and approach Gammon has adopted the following policy on quality:

- Work in partnership with its customers to understand their needs.
- Enhance customer satisfaction by providing a consistently high quality of services and products that comply with requirements, both contractual and regulatory, and meet or exceed their expectations.
- Strengthen its business processes through the effective implementation and continual improvement of its quality management systems.
- Establish quality objectives on projects and at relevant functions and levels to direct staff commitment and effort towards improvement.
- Enhance the ability and competence of staff to deliver quality results by systematically developing their competence and motivation.
- Regularly monitor and review performance and implement actions necessary to address deficiencies and increase effectiveness.

Gammon has formed a Quality Action Committee to help ensure that this policy and the company quality management systems, objectives and targets are understood, implemented and reviewed. This committee is to report to and advise the company executive team on improvement actions necessary.

To be effective risk managers, the company must also have the right people, operating within a clear framework of formal controls. To achieve this, Gammon has implemented quality and environmental management systems in accordance with the requirements of ISO 9000 and ISO 14001, respectively, and it also operates a comprehensive health and safety management system.

# Values

Central to Gammon's Brand Values is their commitment to social responsibility.

The ultimate goal at Gammon is to deliver a high level of quality to customers. By this, the company means not only the quality of its built products and service outcomes, but also the quality of the way in which they are delivered – reliably, safely and responsibly.

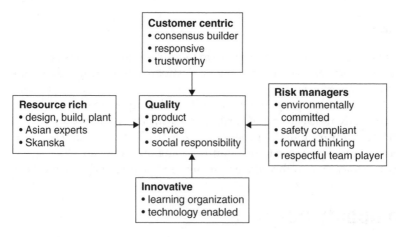

**Figure C10**
Brand Values

Gammon believes that it can best deliver the level of quality to which the company aspire by concentrating on four core values. Gammon is determined to be:

- *Customer-centric* – focusing on the broader needs of customers and not simply on the structures that it builds for them or the services that it provides to them.
- *Resource rich* – combining the company's own extensive local resources with the global resources of its shareholders in a way that puts the company ahead of local competition.
- *Innovative* – being a technology enabled, learning organization that is committed to finding better ways of doing things, making the most efficient and sustainable use of available resources.
- *Risk managers* – having the foresight, attentiveness to detail, and tenacity to anticipate and effectively deal with the problems and challenges that construction inevitably entails.

## Customer-centric

Gammon's focus on the customer, as well as on the structure that it builds, differentiates the company in the market. The company also aims to build 'trust' through its consensus-building approach with clients and its responsive attitude that enables Gammon to anticipate client needs.

Drawing on more than 40 years of experience in a broad range of projects, the company has focused on unlocking value in the design and construction process by applying its specialist knowledge wherever and whenever possible. There is a commitment to focus on the lifetime performance of projects and to delivering solutions that are in the best interests of customers' long-term needs.

## Resource rich

Gammon's wealth of local knowledge and experience, its wide-ranging capabilities in design and build, and its extensive investment in construction plant and support facilities are key competitive advantages for the company. The regional expertise of Jardine Matheson together with the global engineering resources of Balfour Beatty have enabled Gammon to meet the exacting demands that are required to stay ahead of the local competition in building Asia's infrastructure.

To maintain its position as a leader in the industry Gammon has committed the company to continual investment both in its people and its physical resources.

## Innovative

Gammon believes that it has a public responsibility, as well as a commitment to the client, to deliver projects in a way that makes the most efficient and sustainable use of available resources. More than ever before, the company believes that, now, there is a need to be environmentally attuned and compliant. To this end, it has recognized the importance of being a technology-enabled, learning organization. In the construction industry, being ahead means not just acting on new business trends, but setting them.

## Risk managers

Some people consider, in a matter of fact way, that delivering projects to budget and time is simply part of the job. Certainly, it is. But the risks that must be dealt with every day require considerable foresight, attention to detail and tenacity if problems that can add cost or reduce productivity are to be avoided. To ensure the excellent overall quality of everything done in the name of Gammon, effective teamwork is crucial, not only between the company's own staff but also with all other business partners in the endeavour. It is, therefore, a real accomplishment to be a risk manager at Gammon – let alone to excel as one.

# Management processes

## Health, safety and environment (HSE) are core issues for Gammon

Excellence in HSE management is not only essential for ensuring the safety of the work-force and protecting the environment but is also of strategic importance when developing new markets and services, and identifying future risks to the business. Gammon has committed resources to developing management systems based upon current international models. They believe that a high quality, safety conscious and environmentally aware company will be more successful in the short, medium and long term.

## Customer satisfaction survey on HSE perceptions

The company proactively surveys its clients about its performance on HSE and other issues. The latest results have indicated that clients attach great importance to safety and that they also perceive Gammon's safety performance as being strong.

## Safety

In the past few years the company has received the OHSAS 18001 certificate for its health and safety management system in Hong Kong as well as ISO 14001 Environmental Systems Certification. Gammon has developed and implemented a company-wide health and safety management system for its operations: project and design management, civil engineering, underground engineering, foundation and building works, ground investigation, concrete supply, plant maintenance and steel fabrication. The company's commitment to safety management is reflected in its recognition in the HK industry awards.

A health, safety and environment league has been introduced for all Hong Kong projects to encourage further performance improvements. Since 2002, 24 projects have reported no accidents or incidents. Over a four-year period the company's accident incidence rates in Hong Kong have fallen to one third of their previous level and these are now less that one quarter of the industry average.

### Group safety targets

In working towards Gammon's mission of providing a workplace that is without accident or incident, the following goals apply to projects across the region:

- no fatal accident;
- continual improvement in the accident incidence rate;

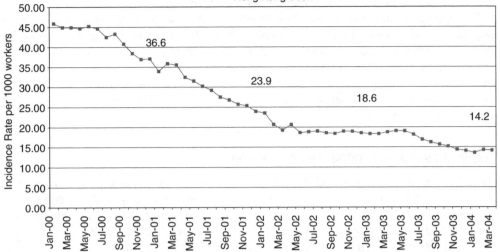

**Figure C11**
Accident statistics for the Hong Kong activities over a four-year period

- continual reduction in the breach of statutory safety requirements;
- continual reduction in the breach of statutory environment requirements.

Furthermore, for Hong Kong operations the following longer-term targets have been set in terms of further reduction of the accident incidence rate; a reduction of breaches of safety legislation to zero and zero breaches of environment legislation.

There is to be an increasing focus on the performance of subcontractors whose accident incidence rate is 50 percent higher than Gammon's and who are involved in 75 percent of safety non-compliances, including damage to utilities.

In order to proactively engage subcontractors, Gammon plans to:

- address HSE issues during tender evaluation and on contract award;
- collate, monitor and disseminate data on subcontractor HSE performance;
- ensure that HSE terms and conditions in the subcontract are fully explained;
- ensure that subcontractor workforce is trained for specific projects.

Achieving the mission to provide a workplace that is without accident or incident requires a concerted effort across the company. Gammon encourages its senior management to exchange views and learn through participation in local and regional associations, forums and conferences.

## Environmental management

Gammon works with its clients to address environmental issues throughout the construction process, helping them to develop solutions that reduce costs and protect the environment. On-site partnerships are developed with staff and subcontractors to meet client requirements, reduce noise and nuisance to nearby residents and to ensure compliance with environmental regulations. The company's improved environmental performance over the last three years is a result of everyone's continued commitment to the ISO 14001 compliant environmental management system.

The company has actively raised environmental awareness in the construction sector and among the general public through sponsorship and participation in environmental events, conferences and community outreach programmes. In recent years Gammon's environmental initiatives and performance were recognized at a number of award ceremonies. These included two Grand Awards (Green Construction Contractor Category) at the Hong Kong Eco-Business Awards for Chater House and Island Eastern Corridor Improvements Projects.

### Gammon's modular construction system

By delivering high quality modular units for erection to the Gammon Technology Park site at Tseung Kwan O, the company reduced on-site concreting and finishing by 70 percent and reduced lorry movements to and from site by 40 percent. This patented system is based on modular units which are fully fitted out to include kitchens, bathrooms and finishes in the factory, all to high quality standards. The unique design features incorporated in these modules enable their ready installation on site using conventional construction techniques. This system can be used to construct entire building structures up to 50 storeys. Construction cost is similar to traditional designs, but the construction is significantly faster, with higher quality, better noise insulation for residents and far less environmental impact on Hong Kong's community.

### Commitment to sustainable buildings

During both design and construction, Gammon has assisted in the development of energy efficient buildings. Examples include several major commercial buildings which have won Hong Kong BEAM 'Excellent' awards and have secured the company's involvement as one of the founders in the INTEGER Hong Kong Pavilion project.

INTEGER demonstrates how innovation in design, construction and management of residential buildings can improve energy efficiency and environmental sustainability. Through active involvement in this project, Gammon has been uniquely placed to develop 'intelligent and green' construction projects across Asia.

## Balanced scorecard approach to drive HSE commitment

The group-wide HSE policy reflects the company's long-term commitment to health, safety and the environment and provides clear guidance to staff at every level. In addition, safety and environment are core elements of the Gammon Brand Values. In 2002, the company introduced two innovative schemes to Hong Kong: the HSE League which ranks sites in terms of HSE performance; and a structured company-wide HSE Bonus Scheme.

The board of directors reviews HSE performance on a monthly basis to ensure that everyone lives up to the HSE principles in the Brand Values. This is actioned through a balanced scorecard, which is one of the company's key performance management tools. HSE targets are included in the scorecards for directors, contracts managers, construction managers and project managers.

Increasingly clients assess the HSE performance of companies and use this information during the tender selection process. Gammon not only attach great importance to these assessments but proactively survey clients about their expectations and perceptions of their HSE performance. The company is committed to the partnering business model and believes that it can improve safety, reduce waste and enhance efficiency by working with clients during the feasibility, planning, design and construction stages.

Gammon also has adopted a policy of proactive engagement with the Labour Department, Environmental Protection Department and other regulatory authorities in Hong Kong to ensure that a responsible attitude is taken during project planning and construction.

Recognizing the critical role of the professional safety and environmental team in supporting the project managers, the company ensures that staff and subcontractors are well trained and encouraged to take an active part in Gammon's incentive schemes. A subcontractor's proposed HSE approach is very seriously assessed during tender evaluation, selection and during the construction phase. The company monitors HSE performance closely through tight supervision, regular HSE inspections, audits and appraisals.

## Quality

Over a three-year period, Gammon has achieved a steady improvement in the quality of its work as measured by independent assessors.

## Student support and staff selection

Established in 2002, the Gammon University Fellowship Programme – targeting second year undergraduates at universities in Hong Kong – is now moving towards its fourth year. The programme has been designed to allow the fellows to gain first-hand experience and learn about current practice in a world-class construction business. In 2003 the programme was extended beyond Hong Kong to the southern part of China, reflecting the company's business development plans.

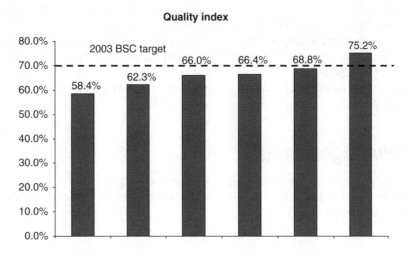

**Figure C12**
Consistent quality improvement over a two-year period

# Recognition for excellence

## Safety

Over a two-year period Gammon has won ten awards in the HK Government Environment, Transport and Works Bureau Considerate Contractors Site Award Scheme.

Building on the momentum of achieving ISO 14001 Environmental Systems Certification, Gammon developed and implemented a company-wide health and safety management system for its project and design management, civil engineering, underground engineering, foundation and building works, ground investigation, concrete supply, plant maintenance and steel fabrication. As a result the company received the OHSAS 18001 certificate for its health and safety management system in Hong Kong.

## Technology and quality

At recent Hong Kong Awards for Industry, Gammon won two accolades – the HKSTP Technological Achievement Award and the Trade and Industry Department Quality Award. The recognition of the company's achievements in both these areas highlights two of the most important aspects of Gammon's work.

The Modular Gammon Construction System received several awards including a Special Merit at the Hong Kong Institute of Engineers Innovation Awards for the Construction Industry and the Hong Kong Science and Technology Park *Technological Achievement Award* at the Hong Kong Awards for Industry. The system also won a Special Merit Award at the Innovation Awards for Construction Industry, organized by the HKIE.

The Quality Award at the Hong Kong Awards for Industry recognizes Gammon's use of a combination of local expertise, innovation, technical competence and global resources to manage risk and deliver services to best-in-class standards of quality management that are underpinned by company-wide ISO 9001 and 14001 certification.

# Review questions

1. List and discuss the advantages and disadvantages of developing the 'management processes' in the way that Gammon has chosen.
2. Critique their approach against the EFQM Excellence Model showing the overlaps and 'underlaps', and provide advice to the management on how their approach could be improved.
3. What is the value of seeking awards in the way that Gammon have done – are there possible disadvantages of doing so?

# Acknowledgements

The authors gratefully acknowledge the assistance of Derek Smyth, Director, and Iain Wink, Group Quality System Manager of Gammon. Most of the text and illustrations are taken from company information.

# World-class project performance from a partnership between IDC and TNC

# Company background

In 1985, CH2M HILL established Industrial Design Corporation (IDC) a high-tech facility design-only firm. Based in Oregon in the USA, the company, with over 1000 employees, has become the world's leading design and project management firm, specializing in the delivery of semiconductor fabrication facilities (FABs).

IDC had developed a close relationship with Intel – providing leading edge high-tech semiconductor facility design service. Intel was in a race to maintain ~90 percent market share over their main competitor AMD. They were in development of the microprocessor that was later named Pentium I. In order to maintain control of the market, Intel needed to build a mass manufacturing plant faster then ever conceived before, it was to be much larger and much more technically complex than any previous facility of this kind.

# Project background

In 1995, Intel interviewed several teams and challenged IDC to create a turnkey design–build company. They needed to build a mass manufacturing plant not only faster than it had ever been thought possible but also much larger and much more technically complex. The new plant would have to be built in less than half the time ever achieved previously – the traditional design–bid–build process was out of the question.

IDC joined with Hoffman Construction to form Technology Design and Construction (TDC) and became the 'umbrella' company which 'hired' IDC and Hoffman to staff the turnkey design build Intel FAB11 project.

Utilizing their then limited understanding of lean construction (LC) principles, IDC and TDC set to work. The practices discussed in this case study describe some of the very early efforts to employ LC. The emphasis on these projects was to shorten the schedule while building to the highest level of quality. While an 'upset' budget was established, these budgets were at the high end of the industry standard for semiconductor FABs. The first project was completed on budget, and in about 30 percent less time than a previous similar facility of somewhat smaller size. The quality end result for the owner was product yields greater than planned, allowing them to have their product in large volume to the market faster than forecast.

On a second project, Intel FAB12, in 1996, the same client asked for the duration of the design and construction to be reduced by a further third and the budget by a quarter. The second project was virtually identical in size, scope and scale to the first. The client's objectives were met on both counts.

Over the two projects this represented a 45 percent reduction in time and a 25 percent reduction in cost against industry standard rates. The projects were delivered defect free on completion.

# Organizing for change

The financial success of these facilities relies on the construction being delivered within a very narrow time-to-market window. Consequently, construction must be very fast and all key deadlines met. Given the accelerated schedules, the construction spend rate is very high and therefore the cost of any errors or delays is similarly very high. Typically the manufacturing processes within these facilities are developed in parallel

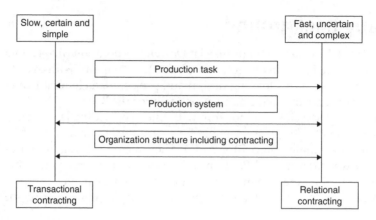

**Figure C13**
Management framework (Miles 1997)

with the building project, further threatening the achievement of the project budget and schedule goals.

While the risks are high and the challenges considerable, the rewards for improved production outcomes are also potentially great. The dramatic increase in competition among the high-tech industries requires that facility construction cost and time be reduced, while maintaining or improving the high quality. The processes on these two projects are characterized by the companies that delivered them as *lean construction*.

While some value engineering in the traditional sense occurred on a small scale, the remainder of the cost reduction and project acceleration was created by improvements in the procurement (supply chains) and production processes. It was found that traditional project organization does not support the demands and goals of such high-tech projects. These projects are quick, complex and uncertain. The project organization was designed to respond to these demands. The fundamental unit on these projects was the Multifunctional Work Group.

The work groups were formed around facility systems and technologies, and they were co-ordinated in the project teams. Each work group comprised representatives of all the organizations involved: constructor, designer, owner, process owner and operator, relevant government authorities, key suppliers and so forth. Each work group was responsible for and 'owned' their respective portion of the project from inception right through construction to the operation of the facility. They were also responsible for assuring that all cross technology co-ordination with other work groups was effective. They were responsible for their part of the project budget and schedule and the interaction of these with the project as a whole.

The closest parallel to the processes that were adopted was the simultaneous engineering and supply chain integration concepts within the Japanese automotive sector. The allocation of responsibilities in these projects was similar to Toyota's practice of setting cost, weight and performance targets for a vehicle and provisionally distributing those targets across the vehicle components. Then negotiations were held with all the component suppliers during the development process in order to stay within the total target cost or weight.

The work group members were empowered with an unusual level of authority to resolve issues and make decisions committing their companies; without such authority, the groups would have been impotent. There were rarely more than two levels of project management above the work groups. In order to avert logjams in decision-making, the members were given time triggers that limited the duration that any issue remained unresolved before it was automatically escalated. This was effective in ensuring resolution at the lowest level and in ensuring smooth operations at all times.

# Communications

Communications were focused within and among the work groups. The use of state-of-the-art tools played an important roll, but these were never allowed to replace face-to-face interaction within and between the work groups. Simultaneous engineering tools such as integrated piping design/construct software were used to create seamless, electronic communications within the project and across organizational boundaries.

Occasionally, it may have appeared that the time committed to meetings was excessive; however, the speed and complexity of such projects require frequent and intensive interaction. These meetings frequently revealed otherwise overlooked or hidden problems that could have seriously undermined the goals of the project.

It was not unusual for the work groups to meet, evaluate and develop a mutually acceptable work programme all within a one-hour weekly meeting. The result was the avoidance of serious and costly rework and delays.

# Project controls

The work groups were also the key to project control: they met every two weeks for one hour to review reports on cost control, the scope of work including change orders and the schedule. Based upon both approved and non-approved but anticipated change orders, design in progress and current schedule status, each work group jointly prepared a forecast.

Within a one-day period, the forecasts of all the groups were rolled up into a project report ready for the project management team the next day. Problem areas were quickly identified and assigned to an 'owner' to resolve. Most importantly, any mismatch between plans and actuals was immediately investigated. The goal was to re-evaluate and reform the planning and management systems in order to improve their reliability and overall project productivity.

# Motivation for change

The motivation for implementing the ideas outlined in this brief case study was the 'desperate need' of Intel to bring its product to market as soon as possible. The massive size, enormous complexity and critically short delivery schedule of the Intel FAB11 project demanded a total rethink of traditional delivery methods. The invention of the specific project processes was facilitated by Intel's familiarity with lean production from Japanese industry practices.

Most of what was done was 'invented' during the project. The key to success was the concept of self-managing teams based around the technological systems of the facilities (such as clean room air systems, high purity water, etc.). Previous methods of centralized project management control simply could not deliver the vast number of decisions required at an extremely high rate.

# Recognition for excellence achieved

High-tech projects were very confidential at the time of these projects. Neither clients nor suppliers wanted any information made public. While there was lots of anecdotal publicity in the trade press about the phenomenal achievements of the projects, no submissions were made to organizations that would have awarded such recognition.

IDC was unable to take advantage of the extraordinary delivery system that its people had invented, the company was severely damaged by the market crash in demand for semiconductors, they did not take their innovation into other markets and consequently lost the very staff who had created this innovation.

# Review questions

1. How did the partnership between IDC and TNC contribute to the improvements in performance?
2. Discuss the issues of measurement in a partnership such as this one.
3. What should the senior management of the organizations involved learn from their experience and what systems and controls could be developed to sustain momentum?

# Acknowledgement

The authors are indebted to the assistance of Bob Miles, a former employee of IDC and the manager responsible for introducing lean construction practices into the company. Today Bob Miles is the Director, Capital Planning and Management, Design and Construction Services at MD Anderson at the Texas Medical Center in Houston, Texas.

# Reference

1. Kirkendall, R. and Miles, R. (1999) 'Application of Lean Design and Construction in High Tech Facility Construction', *International Group for Lean Construction Annual Conference IGLC8*.

# Commitment to total quality at Mirvac

# Company background

This case study demonstrates how senior management's focus on quality can lead a company to becoming a nation's outstanding builder of quality dwellings.

Founded in 1972 Mirvac has grown in a short 33 years into one of Australia's most successful property groups. At the outset, the company saw its niche in the construction of quality homes; the guiding principle of the founding partners from their very first project was that they should never build a dwelling that they would not live in. This has set the standard to this day.

Initially the company built medium density projects and then it branched into housing estates and eventually into the development and management of commercial properties and hotels. The company built the Olympic athletes village for the 2000 Olympics, which has now been converted into a suburb of Newington. The project won a total of 36 building, architectural and design awards.

In the relatively short time since its inception, the company has successfully built an enviable brand image and a reputation for the quality of its buildings that has led to their properties holding their value better than other properties in the market. They have won an enviable reputation for excellence in initial quality and whole of life performance.

It is interesting that all this has been achieved without a formally documented quality system. This case study explores how that has been achieved.

# Industry setting

The period of the company's growth, the last 30 years has also been a time of extremely high construction risk in the newly emerging market sector of high density, high rise housing. Mirvac had many firsts during the period: its *Castlevale* project in 1976 in Willoughby, a Sydney suburb, was the first large scale integrated development in the city; in the early 1980s, Mirvac built the *York*, the city's first luxury inner city high rise housing project.

In the late 1980s *Raleigh Park* comprising 360 apartments and 150 houses was commenced on a 12-hectare site as a fully planned, self contained suburban estate on high value land in a central location within 5 km of the Sydney CBD. Because of the prestige location of this estate high quality, trouble free design and construction practices were critical, success was assured through the regularly review of construction practices and product performance by a team comprising both designers and construction personnel.

During the intervening period, this sector of the construction market in Australia has been relatively unprofitable, with construction related risks eroding most of the profits made by builders. Those who were profitable made their money in the development of projects, not from their construction operations. Typically construction companies had broad supply chains, with cost being the major factor in the selection of suppliers.

Mirvac has been the outstanding exception. Its focus on quality helped it to avoid the losses made by others. When research published in 2002 showed that typically this sector was wasting between 3.3% and 6.4% (Thomas *et al.*, 2002) of turnover on rework within the contract period, Mirvac unofficially reported that its rework was a fraction of 1 percent.

# Managing director leadership

Quality is foremost in every conversation within Mirvac and the MD drives this. There is no quality manager as such, since quality is the responsibility of very person in the organization.

An interesting feature of this case study is that the company has relatively little formal documentation in relation to its quality systems. In fact, it is only relatively recently, as the company has continued to grow and with changes in management, that a more formal development of the company's quality systems has commenced.

How then has the company achieved its exceptional results? It is through the MD's constant focus on quality; it has always been the most important aspect of everything the company does. Quality and service to the customer is the first consideration in everything. As a result, the company has developed systems and processes that reflect this value, despite the fact that their system has not been formally documented (as it seems to be the case with larger companies).

At Mirvac, the consideration of quality-first influences how every decision is made, and how every problem is solved. The most critical issue is the total focus on the end customer. Although the company acts as both developer and builder, the two functions are completely integrated. All decisions, whether they relate to development or to construction are made with the satisfaction of the end user in mind.

# Core values

The core value of Mirvac is customer satisfaction above all else.

This focus of customer driven quality of the overall product throughout its service life has shaped the development of all the company's processes and the everyday culture of the entire organization.

# How the quality focus influences every day practice

## Site selection

Property development starts with purchasing land; the company will not buy a site that compromises the quality of the dwellings that can be built there, even if a considerable sum of money has already been spent investigating the feasibility of the project.

## Dwelling design

Since the early days dwelling design has always been underpinned by market research to assess customer feedback on recent sales and to identify product developments that purchasers would favour. This market research based development of product design and product improvement enabled the company to lead the field in terms of price and reputation.

Everyone including the MD gives a great deal of attention to the design of each dwelling in a development. As with every other decision, the layout of dwellings is driven by a desire to create excellent homes. In a situation where the outlook in certain directions is compromised by a nearby building or structure, even though more dwellings might be fitted on a floor plate to maximize potential income, the company will build fewer, better quality dwellings and avoid the construction of any inferior units.

## Managing design related construction quality risk

The focus on quality has translated into the company taking a lead role in attacking quality problems which have arisen from the introduction of new materials, declining craft skills and the need for new craft skills in the application of new materials in the late 1980s.

When the waterproofing of shower recesses was changing from traditional detailing in metal (copper or zinc) shower trays to polymer based materials, the risk of waterproofing failures suddenly increased. Mirvac immediately invested in laboratory based research and site trials to find the most robust, commonly available solution. Then to ensure this was successfully implemented, standard guidelines were developed and designers, construction supervisors and applicators were appropriately trained. An important part of the company's strategy was to employ and train apprentices to play a key role in the site coordination process. Employing this range of strategies the company avoided the extremely costly problems that most builders encountered arising from water leaks in wet areas.

In a similar way the company took a scientific, risk management based approach to issues such as the selection of natural stone to minimize the opportunity for customer complaints and the development of standard, robust detailing wherever possible to reduce construction risk that arises from unnecessary architectural variety in common areas of detailing. This use of independent laboratory based research to underpin the development of robust detailing and the specification of preferred materials that are known to be reliable was very unusual in the Australian construction sector at the time – though many companies are now emulating these practices some 10 or 15 years later.

## Minimizing supply chain risk

The quality focus led to the company having a narrow supply chain, using the services of a short list of preferred subcontractors at a time when other companies competing in the sector were market driven in their thinking, selecting suppliers substantially on cost in the belief that they could manage the production risks associated with using weak but cheap suppliers though their site staff.

During the period, Mirvac grew relatively gradually, it had a policy of using a limited number of subcontractors who it knew well, who understood the company's quality focus and the way things were done. This strategy helped the company avoid the pitfall of expensive rework arising from poor construction quality.

In recent years Mirvac has grown quickly and it is faced with new challenges, enlarging its supply chain to balance its growth and training new members so that they can build at the quality that is consistent with the company's history and expected by its customers.

## Customer handover

A final inspection of every unit is undertaken immediately before the keys are handed over to the customer. This is done to make sure that each unit is spotless in every way so that the purchaser's first impression is one of excellence. Once this final inspection is completed, the work team fixes a certificate on the door to signal to the sales person handing over the keys to the new owners that all is ready.

If the final inspection certificate does not confirm that the dwelling is not completed to the Mirvac Standard, the new owners are not allowed in, as the company does not want to risk letting them in to view a product that is not perfectly presented.

# Conclusion

This case study presents a brief snapshot of a development and construction company that drives product design and improvement through market research, that details design for construction in a manner that minimizes design related construction risk, and one that has a narrow supply chain to maximize intellectual and social capital in order to minimize quality risk. This thorough, total quality based strategy to its business has helped Mirvac to prosper in a market sector where others have struggled.

## Review questions

1. List and discuss the many different ways in which the commitment to quality has shaped the business of Mirvac.
2. Generally quality systems are formally documented, yet this is an example of a company which during its growth phase achieved excellent quality without the formal systems. List the ways in which this success was achieved.

## Acknowledgements

The authors gratefully acknowledge the assistance of John Carfi, the construction director of Mirvac and retired director Peter A Carter for their assistance in putting this case study together.

## Reference

Thomas, R., Marosszeky, M., Karim, K., Davis, S. and McGeorge, D. (2003) *Enhancing Project Completion*, Australian Centre for Construction Innovation, ISBN 0733420613, pp. 28.

# Triple bottom line reporting as a catalyst for organizational culture change at Landcom

# Company background

Landcom was originally set up in 1975 as the Land Commission of New South Wales, Australia. It is a government-owned property developer, operating on the open market in a competitive environment. The organization's traditional focus has been on the development of serviced residential land for new homes in fringe metropolitan locations and it remains a significant player within this market. In recent years, Landcom has extended its operation into urban renewal or redevelopment projects and to include commercial and industrial projects.

The corporation's vision is to 'take the lead in creating better communities' by adding value to a portfolio of strategic and complex projects that result in high-quality urban design and demonstrate best practice in social, environmental and economic sustainability.

On 1 January 2002, Landcom was established as a state-owned corporation, constituted under the Landcom Corporation Act 2001. Today, Landcom's value to government lies in its ability to:

- take on strategic and/or complex projects in which the private sector may be either unwilling or unable to become involved;
- deliver more positive development outcomes by facilitating creative partnering arrangements on selected projects between government and the private sector; and
- using its trusted position in government and in the development industry to create high-quality urban developments that lead practice in key areas of excellence (such as urban design, sustainability and housing affordability).

As a Government owned developer, Landcom needs to differentiate itself from the private sector players. Leading by example in urban design, social and environmental sustainability and triple bottom line (TBL) performance reporting are just a few aspects of Landcom's differentiation strategy.

# The importance of corporate commitment

In 1999, Landcom appointed a corporate Environment Manager to develop an Environmental Management System (EMS) compliant with ISO 14001. Once the EMS was certified, the Board and senior management shifted their focus to establishing Landcom as an industry leader in sustainability. This led to the adoption of a triple bottom line business model and reporting framework. This decision had the full support and sponsorship of the Board and senior executive and was driven by them.

This was critical as the TBL business model requires a major change in corporate culture and substantial development investment and effort. It required additional resources and expertise; significant changes to the 'business as usual' model; willingness to publicly report non-financial performance; and most importantly, the confidence to face any criticism that may arise as this was the first government sector agency to adopt such an approach.

A corporation's true values can often be determined from its organizational structure. Landcom's sustainability drive began with an environmental agenda and started with a position on the organizational chart containing the words 'Environment Manager'. Since 1999, the organizational structure has gone through several iterations, reflecting management's priorities. The initial environmental agenda has evolved into a holistic sustainability approach and the current structure clearly demonstrates Landcom's commitment, senior level interest and involvement.

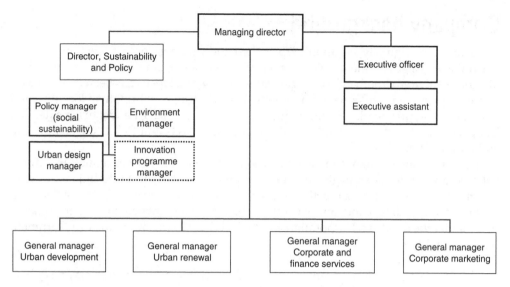

**Figure C14**
Landcom's current executive unit organizational structure – 2004

# Why implement triple bottom line (TBL) reporting?

The foremost consideration for TBL reporting is that it must reflect the core values of the enterprise; it must add value to business decision-making and to the everyday operation of the business. If TBL adoption is motivated by external drivers – something done for someone else, the integrity, consistency and continuity of TBL will inevitably become compromised. The key is to integrate TBL performance values into business processes and procedures, starting with the core business.

For Landcom, TBL was clearly in line with the corporate vision of 'creating better communities'. It needed a clear strategy for delivering measurable public benefits as a way of differentiating itself from the private sector, proving its value as a government owned developer. In addition to achieving increased transparency, and consequently increased trust among its stakeholders, the corporation wanted to enhance its reputation as a good corporate citizen.

Furthermore, the corporation had been implementing various good urban design, environmental and social measures in the planning of its projects for some years, however it had never measured the effectiveness of its initiatives against a consistent set of key performance indicators designed specifically for non-financial initiatives and outcomes. The adoption of TBL essentially underpinned the direction that Landcom had already committed to.

# Integrating TBL into business decision-making

Landcom's corporate plan provides the link between its strategic commitment to sustainability and its day-to-day operations. Sustainable development and TBL performance reporting are among Landcom's 'top ten' priorities and sustainability considerations permeate the business plans for each of its business divisions. Furthermore a culture of accountability has been created at an individual employee level, by specifying TBL indicators and targets in all performance agreements.

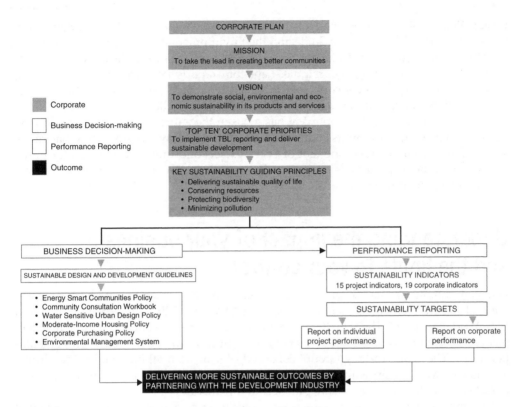

**Figure C15**
Overview of corporate integrated TBL decision-making framework

While TBL reporting alone does not guarantee sustainable outcomes, it can facilitate the integration of sustainability criteria into business decision-making by allowing management to observe the real social, environmental and economic outcomes of their decisions.

Landcom sustainability philosophy follows four key principles:

- To deliver a sustainable quality of life (cultural, economic and social well-being).
- To conserve resources.
- To protect biodiversity.
- To minimize pollution.

Based on these key principles, Landcom has developed TBL performance indicators and targets that are relevant to its core business. Using these, minimum benchmarks have been established for all its projects. The indicators and associated targets are compulsory requirements for all development projects and have been integrated into all stages of every project from acquisition to planning, design and delivery.

The performance indicators and targets are incorporated into all tender specifications and form part of the evaluation of all tender bids. For service providers such as project managers, civil works suppliers and landscape contractors, these criteria are also used during the pre-qualification process, ensuring that Landcom works exclusively with organizations that have compatible values and sufficient sustainability experience and credentials.

The corporation has devised sustainable design and development guidelines for design consultants, developers and builders. These guidelines are supported by specific and practical policies such as its Energy Smart Communities Policy, Water Sensitive Urban

Design Policy, Moderate-Income Housing Policy, Stakeholder Consultation Policy and Workbook, Social Sustainability Guide and the Environmental Management System.

Each project is monitored against relevant TBL performance indicators and targets through the TBL reporting programme. Landcom reports its performance publicly to make the Corporation more accountable, this also allows stakeholders to question results, understand the constraints faced and challenges the organization to strive harder.

Landcom's journey to TBL reporting has focused the entire organization not only on its performance, but also on its commitment to sustainability. TBL reporting has made sustainability an integral component of every project and therefore a key factor in the evaluation of the corporation's performance – as well as that of its consultants, contractors, builders and developer partners.

# Understanding the impact of your business and the limits to your control

Landcom commenced its TBL programme by engaging its own staff through an extensive internal stakeholder engagement process. It established a 'TBL Reference Group' and a 'TBL Working Group'. The 'TBL Reference Group' comprised senior and middle management representatives and its role was to provide direction and advise to the implementation team. The 'TBL Working Group' consisted of staff from all internal business units and its role was to help with problem solving, process integration, ensuring that proposed changes to procedures were practical and realistic.

The impact identification process requires input from a team of internal personnel across a wide range of work disciplines. Landcom used its 'TBL Working Group' to identify impacts and to determine their relative significance.

Following impact identification, the 'TBL Working Group' determined potential indicators for measuring each of the impacts. During this impact identification process, over 90 potential performance indicators were considered. However, as there is no value in measuring performance based on an impact that is outside the control of the organization to change (e.g. due to limited control over its cause), Landcom's control over the cause of the impact was assessed.

The next step was to understand and agree on what sustainability means to Landcom. For Landcom, its commitment to sustainable development means the Corporation strives to have no adverse impact on the environment, economy or the community with its long-term goal of adding net value.

This internally agreed position brought a degree of clarity, ownership and support for the implementation of the TBL programme. Once this was in place, Landcom was then ready for dialogue with its external stakeholders.

# Process for developing the TBL framework

In working towards the development of the TBL framework the TBL Working Group with guidance from the TBL Reference Group went through a comprehensive development process, which included:

- Identifying core business processes.
- Identifying the potential environmental, social, socio-economic and financial impacts of the core business processes.

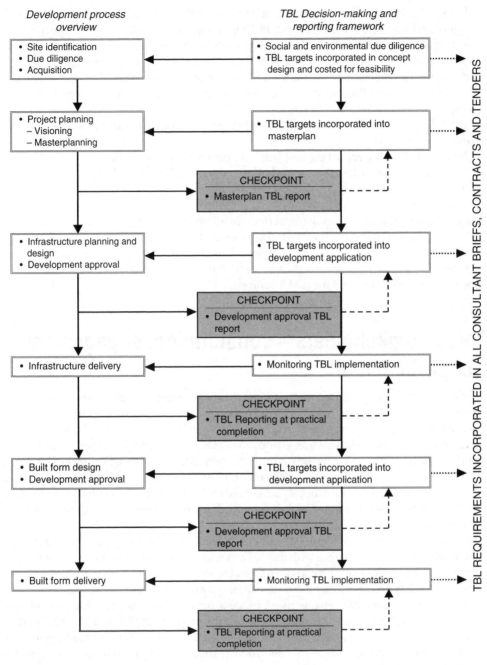

**Figure C16**
Landcom core business processes map

- Selecting meaningful performance indicators that can show performance over time and can be measured with relative accuracy.
- Gap analysis to determine gaps between existing data and processes and those required for TBL performance reporting.
- Setting meaningful and measurable targets.

- Identifying steps within core business processes where performance indicators, targets and appropriate monitoring and reporting mechanisms must be specified.
- Designing suitable data collection, interpretation and communication mechanisms.
- Revising all processes, procedures, standard contracts and other documents.
- Specifying the relevant performance indicators, targets, the recording and reporting responsibilities in all performance agreements for employees, contractors and suppliers.
- Retraining and capacity building for employees, contractors, suppliers and customers.
- Closing the loop by reporting and tracking performance over time.
- Continuous review of relevance and effectiveness of the TBL indicators, targets and reporting framework.

Apart from the first few steps listed above, the development process did not unfold in this order, as most of the steps occurred concurrently.

Stakeholder engagement is not mentioned as a distinct step because engagement with key stakeholders is a continuous process that must run concurrently with all these steps to provide insight and mutual learning.

# Role of stakeholders – consultation, engagement and participation

Most good corporate practices and all sustainability reporting guidelines advocate the importance of engaging key stakeholders. In the absence of genuine community consultation, well-organized action groups tend to move in and hijack the agenda. The political views and objectives of these action groups may have nothing to do with the community agenda and, in many cases, agreeing on outcomes with a vocal action group may not deliver the best outcomes for the greater community. This makes it difficult to both find, and engage with, genuine community stakeholders. To make matters even more difficult, running effective stakeholder consultation in today's time poor society is an ever more difficult task.

Nevertheless, corporations must devise effective methods of, firstly identifying the real stakeholders and then engaging with them. Perhaps the best way of looking at this process is to think of it as an opportunity to learn from the stakeholders and, in turn, to demystify some of the misconceptions that may exist among them about your organization.

The external stakeholder engagement process commenced shortly after this internal consultation and both processes ran parallel for 18 months, until the set of TBL performance indicators was finalized. It should be noted that there was a surprisingly close correlation between the indicators identified through both internal and the external stakeholder consultation processes.

Initially, over forty external stakeholders were identified from Federal, State and Local governments, industry and non-government organizations. The stakeholder engagement process started with a very high level of participation but it proved difficult to maintain continuity of engagement with such a large group. In addition, many of those who were identified as potentially interested parties proved to be either not particularly interested in Landcom or were not directly impacted upon by Landcom.

As a result, it was decided to create a smaller group of key stakeholders who were either directly impacted by or demonstrated a strong interesting Landcom's core business. This smaller external stakeholder group included three development industry partners, two not-for-profit green groups, two not-for-profit community groups, representatives of the National Parks and Wildlife Service (NPWS), the Department of Infrastructure Planning Infrastructure and Natural Resources (DIPNR), and representatives of Landcom's Shareholder Ministers and the Portfolio Minister.

However, even with this smaller number of external stakeholders, participation proved difficult due to competing demands on stakeholders' time and resources. Landcom has subsequently changed its stakeholder engagement programme to a less time consuming and individually customized programme. Part of this programme has included the appointment of an independent consultant who conducts face-to-face or telephone interviews with key stakeholders following the publication of each annual TBL Report.

Landcom offers a nominal donation to not-for-profit organizations for their participation, equivalent to an employee's hourly or daily rate. The donation has been criticized by some as compromising the integrity of the consultation processes. However, the experience has been that donations to not-for-profit organizations have not tempered the stakeholders' keen eye and criticism.

Landcom considers that corporate donations to not-for-profit organizations are justifiable compensation for the time and effort involved and not an attempt to buy or influence opinion.

# Selecting indicators

After identifying the potential business impacts and associated indicators through the stakeholder consultation, Landcom had a list of over 90 indicators which could not be effectively measured given the usual business resource constraints. Therefore, an 'indicator selection test' was designed to test each of the identified indicators against a set of criteria. The criteria were of equal weight and were chosen to ensure that the final indicators would be robust and reliable.

The selection criteria were based on:

- the significance and relevance of the indicator to Landcom's core business,
- whether the indicator enabled Landcom to demonstrate a change in its performance over time,
- whether indicators were 'input' or 'output' indicators,
- the ability of Landcom to exercise direct control or influence over the indicator,
- whether the indicator could be easily understood and communicated,
- whether clear and measurable targets could be developed to illustrate Landcom's performance against the Indicator, and
- the likelihood of data being readily available to enable the reporting of Landcom's performance against the indicator and the ability to retrieve accurate and meaningful data from third-party sources (contractors and builders, etc.).

In setting the criteria, it was considered essential to ensure that the TBL reporting process included indicators that were capable of clearly demonstrating changes in performance over time.

It was also important to select and focus on indicators which describe either inputs or outputs. For example, an input indicator would be the amount of money invested in energy conservation technologies, whereas an output indicator would measure the effectiveness of these technologies in terms of the amount of energy and greenhouse gases saved.

The majority of Landcom's chosen TBL indicators report on outputs. No single indicator can demonstrate sustainability performance – the indicators must be assessed as a set. Furthermore, each indicator can only show a small picture. Interpretation and understanding of the indicators will always be influenced by the perception of the reader and changing external circumstances. For example, benchmarks will vary over time in accordance with changing community, government, employee and management attitudes.

Finally, the indicators selected for Landcom's TBL reporting system are not the only criteria used by the Corporation in carrying out its commitment to sustainable development. Instead, they are intended to reflect the many activities undertaken by Landcom during the year, and they are subject to regular review to ensure that they remain relevant to those activities and to community expectations.

# Data gap analysis

Following the selection of the indicators, a gap analysis was carried out to determine how practical it would be to collect data and report on each of the preliminary indicators. The gap analysis was essentially a search for existing data and a determination of practical data collection processes. The gap analysis identified:

- what type of baseline information was available
- who collected the information
- how frequently the information was collected
- how accurate the information was
- how the information was stored
- how the information was presented
- how much it cost annually to collect or buy the information.

The gap analysis revealed that most indicators did not possess any existing data and that new mechanisms for data collection would have to be funded and implemented. Consequently, Landcom committed itself to expanding its data collection obligations in an effort to meet the requirements of its TBL indicators. These new data pathways applied principally to project reporting and have generated substantial amounts of information about the environmental, social and economic impacts and benefits of Landcom's urban development and renewal programmes.

# Landcom's indicators and targets

Once TBL indicators were selected and following the completion of the gap analysis, it was time to set performance targets. TBL targets were determined through several indicator specific research programmes – investigating current industry performance, current Landcom performance and best known performance. Stretched targets were then set for those indicators where a measurable target was possible.

The following table lists the indicators which were selected to measure performance on Landcom's development projects.

**Table C1** Landcoms' indicators and targets

| Indicator | Target (overall Landcom performance) |
| --- | --- |
| 1. Water Cycle Management | (a) 100% of projects to have project-specific WSUD strategies. (The strategy should be appropriate to the size, scale, sensitivity and location of the project.) For detailed case studies and specific requirements, refer to Landcom's Water Sensitive Urban Design Policy on http:// www.landcom.nsw.gov.au under Environmental Initiatives/WSUD Strategy. |
| • Water conservation | (b) Combination of water efficiency and reuse options – achieve 40% reduction on base case.<br>(c) Public domain irrigation must be from non-potable sources and designed with water efficiency in mind. |
| • Pollution control | (d) 45% reduction in the mean annual load of total nitrogen (TN). Based on EPA best practice guidelines.<br>(e) 45% reduction in the mean annual load of total Phosphorus (TP). Based on EPA best practice guidelines.<br>(f) 80% reduction in the mean annual load of total suspended solids (TSS). |
| • Flow management | (g) Post-development storm discharges = pre-development storm discharges for 1.5 year ARI event. The purpose of this is to minimize the impact of frequent events on the natural waterways and to minimize bed and bank erosion. |
| 2. Moderate Income Housing | 7.5% of Landcom's total product is moderate income housing. Moderate income housing is delivered where commercially viable consistent with existing Landcom's Moderate Income Housing Policy. |
| 3. Community Consultation | (a) 100% of projects have Community Consultation Plans developed and implemented in accordance with Landcom's 'Stakeholder Consultation Workbook'.<br>(b) 100% of identified stakeholder groups being engaged through the consultation and participating. |
| 4. Community Facilities | Targets are to be determined on each project based on the community demographic and needs. |
| 5. Welcome Programme | (a) All projects greater than 200 dwellings must have a welcome programme.<br>(b) Where there are welcome programmes, initial contact to be made within 14 days of residents moving in. |
| 6. Consumer Education on Sustainable Living | (a) Develop educational programme and materials for the project.<br>(b) 100% of projects have marketing material that includes consumer education on sustainable living. |
| 7. Percentage of construction and demolition materials reused on-site and elsewhere | (a) Achieve 95% recovery (reuse and recycle) of total construction and demolition waste materials generated from sum of civil works contracts completed in that year.<br>(b) Achieve 76% recovery (reuse and recycle) of total construction and demolition waste materials generated from sum of building projects delivered in that year.<br><br>Note – target is based on Waste Avoidance & Resource Recovery Strategy 2003 – published by Resource NSW. |
| 8. Energy efficient design (residential and commercial buildings) | For detailed case studies and specific requirements, refer to Landcom's Energy Smart Communities Policy on http://www.landcom.nsw.gov.au under Environmental Initiatives/Energy Smart Communities Policy.<br>(a) All Landcom projects to achieve minimum of 40% conservation of greenhouse emissions, compared with base case. As per the Energy Smart Communities Policy. |

(*continued*)

431

**Table C1** *(continued)*

| Indicator | Target (overall Landcom performance) |
| --- | --- |
| | (b) Landcom projects over 500 dwellings (or projects with commercial component or town centre) achieve more than 40% conservation of greenhouse emissions, compared with base case (including a %age of renewable energy generation). As per the Energy Smart Communities Policy.<br>(c) Average NatHERS (National House Energy Rating Scheme) rating for residential buildings is 4.5*. As per Landcom's Energy Smart Communities Policy.<br>(d) Average rating for all commercial projects is 4.5* under the Australia Building Greenhouse Rating tool. As per Landcom's Energy Smart Communities Policy. |
| 9. Design guidelines for built form (e.g. guidelines for courtyards lots, corner lots, etc.) | (a) 100% of projects to have design guidelines to control the siting of dwelling, garages and fencing and incorporate appropriate building elements which contribute to the streetscape quality and promote casual surveillance.<br>(b) All design guidelines produced by Landcom to include, as a minimum, the following sustainability criteria:<br>• Minimum solar access zones (generally indicates where private open space should be located) in accordance with Solar Access for Lots Guidelines for residential subdivision in NSW.<br>• Energy efficient design – 4.5 star NatHERS rating and gas reticulation.<br>• Water conservation (AAA rated shower roses, 3/6 litre flush toilets).<br>• Construction waste minimization and on-going recycling facilities. |
| 10. Sustainable energy technology use | (a) All dwellings are to be fitted with gas boosted solar water heaters sufficient to meet 60% of annual hot water requirements except where:<br>• No gas supply is available; a heat pump hot water system should be fitted instead.<br>• Solar panels cannot be suitably positioned (for example because of orientation or overshadowing). In which case single dwellings should be fitted with an Australian Gas Association (AGA) registered 5* gas water heater.<br>• A cogeneration system is supplying the hot water.<br>(b) All projects greater than 500 dwellings (or projects with commercial component or town centre) include a percentage of on- or off-site renewable energy supply. As per the Energy Smart Communities Policy. |
| 11. Native vegetation management (net loss or gain). | No net loss for high conservation value vegetation.<br>(a) No loss of length in Category 1 and 2 streams.<br>(b) Category 1 – Environmental Corridors – greater than 40 m riparian corridor on either side (from top of bank). |
| 12. Riparian corridor management (net loss or gain). | (a) Category 2 – Terrestrial and Aquatic Habitat — 20 m riparian corridor + 10 m buffer (from top of bank).<br>(b) Category 3 – Bed and Bank Stability and Water Quality — 5–10 m riparian corridor (from top of bank). |
| 13. Conservation of indigenous heritage (including items, values, and places of cultural significance) | (a) 100% of significant items and places in Landcom projects conserved.<br>(b) Consultation occurs for 100% of projects with indigenous heritage issues.<br>(c) 100% of projects with indigenous heritage issues have Conservation Management Plans. |
| 14. Conservation of non-indigenous heritage | (a) 100% of significant heritage items and places conserved (unless where theirs is safety or contamination issues).<br>(b) 100% of projects with non-indigenous heritage have Conservation Management Plans. |

**Table C1** *(continued)*

| Indicator | Target (overall Landcom performance) |
|---|---|
| 15. Number and nature of non-compliances | (a) 100% of projects achieve full compliance with the Protection of Environment Operation (POEO) Act – Landcom actions (i.e. Penalties issued in Landcom's name). |
| | (b) 100% of projects achieve full compliance with Protection of Environment Operation (POEO) Act – contractors' actions. |
| | (c) 100% of projects achieve full compliance with other environment, OH&S and planning legislation – Landcom actions. |
| | (d) 100% of projects achieve full compliance with other environment, OH&S and planning legislation – contractors' actions. |
| | (e) 100% of Landcom contracts have environmental audits carried out. |
| | (f) 100% of environmental audit scores are greater than 75%. (i.e. every audit score, not the average of all audit scores). |

# Role of consultants

External consultants cannot implement a system or programme that is, firstly, dependent on a shift in organizational culture. To achieve successful cultural change, an organization must understand, appreciate and own the vision and the objectives of the change required. Implementation must be done internally, by involving as many people across the organization as practicable.

The effective implementation of Triple Bottom Line programmes, social and environmental reform and sustainability management systems are concepts that rely on the premise of organizational culture change. Accordingly, while external consultants certainly have a role, this role is not in implementation. Consultants can add value in areas of research, benchmarking, peer review and by providing a wider range of experience due to their exposure to different industries and specialized expertise.

In Landcom's TBL programme implementation, consultants were employed to deliver such services as:

- Initial research on national and international triple bottom line guidelines.
- Research on triple bottom line programmes within other companies and industries.
- Preparing a discussion paper to facilitate internal discussion.
- Facilitation of internal discussion on how to approach triple bottom line performance monitoring and reporting.
- Facilitation of stakeholder engagement programmes.
- Providing peer review during implementation drawing on consultants' diverse knowledge and exposure to other industry sectors, private sector companies and government organizations.
- Benchmarking Landcom's performance against others within development industry sector and against other industries both nationally and internationally.
- Benchmarking Landcom's TBL programme against criteria set by United Nations Environment Programme (UNEP) and Global Reporting Initiatives (GRI).
- Conducting ongoing stakeholder interviews and surveys.
- Independent auditing and report verification.

# Global reporting initiative

The Global Reporting Initiative (GRI) is an internationally recognized group whose mission is to develop Sustainability Reporting Guidelines (SRG) that are capable of universal application. These guidelines are intended for organizations wishing to report on the economic, environmental and social dimensions of their activities, products and services.

Initially developed for use by public corporations, it is expected that government and non-government organizations will increasingly adopt the guidelines. However, the current GRI framework seems to be more suitable for extractive and manufacturing industry sectors and multinational operations.

Since Landcom is a state Government owned corporation with operations only in NSW, Australia, indicators such as 'child labour in third world operations' are not relevant. Landcom's 'product' is creating better suburbs and communities. It is not possible to normalize Landcom's product/service performance relative to a single unit of measurement – it cannot measure its performance based on a tonne of product.

Accordingly, the majority of Landcom's TBL indicators have been specifically developed to reflect Landcom's core business. While Landcom's TBL report is not strictly in accordance with the requirements of the Sustainability Reporting Guidelines, the Corporation has generally followed SRG principles and criteria.

The relevance of Landcom's TBL reporting, in the context of SRG principles, is described below:

## Transparency

Relevant policies, strategies, assumptions, processes and reports are referenced throughout Landcom's TBL reports. Reference documents are available in electronic format on an accompanying CD-ROM or on Landcom's website.

## Inclusiveness

Landcom has undertaken an extensive stakeholder engagement programme in conjunction with the development of its sustainability indicators and reporting regime and continues to engage with its stakeholders by informing them of key decision-making steps in the TBL reporting process. The relevance and applicability of the indicators will be monitored through continuing consultation.

## Auditability

All reported information is methodically recorded and analysed. External auditors have access to all databases, calculations, analysis reports, relevant policies and procedures.

## Completeness

Data for all indicators is reported.

## Relevance

All indicators were selected for their relevance and significance to Landcom's business. They encompass all of Landcom's core business activities.

## Sustainability context

This SRG principle recommends that the reporting regime addresses how an organization's performance affects economic, environmental and social capital formation (or depletion) at the local, regional and global levels. At this stage, Landcom's reporting system is not yet sufficiently developed to enable its sustainability performance to be assessed at a macro level with any reasonable degree of accuracy. Accordingly, this principle will be revisited in future reporting years.

## Accuracy

Considerable effort has gone into ensuring that a high level of accuracy is maintained in the reported data. Accuracy will be further enhanced through proposed mechanisms for capturing reliable details from third-party service suppliers.

## Neutrality

All data is presented without bias. In the first year of reporting there were gaps in the data which was due to the unavailability of data from third-party service suppliers, such as builders and contractors, or because there was insufficient time for Landcom to fully implement data-collection mechanisms prior to the production of the report.

## Comparability

Defining the reporting scope and procedures ensure consistency between projects and reporting years, enabling Landcom's performance trend to be compared over time.

## Clarity

A summarized/simplified version is published in hard copy for the general public. Detailed data tables with calculation notes and other relevant references are available in electronic format for more technical readers.

## Timeliness

Landcom publishes its TBL reports annually based on the financial reporting year.

# Logistics of data collection and reporting

Reporting is not possible without data collection and analysis tools and procedures. To calculate performance against each of Landcom's TBL indicators, several data elements are needed. Landcom collects over 140 data elements to calculate its performance against the 15 chosen project indicators. In terms of its financial indicators, Landcom's existing financial database was adapted to allow TBL data collection and reporting.

Methodologies for calculating certain indicators have been developed where industry standards did not exist. In addition, processes for data collection, verification and quality control have had to be developed and implemented. All standard contracts have had to be revised to specify the provision of data by third party service providers.

# Communicating performance or marketing opportunity?

TBL report, a corporate positioning medium or a sales promotion.

Spin doctoring is the greatest danger that faces an organization in communicating its performance. Just as profit or loss should be reported honestly to protect investor interests, social and environmental performance should also be presented honestly, transparently and without bias. The community has grown sceptical to claims of big business and consumers in general have become more astute and discerning. If a reader detects bias, public reporting can cause more damage to the reputation of the business than not reporting at all.

Therefore, the important message in communicating TBL performance is to leave the reader with the assurance that the reporting corporation takes responsibility for both poor and excellent performance.

The report should be used to communicate not only performance but also the dilemmas, challenges and risks that the business faces in making its key decisions. The actions either already taken or under consideration to address poor performance should be clearly communicated to and help the reader understand the challenges being faced. It is important to admit, wherever relevant, when an organization doesn't have all the answers.

# Conclusion

Corporate sustainability and triple bottom line performance reporting requires significant changes to the conventional business model and it is not a decision to be made lightly. Integrating sustainability in business decision-making requires senior level commitment, significant resources, a willingness to accept responsibility for your business impacts and most importantly a willingness to right the wrongs.

# Review questions

1. Landcom as a provider of land to developers has put itself into the position of influencing the downstream products of developers and builders. Methodically review what is being achieved and discuss the limitations of such an approach, are there some environmental quality issues that cannot be approached in this way?
2. Discuss the relationship between and quality and the TBL reporting system implemented by Landcom in this case study.
3. In the context of this particular case, explain how the managers in Landcom have implemented a TBL system of reporting from scratch.

# Acknowledgement

This case study was prepared by Armineh Mardirossian, Director, Sustainability and Policy, Landcom.

# Best value in Harrogate Borough Council

# Background

'Best value' is a UK Government initiative which places a duty on all local councils and authorities to deliver the most economic and efficient services possible. Councils must report to their public and the Government each year on their performance, in addition to reviewing all their services to identify and achieve continual improvements. In this way the Government has challenged local councils to look at the way they deliver services and raise their quality at a reasonable cost.

This case study looks at the way Harrogate Borough Council in North Yorkshire – John Oakland's own district – has addressed the needs and challenges of Best Value through a five-year review programme of:

- Culture and community safety.
- The local economy; local taxation and benefits.
- Managing the council; access to services.
- Public health and protection; the local built and natural environment.
- Highways and traffic management; housing.

# The vision, objectives, etc.

The council's vision is to 'provide civic leadership and high quality, cost-effective services to fulfil the aspirations of the community, local people and visitors'. In working towards achieving its long-term vision, the council has identified three broad aims and nine key objectives (Table C2).

**Table C2** Harrogate Borough Council's three broad aims and nine objectives

| Aims | Objectives |
|---|---|
| A sustainable environment | To contribute to a transport infrastructure that ensures that people and businesses can travel safely and conveniently |
| | To work in partnership with the health agencies to protect and improve the general health of people in the district by providing a range of environmental health services and promoting individual well-being |
| | To protect and improve the natural and built environment and to promote sustainable development across the district |
| Building local communities | To work with others to build a prosperous and robust local economy. |
| | To work in partnership with the police and other agencies to reduce crime and the fear of crime in the district |
| | To seek the views of local people, to respond to them and to keep them informed through timely and well-presented information |
| | To facilitate the provision of a range of good quality housing appropriate to all ages and income levels in our community |
| | To ensure the provision of a range of leisure, cultural and amenity services which meets the needs of all individuals and communities in the borough and benefits both residents and visitors |
| Delivering services for all | Continue to be a well-managed, responsive authority that meets the needs of all its customers |

# Core values

The council's vision and nine corporate objectives are supported by the following core values:

**Involvement:** We will involve local people in the council's decision-making process through consultation, discussion and engagement initiatives, both corporately and at a service level.

**Fairness:** We will work towards fairness and equality of opportunity for all people regardless of age, culture, disability, economic status, gender, race, religion or sexuality.

**Openness:** We will ensure that the decisions we make are clear, open and honest; that we will listen to people and ensure that people have the right to challenge our decisions.

**Respect:** We will treat people with dignity and courtesy in providing services which reflect and celebrate local diversity, local need and provide choice.

**Sustainability:** We are committed to giving people a better quality of life now, without leaving problems for future generations either here or elsewhere.

# Principles

The council is committed to seven key long-term principles:

**Quality services:** Providing responsive, customer-focused and efficient quality services, accessible to all, which try to meet the needs of all our customers, including vulnerable groups.

**Effective management:** Managing the authority's financial and other resources effectively to achieve its service commitments within agreed budget limits.

**Integrity and accountability:** Maintaining the highest standards of honesty, integrity and accountability and demonstrating fairness and equity in dealing with customers, employees and specific interests.

**Employee development:** Developing employees' potential, their commitment to public services and the contribution they can make to improve the services that the council provides.

**A prosperous economy:** Working to support the development of a balanced local economy with rising prosperity shared by all.

**A quality environment:** Preserving and improving the health and the quality of life by protecting and enhancing the natural and built environment of the district.

**Community leadership:** Providing community leadership and focus so that the community's views and opinions are taken into account by the council's actions whilst working to sustain and enhance pride in the Harrogate district.

# Quality of life

Working together – councils, voluntary sector, businesses, health agencies, Police, etc. – to achieve a sustainable society which has, at its core, the national quality of life agenda. (Department of the Environment, Transport and the Regions – 'A Better Quality of Life').

That agenda has the following characteristics:

## Economic

- Combating unemployment.
- Encouraging economic regeneration.

## Social

■ Tackling poverty and social exclusion.
■ Developing people's skills.
■ Improving people's health.
■ Improving housing opportunities.
■ Tackling community safety.
■ Strengthening community involvement.

## Environmental

■ Reducing pollution.
■ Improving the management of the environment.
■ Improving the local environment.
■ Improving transport.
■ Protecting the diversity of nature.

Harrogate Council has incorporated the above agenda into its priorities, plans, budgets and targets. As both an employer and a provider of services, it monitors and reviews the quality of life agenda both as part of its Best Value reviews and in its approach to every-day management.

# The Corporate Action Plan and Best Value Performance Plan

The Corporate Action Plan sets out the planned actions and targets which deliver the council's corporate objectives and priorities. It enables the authority to look beyond immediate issues and problems and to plan ahead for the longer-term future of the district. The Corporate Action Plan links into both the Best Value Performance Plan and the service and business plans prepared by the council departments to deliver their part of the council's corporate plans and targets.

Some of the actions in the Corporate Action Plan are designed to meet a local need or policy issue while others are to address the council's current performance. All of them are agreed by the council for implementation, following consultation with local communities and partners in the district. The council reviews the Corporate Action Plan twice a year to measure the progress being made in meeting the council's longer-term vision and strategy through the achievement (or not) of service actions and targets each year.

The Corporate Action Plan is divided into action tables – one for each of the council's corporate objectives. The council agrees a number of key priority areas for action to help deliver each of its corporate objectives and these are set out in detail in the plan, together with the actions and targets planned under each priority area, and the links into the relevant service and other council plans. An example part of the plan's details under 'Sustainable Environment' – Highways and Traffic is given in Table C3.

Details of the council's longer-term priorities and targets are set out in a separate 'corporate strategy' document.

The council's budget for the financial year is explained in detail in a separate 'Budget' document and each year the council allocates funding in its General Fund Revenue Budget to enable it to deliver its annual corporate priorities and targets. Details of the council's funding of corporate priorities is set out in the Best Value Performance Plan.

**Table C3** Part of the 'Sustainable Environment' Corporate Plan: Highways and Traffic

**Objective No. 1 – To contribute to a transport infrastructure that ensures that people and business can travel safely and conveniently**

| Action | Target | Responsible officer | Revenue budget ref. | Service ref no. | Other plan refs |
|---|---|---|---|---|---|
| **Priority No. 1.1 – Encouraged use of sustainable forms of transport** | | | | | |
| 1.1.1 Phase in the use of cleaner fuels in council vehicles | Acquire four council vans, which use liquid petroleum gas (LPG) fuel | Name 1 | RB3 | DH21 | HH1, HH10 |
| 1.1.2 Continue to develop the North Yorkshire Concessionary Fares scheme in the Harrogate District | ■ Review the first year of operation of the scheme by September<br>■ Issue travel tokens to eligible residents by June | Name 2 | RB8 | DT12 | TS1, TS2 |
| 1.1.3 Influence the draft North Yorkshire guidance on parking, transport assessments and travel plans | Work with other agencies to influence the guidance | Name 3 | RB4 | DT03 | TS1, TS2 |
| 1.1.4 Input into the local transport plan | Make representations by 31st March next year | Name 4 | RB4 | DT03 | TS1, TS2 |
| 1.1.5 Encourage more people to make use of public transport and encourage more walking and cycling | ■ Work with others to complete the Harrogate bus station<br>■ Undertake further studies into providing more rail halts<br>■ Implement the Harrogate and Knaresborough Cycling Strategy | Name 5<br>Name 4 | RB8 | DT10 | TS1, TS2 |

A Best Value Performance Plan is generated for each coming financial year. This provides a snapshot of the council's performance and achievements for the previous year – what worked/what did not – and looks forward to what the council needs to do to meet its commitment to provide high-quality, cost-effective services which meet the needs of the people of the Harrogate District.

The objectives and priorities are stated together with the long-term issues facing the district. The council's performance has improved in a number of areas and, where it has not improved, the council has taken action to address this. On the Government's national top 11 indicators for District Councils, Harrogate's performance in the year of the case study preparation was in the top quartile on five indicators, average performance on three indicators and below average performance on three indicators. Over 70 percent of people living in the district were satisfied with the overall service provided and the council met over two-thirds of its performance targets and 'almost met' a further 7 percent.

# Performance indicator support pack

A document on 'Best practice guidance for staff working with performance indicators (PIs)' has been issued to address the users and uses of performance information. This contains the following information:

### Performance indicators

- What they are for.
- What they do.
- What makes good PIs.

### Developing new PIs

- Who the PI is for.
- How the PI will be used.
- The importance of PI focus and balance with the 'bigger picture'.
- Robust PIs.

### Documenting PI calculations

- Audit trail.
- Support/guidance.
- Evidence-based.
- Transparent/replicable.
- Sign posts to evidence.

This excellent document points out that PIs indicate how well an organization is performing against its aims and objectives, they are not a means to an end but:

- Measure progress towards achieving corporate objectives and targets.
- Promote accountability of the service providers to the public and other stakeholders.
- Allow comparisons of performance to identify opportunities for improvement.
- Promote service improvement by publicizing performance levels.

The council recognizes that good performance information helps identify which processes/policies work, and why they work, and is the key to effective management including service planning, monitoring and evaluation. Clearly in this public sector environment performance information is important externally as it permits greater accountability and allows members of the public and stakeholders to have a better understanding of relevant issues and to press for improvements.

The Audit Commission in the UK has a set of five-point guidelines on good practice for performance information: 'Councils should try to develop and use a range of performance indicators that measure five aspects of their service':

- *Its aims and objectives* (why the service exists and what it wants to achieve).
- *Its inputs and outputs* (the resources committed to a service and the efficiency with which they are turned into outputs – cost and efficiency).
- *Its outcomes* (how well the service is being operated).
- *Its quality* (the quality of the service delivered explicitly reflecting users' experience of the service).
- *Its accessibility* (the ease and equality of access to services).

Services will need to consider over time the set of the PIs that they have in operation (national PIs, local PIs and management information) and judge whether they need to adopt new PIs to fill in gaps or cover any new work areas. This can only be done once councils consider what they currently monitor and its usefulness, the department/service aims and objectives and where they want to take the service in the future.

# Performance management corporate arrangements

Harrogate Borough Council has prepared information and advice on the authority's corporate performance management arrangements.

Each department has its own performance monitoring arrangements which cover, at a service level, setting objectives and targets together with the reporting of performance to both officers and members. The Business Unit Manager's Handbook issued to the Authority's managers sets out the council's policy on the management of a monitoring/reporting at a service level. The focus of this document is on the arrangements to manage the authority's performance corporately through both the Corporate Management Team (CMT) and the cabinet.

## The leader's annual statement

The leader produces an annual statement of the Political Administration's key aims and objectives, policy targets, etc., for the next financial year and coming years. The annual statement is the key outcome of discussions in the spring each year between CMT and the cabinet.

## Corporate action plan

The Corporate Action Plan (CAP) is part of the Best Value Performance Plan (BVPP). Its purpose is to summarize how the council intends to deliver its long-term corporate objectives in the coming year. The BVPP and CAP are adopted by council in early March and published by the 31st March each year. A separate summary BVPP is printed and distributed to every household, to local businesses, to the voluntary sector, etc., in early March each year. The CAP is reviewed mid-year and at the end of the financial year and the results are reported to CMT and the cabinet.

## Business units

The Business Unit Manager's Handbook requires all council service managers to prepare a business and/or service plan each year, setting out the planned actions, targets and accountability arrangements for the coming year. The handbook also requires all council service managers to agree with their relevant chief officer and cabinet member the frequency of reporting of service or business performance. (The minimum reporting requirement is twice a year.) In addition, each service manager must prepare and agree a work plan to deliver the agreed business and/or service plans.

Each business unit's performance monitoring report is based on the agreed business and/or service plan and includes links to the relevant corporate objective and action in the BVPP and CAP. Chief Officers and senior managers meet regularly to review the performance of their services including benchmarking with other providers (public, private sector, etc.), comparing performance over time, etc. Service-related performance is reported to CMT only where there is a corporate issue to address.

## Performance appraisal

At their annual appraisal interview, conducted by the Chief Executive, the Chief Officer is required to account for their own performance and that of their department. Their main service and departmental objectives are reviewed during the Chief Officer's performance appraisal.

A Chief Officer's performance is reviewed half-yearly with each Chief Officer reporting to the Chief Executive on progress and changes in their objectives over the last six months. Performance appraisal looks back, looks forward and sets objectives which reflect the chief officer's priorities and the Chief Executive's priorities. The relevant committee cabinet members are present at a Chief Officer's performance appraisal. They are also present where appropriate at the appraisal of executive officers. In order to report effectively to the Chief Executive on their policy priorities and performance, each Chief Officer must review the performance, each of their own services through ad hoc reviews, department management team meetings and individual performance appraisals of senior departmental managers. The Chief Executive's performance appraisal is a report back to members on the Council's priorities and a look ahead to set policy priorities for the coming year. Within each department, the Chief Officer conducts regular performance reviews with their senior officers. Once a year, the performance of each employee is appraised by their line manager as part of the Council's Staff Appraisal Scheme.

## Corporate planning process

The corporate planning process requires stakeholder consultation to take place, at both corporate and service levels, as part of the development of plans, targets and performance indicators. This requirement is set out in the Business Unit Manager's Corporate Handbook.

The corporate planning process and the departmental/service planning processes are linked requirements for the administration's Annual Statement of Objectives to inform service and business planning and for draft service plans to be summarized in the BVPP. Service and business plans must explain, in detail, how service actions will deliver the council's corporate objectives and priorities each year. They also explain, in detail, how each Business Unit Manager will monitor the delivery of corporate objectives and priorities at a service level, what targets they will use, etc.

## Monitoring performance

Every week the **Chief Executive** meets the authority's political leadership (the leader meeting) and every month has a joint meeting with the leader and the leader of the opposition group. Reports on progress and performance form part of the discussions. Chief Officers and Senior Officers attend the leader meeting as necessary to discuss service and corporate performance.

The **Corporate Management Team** (CMT), comprising the authority's seven Chief Officers, meets fortnightly. The team's composition is:

- Chief Executive.
- Director of Finance.
- Director of Administration.
- Director of Leisure and Amenity Services.
- Director, Harrogate International Centre.
- Director of Technical Services.
- Director of Health and Housing.

The CMT's agenda includes reports from corporate projects and groups on a pre-agreed frequency (the minimum reporting frequency for a project or group is once a year). CMT also receives reports on corporate performance or issues (as necessary), either through the standing CMT agenda item 'Information Exchange' or specific agenda items/reports. The CMT receives regular financial monitoring reports on the authority's revenue and capital budgets. It also receives regular monitoring reports on the authority's corporate performance against national and local performance indicators, the district audit's BVPP action plan, etc.

The **Chief Executive** carries out a variety of reviews during the year. Some are ad hoc, asking for information, and some are planned as part of an annual review programme, including regular meetings with the Director of Administration, the Head of Environment (on environmental health issues), the Audit Manager, the Head of Human Resources (on training), the Borough Administrator (on political management), the Assistant Director of Technical Services (on community safety), the Head of Museums and Arts, the Director of Health and Housing (on housing issues), the Chief Estates Surveyor (on property issues) and the Head of Planning Services.

The Chief Executive attends a Departmental Management Team (DMT) meeting in each department twice a year to explain the council's approach to budgeting and other major issues such as the New Political Framework. It is also an opportunity for Senior Officers in departments to raise issues or ask questions.

Each **Chief Officer** is responsible for monitoring and reviewing the performance of their services. A Chief Officer will report on service performance to the Chief Executive or CMT (or both) on an exceptional basis.

Each Chief Officer is responsible for reviewing the performance of their services and budgets during the year using **Best Value Performance Indicators** (BVPIs), local performance indicators and targets. Chief Officers need to ensure that they compare and benchmark their services with other providers (public and private sector) on a regular basis. Each Chief Officer is responsible for reporting the performance of their services and budgets to the relevant cabinet member on a regular basis.

The **cabinet** meets once a month and comprises eight members:

- Leader of the Council.
- Deputy Leader of the Council.
- Cabinet Member (Planning Portfolio).
- Cabinet Member (Housing Portfolio).
- Cabinet Member (Leisure and Amenity Services Portfolio).
- Cabinet Member (Environmental Health Portfolio).
- Cabinet Member (Public Works Portfolio).
- Cabinet Member (Opposition Member without Portfolio).

The cabinet receives regular financial monitoring reports on the authority's revenue and capital budgets. It also receives regular monitoring reports on the authority's corporate performance against national and local performance indicators, the District Audit's BVPP Action Plan, etc.

In addition to this each Chief Officer is responsible for reporting the performance of their services and budgets to the relevant cabinet member on a regular basis.

## Cross-cutting issues

The **Council's budget** has its own annual process which involves central corporate analysis and review by CMT and an established corporate timetable and reporting programme, including a budget seminar for members.

## Economic indicators

Information is picked up through the council's Economic Development Strategy. It includes consultation with other organizations such as major local employers and the Chambers of Trade. The information on economic indicators is fed back through the political leadership into the authority's policy-making process.

The authority's Medium-Term **Financial Plan** is rolled forward each year. The roll-forward involves extensive consultation.

The authority's Capital Initiatives Strategy is rolled forward each year. The roll-forward involves a corporate review of need and resources. The review, etc., will form part of the authority's Asset Management Plan.

The authority has carried out several **Risk Assessment** exercises in recent years, relating to risk management on revenue budgets, on capital budgets, forward planning and on planning for high percentage budget reductions.

The Strategic Management Officer is responsible to the Corporate Management Team for reviewing the authority's **Corporate Performance**, on a six-monthly basis, in six key areas of corporate management:

1. The council's Corporate Action Plan.
2. The implementation of the council's agreed Service Improvement Plans and/or Best Value Inspection reports.
3. The implementation of the District Auditor's BVPP Action Plan.
4. The authority's audited performance against the national BVPIs and targets.

5. The overall performance of council services against last year's targets and the current year's targets.
6. The preparation of draft service and business plans for next year.

There is an annual corporate performance monitoring timetable to meet the above corporate performance requirements. The outcomes of each of the six corporate performance reviews are reported to CMT and the cabinet for each to challenge and agree. In addition each Chief Officer is responsible for comparing, monitoring, reviewing and reporting the performance of their services and functions.

Each Chief Officer is responsible for their department's performance management arrangements through their line management structures and processes. These arrangements must enable the Chief Officer to monitor each year whether the corporate objectives and targets set out in the BVPP are being delivered by service actions and expenditure in their department. The annual work programmes of the authority's corporate groups are discussed and agreed by CMT in advance of the start of the year to which they apply.

# Conclusions

Harrogate Borough Council's performance management arrangements reflect the devolved service culture in the authority. They are supported by a management culture of delegation and accountability at a service level.

The developed management culture includes appropriate checks and balances, together with 'incentives' which encourage effective business unit management. The incentives are set out in the Business Unit Manager's Handbook and cover the treatment of budget surpluses and losses, virement and internal trading relationships.

The authority's approach to performance management is 'hands on', whether at a service or corporate level. This provides an open management environment in the authority where problems, failings and successes are reported upwards on a regular basis.

In a value for money study on the authority's service and financial planning arrangements, the main conclusion was 'The council has a well-developed corporate, service and financial planning process.'

# Review questions

1. Explain the principles behind 'Best Value' giving clear statements of what it is trying to achieve in the delivery of public services. What sort of organizations may usefully adopt these ideas and how may the deployment have to change to accommodate a particular situation?
2. Evaluate Harrogate Borough Council's approach to 'Best Value', indicating strengths and areas for improvement.
3. In terms of performance measurement and management, how does this approach compare with other perhaps simpler arrangements and what could be done to streamline and improve the application?

# Acknowledgement

The authors are grateful for the contribution made by Rose Johnston and Ben Grabham in the preparation of this case study.

# Takenaka – 400 years old and a commitment to innovation and excellence ensures the future

# Company background

Takenaka Corporation is a privately owned family company with a long history and rich tradition. It is Japan's oldest architecture, engineering and construction firm – a shrine and temple master builder started the business in 1610. The present company was officially founded in 1899 and in the following 60 years it grew to achieve national prominence. The company steadily increased its capital, moved its head office a number of times and by 1938 it had established a national network of branch offices. By 1953 it had established technical laboratories and in 1958 it had built Tokyo Tower, the tallest communication tower in the world at the time.

In the two decades following 1960, the company turned its attention to international markets, establishing several offices in the US, several within Europe and South East Asia and in Brazil. In 1979 Takenaka became the first non-manufacturing company to win the prestigious Deming Award for Quality in Japan for its successful application of the principles of TQM.

In 1992, Takenaka won a second significant national quality award, the Japan Quality Control Medal, a major achievement, the award being made for the further development of their quality system. A major specialization of Takenaka is the construction of sport stadia: they have built the majority of these structures in Japan.

Today Takenaka has yearly sales of US$9 billion. It has 20 overseas offices, the largest construction R&D laboratory in the world and some 8000 employees, two thirds of whom are professional architects or engineers. Takenaka offers comprehensive services, worldwide, across the entire spectrum of space creation: from site location and planning to design and construction as well as post-completion services such as building maintenance. At any one time, Takenaka employees are delivering as many as one thousand projects around the world: three quarters of these in Japan and one quarter are international.

At Takenaka the goal is to consistently improve the quality of all works, with the overall aim of giving customers confidence that their project is completed to the highest standards throughout the contract period and beyond.

## Policy development and management

The company has a formal process of consultation that feeds into its the annual policy and performance review. The process is conducted over a six-month period, from March to September. Starting at the division manager level, formal consultations are conducted across the organization right up to the president.

At the top level, policy is formulated in terms of four primary issues:

- long-term strategic vision;
- market and environmental data;
- strategic objectives;
- past business records.

This process results in a three-year plan, which outlines planning and management objectives and articulates the current year's goals and policy. There is then a formal interpretation of this plan for action at the division and departmental level.

# Total quality management at Takenaka

Takenaka's corporate philosophy is to contribute to society by creating quality works that will serve current and future generations. The built environment is conceptualized

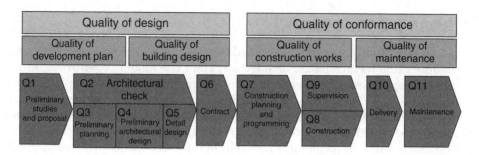

**Figure C17**
Job-specific quality assurance flowchart

as being for the conservation of life and wealth, while, at the same time, it is also a vehicle for the transmission of culture from one generation to the next. Takenaka sees its projects as a complete and integrated solution, stretching from conception to final production through to operation in use. With this in mind, the design–build integration system is considered to be the best way of ensuring that the projects provided by the company are of the highest quality.

1. The design quality of buildings are characterized as having:
   - a timeless quality that anticipates the needs of the next generation;
   - functionality, economic viability and beauty through the whole life of the building.
2. Building in quality:
   The process of building in quality is defined in several logical stages:
   - gaining a clear picture of customers' needs;
   - synthesizing those needs through 'quality of design' documented in drawings and specifications;
   - building in this design quality in the workplace and assessing it as faithful 'quality of conformance';
   - implementing quality assurance activities through after-sales service.

Takenaka's TQM activities are positioned within the company as a means of promoting vision management and improving the quality assurance of all work, with the ultimate aim of improving customer satisfaction.

## Progressive development of TQC/TQM at Takenaka

In the early 1970s, motivations for the company leadership to embrace quality as a core business value came from two directions: the need to improve technical reliability and the need to survive in an unexpectedly less stable business environment. Technical challenges for Takenaka included the collapse of temporary shoring in an excavation in Okinawa and the use of understrength concrete on another project. At about the same time, the OPEC oil shock of 1973 created instability in prices, resource flows and demand.

Table C4 traces the progressive development of the company's commitment to total quality over a period of almost 30 years. It characterizes both the evolution of the company TQM philosophy and the necessary staged implementation that is a must if a fully effective TQM system is to be achieved.

452

**Table C4** Progressive development of TQC/TQM

| | |
|---|---|
| 1976 | TQC introduced – the focus is on zero defects |
| | TQC central promotion committee inaugurated |
| | TQC advisory committee set up |
| 1977 | Quality assurance manual enacted – the vision is broadened to embrace the values of total quality, though it was called TQC |
| | QC circle activities started |
| 1977 | Demand drops significantly |
| 1978 | Decided to improve the level of quality assurance as the basic policy of promoting TQC |
| 1979 | Deming Prize received – visit by Mr Deming |
| 1982 | Takenaka Quality Control Prize (T Prize) initiated – for outstanding subcontractors. Demand picks up (during the 1980s) |
| 1988 | Deming Prize received by Chairman Takenaka |
| 1991 | Japanese economy hit by recession again – focus on reducing overheads and strengthening financial base |
| 1992 | Environmental approach initiated – E1 |
| 1992 | Japan Quality Control Medal – received |
| 1992 | Takenaka Environmental Charter established – defining target areas for focus, goal setting, measurement and improvement |
| 1995 | TQC to TQM – redefining of terminology |
| | Further strengthening price competitiveness |
| | IT-based technologies adopted |
| | Strengthening client focus – vision management, adoption of long-term strategic vision of products through their entire life cycle and concurrent strengthening of community focus |
| | TQM process expanded – to include E2 – ethics, E3 – ecology and E4 – employees |
| 1996 | Japan Quality Control Medal won (second time) on the basis of a three-year survey of system performance |
| 1998 | ISO 9001 and 14001 certification obtained |
| 2001 | TQM, basic policy for promotion – reviewed |
| | Quality of management improved – by promoting vision management, improving the quality of products and service, and the quality of operations |

# Development of the TQM approach

Between 1976 and 1995 the conceptualization of quality management at Takenaka was based around the concept of *quality first* and the decision that the *design-build integration system* was the best way to achieve this outcome, with a broad focus on all aspects of the production process through the *QCDS* (quality, cost, delivery, safety). This approach is described in the model in Figure C18.

In 1996, with a shift in the customer focus, this conceptualization was altered to consider the entire life cycle of a facility and simultaneously to more clearly articulate the company's commitment to the environment as an expression of its genuine contribution to the community.

The shift in focus can be clearly seen in the differences between the two process diagrams, the latter, described in Figure C19, clearly articulating a more complex vision that embraces both a longer-term view of the customer relationship and a response to socio-economic factors in its commitment to the community. This latter issue is also reflected in the addition of the environment to the *QCDSE*.

The starting point for TQM at Takenaka is to implement all the quality assurance activities that enable a client to place an order for any project with confidence and to

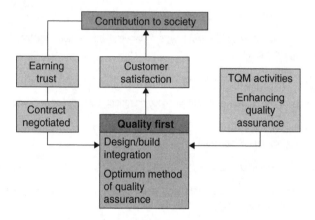

**Figure C18**
Initial conceptualization of quality in 1976

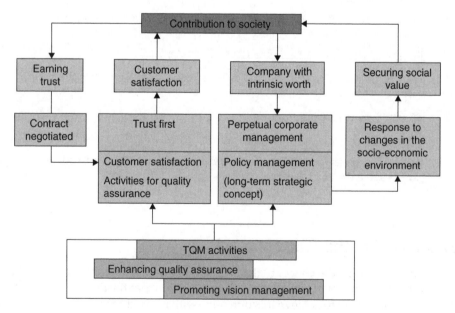

**Figure C19**
More complex reconceptualization of quality in 1996

remain reassured and satisfied throughout the many stages from planning to completion that the agreed brief will be achieved in every regard.

At the end of a project, TQM is to ensure that the client takes delivery of work at a guaranteed quality standard. To achieve this goal, Takenaka have developed a 'Job-specific Quality Assurance System'. This is a framework that sets out the quality assurance procedures at every stage of the project. The company's quality system incorporates the best features of ISO 9001 and ISO 14001.

Takenaka's focus on creating and managing to a *vision* delivers a customer-oriented focus throughout the entire procurement process. In response to the rapidly changing business environment, the quality management system is designed to respond to the specific needs of customers in terms of the key parameters of Q (Quality), C (Cost), D (Delivery), S (Safety) and E (Environment). TQM activities are used as an effective means of developing, maintaining and improving customer satisfaction, and while Takenaka is utilizing a variety of different methods in its commitment to TQM, the primary focus is on customer satisfaction.

## Focusing on customer satisfaction

Takenaka commits significant resources to ensuring that it is aware of its customers' needs and satisfaction as well as market expectations. This is achieved through regular client meetings, customer satisfaction surveys, customer expectation surveys and more general market surveys to establish end-use expectations. Customers are routinely surveyed 12 months after project completion. Currently most survey work is in-house; though, consideration is being given to using second parties to ensure the process is objective and unbiased. The aims of the company are to clearly understand how the customer feels and to quickly disseminate this information throughout the company.

## Training at all levels throughout the company

Training is implemented using internal training programmes as well as external providers. The training of managers follows a clear philosophy. It addresses the following hierarchy of issues:

- learning social and cultural values;
- learning to lead and manage;
- learning management and problem-solving skills;
- learning deeper knowledge of specialist skills;
- learning traditional techniques and attitudes.

To build a solid base of technical experience in their employees, graduates younger than 30 are rotated throughout the company to gain experience in research and design before they specialize in construction and project management.

The TQM training programme includes subcontractors; they are encouraged to form and to participate in quality circles and the Takenaka QC prize for the best subcontractor motivates subcontractors as a part of the team. All employees have to undertake a two-day TQM programme as part of their induction into the company.

There is also an annual TQM convention to review challenges and achievements and award prizes for excellence. Takenaka people as well as their subcontractors attend this convention.

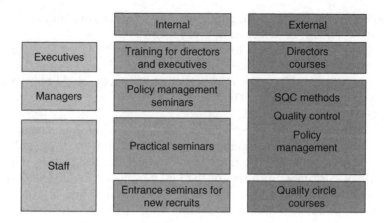

**Figure C20**
TQM training

## Quality control process – approach and method

There is a difference in the approach used within the quality process between maintenance and improvement on the one hand and development on the other. Both are based on feedback provided through a typical PLAN–DO–CHECK–ACT cycle. While they draw on different skills and information, the process in each case is one of implementation, evaluation and improvement to close the gap between expectation and reality.

A constant review of past performance produces the information that drives quality activities in relation to maintenance and improvement. This is done through formal data collection and analysis to check actual outcomes against targeted performance and to focus action on closing identified gaps. Data is collected, analysed, interpreted and discussed in visual formats using a range of standard quality techniques. These include numerical recoding systems, graphical representation of data and analyses, cause and effect relationship analysis, relational diagrams and decision matrices. The aim is to prevent the recurrence of errors and to avoid errors before they occur.

The challenge of new development is different because information and motivation cannot be drawn directly from past experience, yet quality here is just as important as the maintenance and improvement part of existing processes. Development themes are drawn from projections based on the analyses of customer expectation and technical opportunity as well as the corporate vision process, which defines long-term strategic direction. The aim of development is to create new, exciting improvements in products and services, attractive aspects of quality to excite customers. This is all developed on the basis of projection rather than learned lessons from past experience.

Currently there is a strong focus on implementing IT more effectively both in terms of generating real time information and in terms of implementing an IT-enabled knowledge management (KM) system. Within the KM system there are two separate foci, improving KM project-based data as well as improving KM for areas of specialist knowledge.

## Outcomes achieved

Figure C21 plots the cumulative number of awards received by the five major construction companies in Japan, Takenaka and its four main competitors, annually, since 1960. Each company has approximately 4 percent of the national market and hence company size is not a differentiator. The graph shows that each year Takenaka wins more awards as its next best competitor.

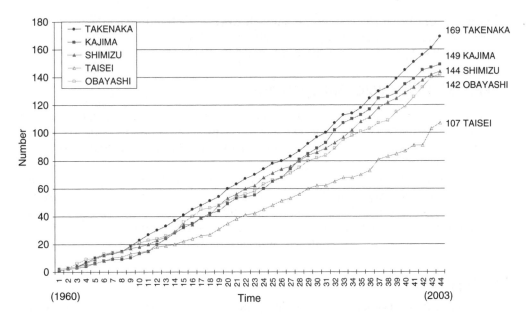

**Figure C21**
The cumulative number of BCS (Building Contractors Society) Prizes awarded to a construction contractor

# Takenaka's commitment to the environment

Perhaps the best demonstration of Takenaka's commitment to being a responsible corporate citizen is its commitment to improving its environmental practices through a very thorough and comprehensive strategy. In 1992 the company first developed its environmental charter. In this statement they set out the extent of their commitment to environmental preservation. They set their sights on reducing a very wide range of environmental impacts and, since then, each year they establish new targets for improvement during the forthcoming year. They have targeted the improvement of design, construction practices and the operational phase of the facilities as well as the renovation and demolition phases through the adoption of action plans and improvement targets on an annual basis.

For example, in relation to design, the company has targets to:

- landscape building roofs to reduce the urban heat-island effect and to blend gardens throughout the built environment. They have also extended the use of trees and natural vegetation for shading and wind protection;
- improve the energy efficiency of designs by expanding the use of solar energy and natural lighting; using geothermal energy wherever it is available; energy storage systems in buildings; reusing urban exhaust heat; and using energy efficient materials and equipment. PAL (perimeter annual load) and CEC (coefficient of energy consumption) targets are set and audits of actual energy consumption are used to ensure that design objectives are actually achieved;
- recycle water;
- reduce waste from building sites and promote recycling. In the 10 years from 1992 to 2002 waste generated from construction sites has been reduced by 35 percent while the recycling of materials has been increased by 125 percent. Several major prizes have been won for recycling activities on projects;

- reduction of $CO_2$ emission over the whole life cycle of a building and waste CFC recovery from the renovation and demolition of buildings;
- an active R&D programme with in excess of 150 current projects in 2002.

## Takenaka's life cycle management

Takenaka supports improvements to the value of customers' assets with a variety of techniques that take into account the life cycle of the facility. The company adds to the value of customers' assets from the perspectives of ecology, economy and quality in all processes over the life of a facility.

The company applies its technical and management capability to life cycle design, this includes financial management planning to improve investment returns over the whole life of an asset, medium- to long-term maintenance planning as well as facility management and operation planning. The company has developed a range of diagnostic tools for the assessment of assets as well as techniques for their improvement.

## Ecology

A wide range of technologies has been developed to reduce the whole life-cycle burden of facilities on the environment. These include greening technology, forecasting simulations of the burden on the environment, purification of polluted soil and water, energy saving and recycling of construction by-products.

## Economy

Takenaka has developed a range of strategic initiatives to reduce the whole of life costs of facilities: business assessments, assessment of economic value at the planning stage, life-cycle cost assessments, computer-assisted facilities management (CAFM), and IT-based technologies.

## Quality

In order to provide support to customers during the whole life cycle of their facilities, and to maintain and improve the function and utility of these facilities, Takenaka has developed high durability, long-life materials, as well as building performance diagnosis and renewal technologies.

# Review questions

1. Describe the 'quality journey' in Takenaka drawing a chart of the activities which a similar company could follow if it was beginning the implementation of TQM.
2. Evaluate the approach taken in this case and offer constructive criticism and suggestions for improvement.
3. What role has the commitment to the environment played in securing the future for Takenaka?

# Acknowledgements

The authors gratefully acknowledge the assistance of Sugjiyama Satoru, General Manager of the Audit Departments of Takenaka, and his staff.

# Business improvement strategies in the Highways Agency

# Background

The Highways Agency is responsible for England's strategic road network, a network that consists of 9400 km (5481 miles) of motorways and trunk roads and carries a third of all road traffic and two-thirds of all freight traffic in the country. This equates to an annual total of around 153 billion kilometers travelled. As an Executive Agency for the Department for Transport the Highways Agency priorities are to:

- Continue to maintain the network in good condition to ensure that it is safe and available for use.
- Maximize performance from the existing network.
- Improve the network where necessary.

At the end of the 1990s the UK Cabinet Office introduced a Better Quality Services Review (BQSR) programme in response to the Government's white paper 'Modernising Government'. As part of that programme the Highways Agency undertook a review of the performance of its activities with a view to considering one of five options:

- Abolition
- Market testing
- Contracting out
- Privatization or
- Internal improvement

The Highways Agency Management Board realized the potential of the BQSR proposals, particularly the opportunity to incorporate a long held policy to improve the management and operation of the organization. The first step was to set up a small team to consider how such a programme could be delivered and by April 1999 this 'Business Improvement Team' presented a paper to the board detailing a potential improvement strategy, which linked the need to implement a programme of better quality service reviews to a structured approach to improving the business.

The proposed BQSR programme was authorized by the Highways Agency Management Board and was due for completion in April 2004. Alongside the authorization of the programme the board approved the establishment of a Business Improvement Co-ordinator in each of its 23 divisions. This role was to be supported by the Business Improvement Team who were charged with facilitating programme delivery.

# The approach

The approach that was adopted by the Agency was to use a small team of internal consultants supported by external expertise to identify good business practices and assist the functional directorates of the Agency to analyse the services provided using the BQSR criteria, identify areas for improvement and implement any improvements. Each of the services identified would then be subjected to a comparative benchmark with a view to aiding the final BQSR decision-making process. Figure C22 is a visual representation of the approach that each directorate used.

The Business Improvement Team worked with the management teams in each area to decide on the service strategy for each of the previously identified directorate key activities. This was effectively the first BQSR analysis designed to identify those services which could be abolished or where there was already consideration of outsourcing.

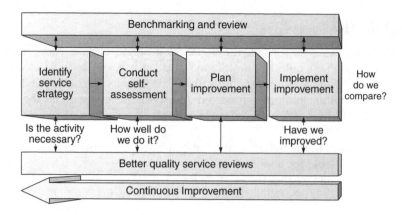

**Figure C22**
BQSR framework

This exercise provided a strategic review helping directorates to clarify their purpose. The end of 1999 saw the completion of this part of the programme with a limited number of services put forward as having the potential for outsourcing. Those services that were put forward were subjected to a comparative benchmark and a subsequent improvement programme was devised.

The Business Improvement Team then facilitated the delivery of the second phase of the framework – 'Self-assessment using the EFQM Excellence Model'. Several of the divisions were already using the EFQM Excellence Model to identify and plan improvements, so it was decided early in the programme that self-assessment against the Excellence Model was likely to be the most effective way to identify, in a holistic sense, the areas that each functional directorate of the Agency should consider for improvement. The Business Improvement Team assessed a number of methods for self-assessment and eventually settled on two. The first of these would be a simple questionnaire-based self-assessment, designed for use with the smaller directorates, where all staff would participate. The second form of self-assessment, to be utilized by larger directorates, would involve training a small group of staff as EFQM assessors who would then gather evidence of business practice, assess that evidence, identify areas for improvement and plan the implementation.

By mid-2001 over 90 percent of the organization had undergone one of the two forms of self-assessment and a clearer picture of the key areas for improvement was evident. Indeed a number of directorates had already agreed improvement action plans and were well into the delivery of improvement.

# Achievements

The improvement strategy has had an impact on the way the Highways Agency manages and operates its day-to-day business. For example, work completed in the financial payments division has led to improvements in key performance, particularly in terms of the handling time of invoices for payment (down from an average of over 15 minutes to less than four minutes per invoice). This has led to an improvement in the prompt payment initiative targets, from less than 75 percent being paid within the mandated 28 days to greater than 95 percent being paid on time, with fewer staff employed in the process.

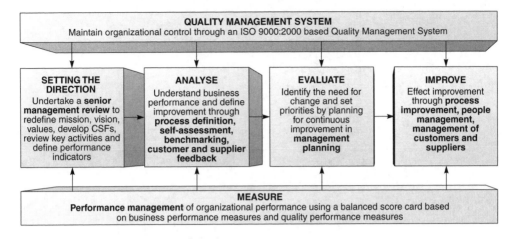

**Figure C23**
Directorate improvement frameworks

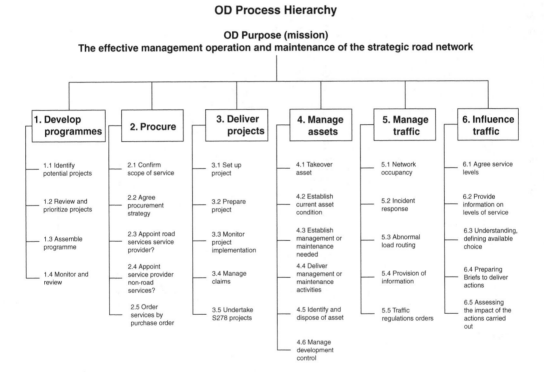

**Figure C24**
Example of a directorate process hierarchy

The planned improvement is led by the customer facing and programme delivery directorates, as they are responsible for the Agency's key or core delivery activity, working to a directorate improvement framework (see Figure C23).

From their clarified purpose (what they are there to achieve), both directorates have identified their key activities (what they are there to do), and their key processes (how

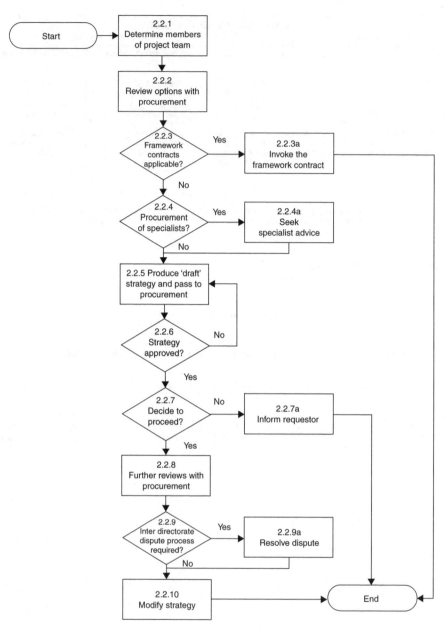

**Figure C25**
Example of a process flowchart

they will operate). This was developed using ICOR (inputs, controls, outputs and resources) techniques and has enabled the production of agreed consistent practices, recorded as flowcharts and working procedures in the form of process tables (see Figures C24, C25 and C26). These enable anyone in these directorates to understand their contribution to the overall work of the Highways Agency and provide clear practical working guidance. These delivery processes, held in an electronic web format, are continuously improved and developed alongside their support processes for customer management, people management, supplier partnerships and management planning, following ISO 9000:2000 principles.

| Item Ref | Task/Gateway | Approach | Forms/Ref Doc's Used | Records Kept | Performed by | Performance standard |
|---|---|---|---|---|---|---|
| 2.2.1 | Is Procurement Division's involvement required? | Determine type of value of project/service required<br><br>Refer to HA Procedures Manual – Procurement Section | HA Procedures Manual – Procurement Section | Record of decision kept on file | Project Sponsor in liaison with Procurement Division | |
| 2.2.1a | Is request for procurement of Utility Services? | Determine whether utility equipment is involved | New Roads and Streets Works Act 1991<br><br>Code of Practice | Record of decision kept on file | Project Sponsor | |
| 2.2.1b | Perform tender process for services | Procurement Manual? | | | | |
| 2.2.2 | Determine members of project team (inc. virtual team and/or DBFO Transaction Team) | Contact potential team members and arrange meeting | Commissioning forms for SSR | Record of Commission-ing form | Project Sponsor | |
| 2.2.2a | Undertake Tender Process for Non-Roads Services | Procurement Manual? | | | | |
| 2.2.3 | Review options with Procurement | Arrange to meet with Procurement Division regarding options | HAPM4 | HAPM4 | Project Sponsor in liaison with Procurement Division | |
| 2.2.4 | Are Framework Contracts applicable? | Consider advice from Procurement Division to determine whether service can be procured from Agreements inc. Framework Contracts | Minute/Email from Procurement | Minute/Email from Procurement | Project Sponsor in liaison with Procurement Division | |

**Figure C26**
Example of a process table

465

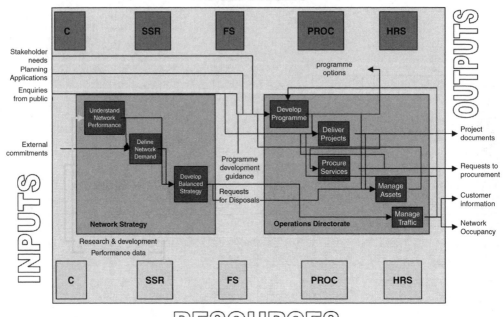

CONTROLS

INPUTS

OUTPUTS

RESOURCES

**Figure C27**
End-to-end Highways Agency process ICOR

This delivery process development work has enabled the picture of the Agency's overall delivery process to be developed, as shown in Figure C27.

This is enabling the wider process picture and to be addressed supporting the Directorates in clarifying their contributions. Measures from these key delivery processes feed forward into the Agency's balanced scorecard.

Other areas include the identification of potential improvements to project delivery areas that, once realized, should improve control over delivery processes, reducing wasted effort through failures, etc.

Work in developing customer satisfaction and management systems allow the line managers to identify key areas for improvement, based on the needs and expectations of the customers, as well as clearly measuring, in both lead and lag terms, how the organization is performing in meeting those needs and expectations.

There has also been a number of improvements identified as the result of detailed benchmarking studies conducted with a range of public and private sector partners and it would appear that the impact of these has been mainly positive.

A notable achievement following the adoption of the Highways Agency business improvement strategy is the greater awareness and desire by all staff to embrace business improvement as a way of resolving problems and delivering improved services to customers.

# Review questions

1. What are the main issues a public sector organization like the Highways Agency faces when designing and implementing a business improvement strategy?

2. Evaluate the BQSR framework (Figure C22), the directorate improvement frameworks (Figure C23) and process hierarchy (Figure C24) and offer constructive criticisms and suggestions for further improvements.
3. The Highways Agency has to deal with contractors and subcontractors in the construction industry. What particular difficulties are there in this sector which might impact on business improvement activities, such as those described in the case study?

# Acknowledgement

The authors are grateful for the contribution made by Dick Tyson and Barry Westwood in the preparation of this case study.

# Case study 10

# Sekisui Heim

# Company background

The Sekisui Chemical Co. Ltd was established in 1947 to manufacture PVA and PVC products. In 1953, the company started to manufacture conduit and pipe for the construction industry; within two years it was making pipe fittings and a year later PVC rainwater guttering for buildings. In 1963 the company started to manufacture FRP bathtubs, septic tanks, and toilet and bath cubicles.

In 1960 the trial manufacture of the 'Type-A House' was undertaken and the Sekisui House Industry Co. Ltd was established. This was the company's first foray into housing manufacture. Sekisui House was established as a separate commercial entity in which Sekisui Chemical Co. still holds some equity. In 1968 Sekisui started to plan its second housing venture. A team of specialists was assembled to develop a premanufactured house that could be mass produced, precision manufactured and would be economical to build. The Sekisui Heim brand was established 1971 and its basic product 'Heim M1' was an extremely robust, fully premanufactured, modular, steel framed house with 80 percent of the work being completed within the factory environment.

In 1977, the company set about developing a timber product based on the combined use of four by two timber frames and prefabricated module technology: the basic structural system having known seismic performance and construction efficiency (its experience proven in the US market). In 1982 the company launched the 'Sekisui Two-U Home'.

A third product line, Crastina, was introduced in 1999: a high end-panel-based timber housing system for the prestige end of the market. Finally, in 2002 the company introduced a further new product system, the Desio GT. This is a system using light, high-strength steel sections that permit larger spans and, hence, greater design flexibility.

The market for each of these is slightly different, Heim appeals to technically minded people. It is built of highly accurate, very robust steel modules, fabricated to three sigma levels of production. Sekisui Two-U Home, being of timber, has been designed to appeal to those preferring a warmer finished product. Crastina, a more customized product for expansive homes, has a greater flexibility in design layout than the other two. In terms of market share, the company builds some 15 000 dwelling units a year, of these Heim represents two thirds, Two-U Home one third and, in 2003, 50 units of Crastina were built.

Since the development of the original Heim M1 in 1970, Sekisui has constantly focused on improving the quality of its modular houses. Factory-built housing permits productivity to be maximized, the construction period to be shortened, costs to be reduced, energy to be saved in assembly, painting and other production processes and the waste of building materials to be minimized during construction. These are all essential requirements if the development of housing is to have a minimal impact on the environment.

By 2000 the company had built more than 400 000 dwellings. Today, Sekisui Heim represents some 50 percent of overall Sekisui Chemical Co. Ltd turnover. It has 36 sales companies and 32 Fami-S maintenance companies and it manufactures some 15 500 dwellings a year from eight factories.

# Central goal

To supply environmentally friendly houses which will provide family comfort and safety for at least 60 years; and to develop a long-term relationship with purchasers for the maintenance, refurbishment and ultimate recycling of Sekisui dwellings.

# Core values

## Contribution to society globally

A focus on reducing and minimizing environmental impacts, on meeting the needs of end-users throughout the life of their dwelling through innovation and quality service and on ensuring the safety of all workers.

## Maximizing the potential and performance of employees

Developing empowered employees whose skills and knowledge grow throughout their working lives, and who achieve improved outcomes in terms of *speed* and *quality* and who overcome new challenges through individual effort and teamwork.

## Maintaining the confidence of all stakeholders

Winning and maintaining the confidence of customers through the excellence of products and service, effectively communicating with customers, dealers, stockholders and local communities and disclosing corporate information correctly and in a timely manner.

## Observing the spirit of all legal obligations

Observing fair-trading practices in accordance with all relevant laws and internal company rules, maintaining proper relations with political and statutory authorities and, as a responsible corporate citizen, focusing on the creation of benefit to society.

## Preserving the global environment and contributing to society

Commitment to all actions that will improve and preserve the physical environment, both within the company's practices and in the wider community; and respect the rights, culture and customs of local communities wherever they may be.

# Sekisui's approach to quality

## Quality divers

From the outset, Sekisui approached the question of quality management from the perspective of a manufacturer rather than a contractor, thus, making this case study rather unique. The three businesses are managed as a group with a common approach to quality management. All dwellings are designed and built to last at least 60 years subject to proper maintenance and the company provides a 20-year guarantee, once again subject to inspection and maintenance at 10 and 15 years. Sekisui sees as a substantial business opportunity the ongoing maintenance, refurbishment and ultimate recycling of the dwellings they sell.

The drive for quality comes from two directions, initial customer satisfaction and excellent service life. The R&D division collects the latest information on dwellings and lifestyles, co-operating with specialists both inside and outside of the company. The division's network includes academics, planners from outside the company and homeowners.

All products are guaranteed against waterproofing and structural failure for 10 years. During this initial period the company has a contractual right to inspect at six months, 12 months and two years, further non-obligatory inspections are offered at five and 10 years. Currently 70 percent to 80 percent of customers permit the five and 10-year inspections and the target is to get this to 100 percent.

However, if purchasers join the company's maintenance scheme, the building is inspected and maintained every five years and customers receive a 20-year warranty. All customer complaints are logged and immediately rectified. Defects and problems identified from these inspections drive quality improvement from the customer perspective.

## Quality processes

The second focus is process driven: this involves the company constantly looking for ways to improve both the performance of the product and the production process. A central development arm creates the product and process design. This includes a research centre and two product development centres. Between them, these head office groups are responsible for product and process development and improvement. They are actively involved in market research, research of customers and their preferences. And through the ongoing maintenance relationship with former customers, they are able to identify both changes in demand and performance improvement goals.

The company CS (customer service) and QA Department are responsible for ensuring that the products perform satisfactorily for their planned 60-year life. They conduct QA inspections at five stages: contract completion, occupancy, and at one, five and 10 years after occupancy. At each stage, a formal questionnaire process is used to gather data. Initially the sales companies handle customer claims and complaints, filtered information is then passed on the CS and QA Department. Structured feedback from both sources is provided to the design and development teams in the research institute. They review and improve product design to ensure enhanced performance. The CS and QA Department also have oversight of the factory quality systems.

Innovation is initiated in the Housing Technology Institute (HTI). HTI staff are responsible for the development of product and process design and for setting workmanship standards. Whenever a new product or detail is introduced, it is thoroughly evaluated at the HTI, before it is allowed trial implementation in a factory. HTI staff, factory staff and CS and QA staff collaborate to ensure smooth implementation and reliable performance.

Prior to the adoption of ISO 9000 each factory had its own quality systems; however, since the international standard was introduced, uniform quality systems and standards are deployed company-wide. These are developed through head office processes that involve HTI, CS and QA and factory quality staff. All quality problems are reported back to the CS and QA Department and are addressed centrally.

The company's central quality council is co-ordinated by the CS and QA Department and meets monthly to review feedback and quality issues that have emerged from anywhere within the company's product areas and operations. This committee combines key people from head office (Housing Technology Department, Production and Purchasing Department, CS Promotion and Quality Assurance Department and Living Environment Division). Its subcommittee includes key people from the eight factories and the seven sales branches.

## Product quality

As a result of its focus on customer satisfaction, Sekisui has paid close attention to those aspects of dwelling performance that impact directly on user health and comfort. This has led to innovations such as a ventilation system that changes 100 percent of the air in a dwelling even when it is fully sealed. The development of fabrics that reduce dust mite and the use of materials that meet the strictest standards of chemical off-gassing. Every dwelling is tested for air quality, formaldehydes and air tightness on completion. Some 70 percent of customers also request acoustic testing.

Other critical performance attributes are engineered into these dwellings. Earthquake resistance was proven in the Great Hanshin-Awaji earthquake in 1995: 180 000 houses collapsed, yet, of the 7700 Sekisui Heim and Two-U Home houses within the earthquake zone none were seriously damaged. Security, fire resistance and water resistance are also designed in.

# Environmental performance

Both nationally and internationally, the company has taken a leadership position within the industry by building to a 60-year performance life standard, and by investing in the development of technologies that optimize the performance of dwellings over that life.

This has led the company to focus on three key areas when constructing houses: the environment – must be protected, therefore the project must have a low load; resources and energy – must not be squandered; and the community – must be respected, therefore the development must be designed to complement it. Sekisui therefore builds dwellings that:

- have low environmental impact and use durable materials with a long life and a low environmental load and utilize natural energy, an area targeted for improvement is the reduction in $CO_2$ emissions;
- save on resources and energy, use solar energy and equipment that is energy efficient, and use recycling and waste reduction in manufacture. Rainwater recycling for garden watering is optional;
- are designed to fit harmoniously into local community environments.

Sekisui has invested in ventilation technology, efficient energy utilization, solar energy collection and insulation, the use of materials with low environmental impacts and enhanced durability of finishes, to name a few key areas. Thermal load is calculated and modelled for every dwelling, taking account of customer choices.

In assisting customers to make design choices, Sekisui has developed modelling to help them give advice on both running and operating costs and the impact that alternate design choices have on them. For example, the use of tiles externally is expensive; however, they last 60 years and therefore the life-cycle cost is low. Furthermore, the cost of placing the tiles in the factory is 50 percent of laying them in situ. Such detailed information on whole-of-life costing helps customers make informed choices from the outset.

In order to be confident about the durability of external cladding systems, Sekisui sends sample panels for full-scale testing to an environmental exposure test site in Arizona.

Achievements to date include:

- ISO 14001 certification by all business units;
- the attainment of zero emission by all factories;
- the availability of totally insulated houses and houses equipped with photovoltaic generation systems;
- the 2000 winner of New Energy Award for Photovoltaic/Thermal Hybrid System;

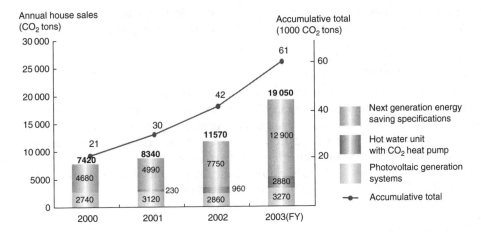

**Figure C28**
Annual contribution to the reduction in $CO_2$ emissions in tonnes

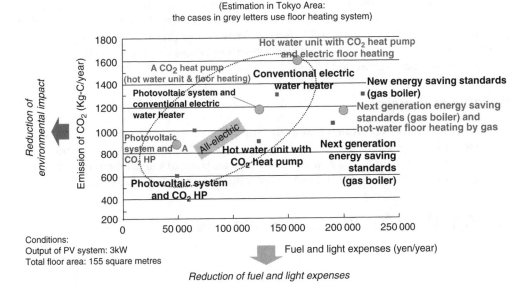

**Figure C29**
Effects of PV systems and efficient equipment

- the 2001 winner of the Grand Prize for Energy Conservation for Hot Water Unit with $CO_2$ Heat Pump from the Japanese Minister of Economy, Trade and Industry;
- the company had sold 20 000 dwelling units with photovoltaic systems installed by March 2003.

# Engineered quality

By adopting a factory-based approach to the construction of its dwellings, Sekisui controls three critical variables in the construction process:

- weather;
- builders' skills;
- materials.

In addition to managing these issues, by adopting a factory-based process, Sekisui can complete a dwelling in 40 days – a third of the time taken using traditional processes.

## Variable weather

Over 80 percent of all construction work is completed at the factory. The length of the on-site construction period is around 40 days. This is shorter than any other prefabricated housing construction methods in Japan. Assembling the modules and erecting the roof in a day protects the dwelling from the effects of bad weather during construction. Factory production also eliminates the effects of dust on painting and finishing, and other factory processes such as welding and gluing achieve their designed strength because they are completed within a controlled environment.

## Variable skills

Skills vary from operator to operator, tradesman to tradesman; however, specialization and work control are easier in a factory environment. The access to work can be radically altered in a factory compared to on site; for example, work that has to be done overhead from a ladder on site can be done on a table in a factory. This has the potential to improve quality, reduce health and safety hazards and improve productivity. Furthermore, automation is much easier in a factory setting where the work can be structured and the environment controlled.

## Variable materials

By purchasing quality controlled materials centrally, Sekisui overcomes the challenge of traditional building where the trade contractors are single-lot purchasers of materials for every home they build. As a manufacturer the company can implement higher standards of quality control both from its suppliers and from within its factories.

# Recognition for corporate excellence

| | |
|---|---|
| 1978 | Okouchi Prize |
| 1979 | Deming Prize |
| 1982 | Science and Technology Agency Directors Prize |
| 1993 | Sekisui Heim *Serano* MITI Good Design Mark |
| 1995 | Sekisui Heim *Desio* MITI Good Design Mark |
| | Nikkei Most Excellent Award for Superior Products and Services |
| 1996 | Sekisui Two-U Home *A-II* awarded MITI Good Design Mark |
| 1997 | Sekisui Two-U Home *Earthia* awarded MITI Good Design Mark |
| 1998 | Science and Technology Agency Directors Prize |
| | Science and Technology Distinguished Service Award |
| 2000 | Awarded Minister Economy, Trade and Industry Prize |
| | New Energy Award for Residential Complex PV-Solar Heating System |

# Review questions

1. Show the relationship between quality and environmental management systems, in the context of this particular case, and explain how the managers in Sekisui have embraced these two vital areas of their operations.

2. In the case study it suggests 'Sekisui approached the question of quality management from the perspective of a manufacturer rather than a contractor'. Explain what this meant for the company in operational terms and evaluate the results derived from this approach.
3. Carry out an 'audit' on the environmental results presented in the case and suggest areas for improvement that could give rise to enhanced performance in this area.

# Acknowledgements

The authors gratefully acknowledge the assistance of Toshiro Takeda, General Manager CS and QA, Sekisui Heim. The information for this case study has been taken from company literature and from an interview with Mr Takeda and his colleagues.

Sekisui Heim

# Boldt leverages lean thinking to drive efficiency and growth

# Company background

Today the Boldt Company is one of the leading exponents of lean construction practices in the world, the company is made up of three separate divisions: Boldt Consulting Services, Oscar J. Boldt Construction and Boldt Technical Services.

Founded in 1889, Oscar J. Boldt Construction is a fourth generation, family-owned construction firm. Operating throughout the United States with a workforce of more than 2000, it provides strategic planning, design management and construction solution services to customers in a variety of industrial, institutional and commercial markets. Oscar J. Boldt Construction is Wisconsin's largest and among the top-ranked general contractors in the US.

Over the 116 years since its inception, the company has grown from housing to commercial construction, heavy industrial construction serving the pulp and paper industries, healthcare, education and power generation. Today Boldt is ranked in the top 5 US contractors in the industrial sector. It has been ranked as one of the largest *at-risk* construction management companies and among the small group of *top contractors* according to Engineering News Record.

Boldt Consulting Services combines a thorough understanding of operations with the disciplines of planning, real estate, construction, finance and law, while Boldt Technical Services provides a single, integrated resource for analysis, ideas, solutions, budgets and plans. They offer value analysis, architectural, project development, value engineering and commissioning services.

Since 1999 the company has embraced the concepts of lean production and they have applied lean thinking to more and more of their operations.

# Corporate values

## Vision

To be the leading national resource which provides consulting, technical, and construction solutions for their customers.

## Boldt promise

To deliver services and facilities which enable their customers to gain maximum long-term value from their operations.

## Strong and lasting values

*Boldt Builds* – Honesty, fairness, hard work, performance and *love of construction* are what are valued.
*Boldt Consulting Services* – Honesty, fairness, hard work, performance and *love of creative solutions* are what are valued.
*Boldt Technical Services* – Honesty, fairness, hard work, performance and *love of technical innovation* are what are valued.

# Implementing lean project delivery at Boldt

Boldt found in the late 1990s that it needed to improve its productivity, and as a part of a search for a set of beneficial strategies and techniques they investigated the ideas of

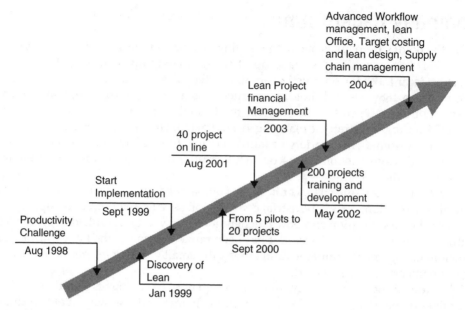

**Figure C30**
Implementing lean project delivery at Boldt

lean construction. They become a corporate member of the US Lean Construction Institute in 1999. The figure below describes increasing commitment the lean production as the company has moved from early adoption of the Last Planner™ technique to a broad based adoption of the ideas of lean production.

Boldt reports that it is not uncommon for projects applying lean principles to experience performance improvements in productivity and schedule acceleration in the range of 10 percent to 20 percent. In many ways the lean process systematically applies and improves on the best practice of the past. More significantly lean fundamentally changes the way work is done throughout the project delivery process. Lean construction principles are applied to capital projects to reduce cost, accelerate delivery and improve both quality and safety.

# What lean delivery means at Boldt

## Cross-functional teams used in product and process design – component standardization

Cross-functional teams are the basic organizational unit for all phases of the lean delivery system at Boldt. Team members are co-located, simultaneously designing the product (that is to be built) and the process (how it is to be built), using 3D modelling to assist in evaluating and optimizing both the product and the process. If the teams cannot be physically collocated, they are linked through virtual web-meetings and project websites.

Labour productivity improves when what is to be manufactured and installed is standardized in configuration and composition. Boldt has developed design product standards to minimize on-site labour effort, facilitate fabrication, installation and maintenance throughout the product life cycle.

## Target costing and value engineering to lower costs and increase value

The company establishes target costs at the start of projects to ensure the budget is an influence on design and decision-making, rather than an outcome of design. This creates the foundation for effective financial management at the outset: at a time when decisions have the greatest influence on the allocation of resources to meet the clients objectives and when cost savings or cost reallocation can be most effectively accomplished. Value engineering is also used to create a high value design, focusing energy on analysis and creativity at the front end to develop solutions that create the highest value for the customer based on functionality and cost.

## 3D CADD as an integrating model for product and process design for the life cycle

Boldt uses 3D CADD to display alternative concepts and ensure interface of systems, subsystems and components. This allows installers to design the assembly process and allows operators to design operations needs and support systems well in advance of construction. This approach enables the designers to realistically plan for the logistics, assembly, and commissioning and operational phase of a facility.

## Detailed design by suppliers and modularization and pre-assembly in delivery

Designers often create details that are inefficient to produce. Later in the process, vendors and specialist contractors are typically required to produce detailed design and fabrication documentation for fabrication and installation of materials and components. Boldt uses the simple strategy of involving the fabricator early and integrating their input with the design at the earliest possible stage of the process.

In addition to detailed design being shifted to vendors and specialty contractors, modularization and pre-assembly are used to shift labour off site. This restructuring of supply chains reduces lead times for material, components and equipment.

Labour productivity improves in the controlled conditions of the manufacturing shop. Jigs and other labour saving devices become possible and the company takes full advantage of the enhanced learning achieved under such conditions. The product is delivered in modules or work packages to optimize assembly on-site. Work packages include everything that is required to perform the work thereby reducing craft delays.

## Structured supply chains to enable JIT delivery and reduce on-site inventories

Costs are reduced when suppliers have an assured and reliable demand. Pricing can be based on volume purchases even though deliveries and inventories on-site are based on small batches. Product is delivered just when it is needed to minimize accumulation of inventories on-site, avoiding re-handling.

Engineered products such as switchgears, pumps, fans and fabricated piping and ducting are often delivered late or wrong. Application of lean techniques such as co-located teams, 3D design, value stream mapping and other supply chain methods can reduce the lead times for engineered products enabling scheduling compression and avoiding premature design decisions.

## Lean production management techniques and tools

The key issue in production control is to match load and capacity. Lean production management is used to avoid workers waiting on work as well as work waiting on workers. Lean techniques such as the Last Planner™ System of Production Control and Interactive Team Scheduling are used to increase the reliability and performance of on-site production.

These techniques allow process reliability and work continuity to be increased. The Last Planner™ production management system allows decentralized planning and control and promotes high reliability in the commitments and performance of work. The Last Planner™ is a systematic applied process that:

- guarantees collaborative planning at the production level;
- releases the right work, to the right people, at the right time; and
- promotes continuous improvement through accountability and learning.

Pull scheduling and planning advances work only when the next station needs it. Pull technique results in higher productivity due to consistent crew flow and reliable workflow.

## Formal methods for designing installation methods to optimize performance

First-run studies are used as a lean tool to design installation methods that meet safety, time, quality and cost criteria. A practice of mock-ups is a version of a first-run study. A detailed level of collaborative planning among those responsible for the work develops best practices and assures continuous improvement.

# The ongoing process

Lean implementation started with a focus on improving process reliability with the use of the Last Planner™ process, however, it is interesting to see how the journey to lean production has changed the way everything is perceived and done. The organization learns to look at its performance through a different lens: one that challenges the traditional way of seeing things and doing things. Thus each year Boldt has redefined what lean construction means within their organization. It is an ongoing commitment to look for waste and develop strategies for its reduction.

# Review questions

1. Map and discuss the ways in which the commitment to the principles of lean production has spread through the Boldt Company's operations. Postulate some of the obstacles that might have been encountered in this development.

2. Identify and list the areas where waste reduction has been achieved, identifying the parties in each instance who are the beneficiaries of the specific innovations.

## Acknowledgement

The authors gratefully acknowledge the assistance of Paul Reiser, Corp. VP Productivity and Quality for his assistance in putting this case study together.

Boldt leverages lean thinking to drive efficiency and growth

# Re-engineering timber floors in the Australian housing sector: an example of process innovation

# Project background

This case study is about process innovation, in contrast to product innovation, in the housing sector in Australia. By this distinction we refer to innovative changes in *who* does certain work and *when* it is done rather than any fundamental change to *what* is done. This type of change is harder to conceive and more difficult to implement than straightforward product innovation, which is generally entirely within the control of a single company. It is also hard for the initiating company to capture the financial benefits of construction process innovation and, hence, the motivation for process-based investment is weak. Yet fundamental changes in the way that we procure buildings require a rethink of both the products and the processes of the sector – simple product innovation simply alters the parts we build with but not the entire sequence or process.

The timber products division of CSR, an international construction materials group headquartered out of Sydney, Australia, decided in the early 1990s to improve the competitiveness of their timber floor system against the main competition, concrete slab on ground.

At that time both the timber and concrete industries in Australia were in an advertising war, each was taking out one-page advertisements in the major newspapers to convince the public that their particular flooring solution was superior.

CSR went to the ACCI, a construction research group at the University of New South Wales under the leadership of Marton Marosszeky, and asked the group first of all to establish the truth regarding the costs of the various flooring systems and to help the company develop a more efficient solution. The company simply wanted to increase the market share of timber floors in the Australian residential market against concrete raft-slabs.

One focus of the case study discusses the factors in construction that are essential for process innovation to be successful. The intellectual and financial capital necessary to re-engineer the supply chain is examined along with the challenge for organizations to capture the benefits of innovation.

Since the project CSR has sold its timber business; however, the development project spurred a number of similar innovations and there are several alternative techniques for building timber floors in the market which use the logic of the innovation described in this case study.

# Housing construction in Australia

In Australia, brick veneer houses comprise structural timber frames and trusses that are clad with an external veneer of brickwork. Normally, they have ceramic or concrete tiles on the roof. The ground floor is either of suspended timber construction or a concrete raft slab on ground. The traditional and still most common way of building a suspended timber floor is to construct brick piers, engaged to the external brick veneer and free standing internally, at approximately 1.8 metre centres in both directions throughout the subfloor area. Hardwood timber $100 \times 75\,mm$ ($4 \times 3$ inch) bearers are run first, then $100 \times 50\,mm$ ($4 \times 2$ inch) hardwood joists, and finally the frame is sheeted with 18 mm (3/4 inch) particleboard sheeting. The competing concrete slab construction consists of a 100 mm (4 inch) slab with thickened ribs around the edge and under internal walls.

On a sloping site, in the case of traditional timber floor construction, the site is left unaltered and the external subfloor walls and internal brick piers create a flat platform for the floor construction. In the case of the concrete slab, a dozer creates a flat platform with cut and fill, and the house is partly built on the ground and partly on piers that pass through the fill to the same stratum as is exposed on the high side of the cut.

This case study is an example of process innovation in the construction of timber residential floors. The ideas were developed specifically for brick veneer houses on sloping sites; however, they are applicable to this type of construction on any site. The construction supply chain in the cottage construction sector in Australia is highly fragmented. Before the timber floor of a cottage has been completed, at least six different parties have already been involved in the process.

A surveyor sets out the reference marks and the builder sets profiles for the actual wall dimensions. An excavator digs the trenches and a subcontractor fixes the reinforcing steel and pours the concrete footings. Subcontract bricklayers build the brickwork up to the floor level and subcontract flooring-carpenters build the timber frame and fix the timber floor sheeting. These processes normally take between 1.5 to 3 weeks to complete.

By comparison, the competing concrete raft-slab is a simple arrangement between the builder and subcontractor – a concreter. This subcontractor arranges for the levelling of the site and the drilling of pier holes, erects edge formwork, places a plastic sheet moisture barrier and reinforcement and pours and finishes the concrete slab. This process is faster and cheaper than the fragmented timber alternative – one subcontractor completes the entire work package. The traditional process takes proximately 1 to 1.5 weeks to complete.

Normal product innovation in building construction is ubiquitous and relatively well understood. A manufacturer conducts market research to test the feasibility of a new product idea and invests in an innovation process. The product is then developed and manufactured. If it finds a successful niche in the market, and performs well financially, the product is deemed a success.

Building construction as a whole, however, is highly fragmented and each party is involved in improving their particular segment of the process. The head contractor increasingly acts as a broker, organizing the work through a chain of subcontractors who are expected to take full responsibility for the development, improvement and delivery of their segment of the whole. Under this arrangement, no one looks at the optimization of the entire process.

There is also a major disincentive for contractors to invest in process improvement. Improvements gained by them cannot be quarantined from their competitors because of the fragmented nature of the supply chain. Any competitive benefits gained are almost immediately lost as competitors on competing projects copy any innovation. Hence process innovation has to come from the manufacturing sector.

Overall the research and development created a process for building a timber floor in normal brick veneer construction that was more than 20 percent faster and cheaper than the traditional alternative. To introduce the new process to the market was as much of a challenge as the process development itself, because it involved forming an organization that could deliver the entire work package. The new entity was required to act as a change agent in the market till the new processes were adopted industry-wide.

# Innovation plan

Step one:  Interviews were conducted with 50 tradesmen and builders to identify their perceptions and experiences regarding the differences between the timber and concrete floor construction.

Step two:  A detailed field research to identify all the factors that impact on the efficiency of individual tasks and the flow of activities in traditional timber construction as well as of the competing concrete slab-on-grade process was carried out. This was conducted using detailed work-study methods, which involved documenting observations and taking measurements on six timber and six concrete slab sites.

Step three: An international review of timber floor construction techniques was undertaken to discover solutions that had been developed elsewhere for the problems observed.

Step four: Based on findings from the three preceding steps of the research, a novel method for timber floor construction was developed and validated through discussion with builders.

Step five: A full-scale prototype was constructed and every aspect of the process was evaluated.

Step six: As the findings were positive, the company developed an implementation strategy.

# Expected outcome

The development process that led to the new timber construction system was informed by a number of key findings.

1. The time difference between the timber and concrete flooring processes was due to periods of tradesmen inactivity in the timber construction process, primarily because of the number of parties involved. *This finding led to the idea that a single point of responsibility was needed for the entire timber floor work package.*

2. It was identified that waste in reinforced concrete, when footings are cast in stepped trenches on a sloping hillside, is of the order of 43–66 percent. It was also found that survey set-out tended to be inaccurate on steeper sites. Though it was perceived that flattening a sloping site for a concrete slab had some negative consequences (the need for retaining walls and drainage provision resulted in a lower profile for the building). *Hence these findings overall led to the idea that sites should be benched, perhaps in about 1.2 metre (4 ft) benches.* This had the positive benefits of reducing waste in footing construction and reducing the subfloor brickwork while minimizing the negative consequences of fully benching a site.

3. The fact that the brickwork in a timber floored house is constructed in two stages and the bricks are delivered in two lots. On most sites there is insufficient space to store all the bricks for the building at the outset, especially on sloping sites. This has led to the practice of having the first delivery of bricks simply for the subfloor, and later, after the frame is erected, another for the walls. Problems occur due to colour differences in the batches of bricks delivered at different times, even though they are from the same brickyard. These problems simply can't be avoided in a clay brick manufacturing process. *This led to a preference for leaving all the brick construction for a single stage, after the floor, frames and trusses had all been erected.*

The solution involved benching the site, cutting accurate footing trenches to avoid concrete waste, then casting concrete strip footings and pads from a stiff concrete mix and standing precast concrete piers into the wet concrete. This process had precedence in Melbourne where experienced workers could stand the piers vertically to accurate levels in the wet, footing concrete. The piers were temporarily braced to the footings at the high side of the site and timber bearers, joists and flooring were then assembled on the piers. All this work was undertaken by a single organization.

# Benefits/results

A number of interesting lessons were learned from this case study.

In a single development exercise more than 20 percent was cut from the time and cost of a multi-disciplinary work package. This gives one indication of the scope of benefits that can be captured through a simple process re-engineering exercise.

In the traditional supply chain, the participants are usually only capable of improving their part of the process. Overall process change normally lies beyond the scope and ability of the single, traditional process participants. Such change requires the involvement of organizations, which are first of all able to make the necessary investment and then to capture the benefits. This project was conducted for a major international manufacturer that had the intellectual and financial capital to conceive of and lead the project. Their benefit is in increased sales of their product.

Increasingly head contactors act as brokers in the construction process, co-ordinating other parties to plan and execute the work. If they engage in the kind of innovation described in this case study, they cannot capture and hold the benefits for any significant time as their innovation rapidly becomes common knowledge, spread through an extremely fragmented supply chain which enables their competitors to copy their innovations. The industry becomes more efficient, prices fall but the benefit for the individual contractor in terms of increased margin is limited and very short term.

However, in an international context, actors in any market cannot afford to be complacent. In a global market, head contractors and manufacturers in other, more competitive markets can simply move into less efficient markets and take significant market share before the home industry can catch up. There are ample precedents for this in other industries.

A major challenge for the construction industry is to develop models of process innovation that enable innovators to capture the benefits on their investment.

# Review questions

1. Evaluate the 'innovation plan' used in this case study and explain why it is particularly process rather than product based.
2. Suggest other areas of the construction industry in which this approach could be applied to develop processes.
3. How does the approach described in this case study compare and contrast with 'business process re-engineering' and 'continuous process improvement'?

# Early adoption of TQM values at Babtie Asia pays off

# Company background

While this case study is about Babtie Asia, a part of the Babtie Group, some references are made to the parent Babtie Group. The Babtie Group, founded in 1895, is a 3500 strong multidisciplinary civil engineering consultancy, centred in Glasgow, Scotland. Since 1956 it has grown from 120 to 3500 people, with offices in five countries in the SE Asian region: Hong Kong, China, Singapore, Malaysia and the Philippines.

Babtie Asia, employing some 270 people in Hong Kong, is a multidisciplinary engineering firm that grew out of the fusion of two competitors – Babtie (60 staff) and Harris and Sutherland (80 staff) – in 1997.

Babtie Asia in Hong Kong is organized in four sections, the Environment Department, representing approximately 50 percent of the income stream and employment, is by far the strongest. This department includes marine, geotechnical and water-related activities. It also includes the very significant activity of project managing the HK Government's school refurbishment programme.

# Motivation and implementation

In the early 1990s, when ISO 9000 was first mandated, Harris and Sutherland realized that it needed to formalize its procedures. In 1992 it was the first construction sector consulting practice in HK to achieve ISO 9000 certification and the group subsequently assisted several other consulting practices to also achieve certification. Harris and Sutherland had always had well-developed and well-documented procedures; however, the introduction of ISO 9000 was grasped as an opportunity to formalize both the procedures and their implementation.

Though Harris and Sutherland's strategy for achieving ISO certification was unusual, it turned out to be very effective. They employed a certified quality inspector to join their staff to develop their quality system in the first instance and then to take on the role of quality manager. Right from the outset, therefore, they were able to develop their system from within. They were also able to reflect the actual needs of the business without an external quality consultant, who may have had little knowledge of their business or of the sector. Furthermore, this experience enabled them to assist other firms to achieve accreditation.

In 1998 Babtie Group reviewed the newly melded Babtie Asia quality system and considered it to be too prescriptive. After careful analysis, however, it was recognized that this merely reflected the more prescriptive regulatory and business environment in Hong Kong compared to the UK.

# Overall aim and strategy

## Aim

The Babtie Group's aim is to be a world-class technical and management consultant, delivering sustainable profitable growth by becoming a leader in their sector through the quality of the service that they provide to their clients.

## Strategy

Their strategy is to focus attention on meeting the needs of their clients by organizing their activities around the strengths of those of their people who are best able to deliver, and by working with their clients in a mutually beneficial, quality relationship.

# Core values and specific aims

In striving to achieve their growth, the Babtie Group will remain true to their values. They will continue to change and embrace new people and new cultures, to refresh and refine their skills, to meet new challenges and to strive for sustainability and excellence in all that they do.

Their key strategic aims over the next three years are:

- People centred organization.
- Client focus.
- Quality service delivery.
- Sustainable profitable growth.

In this way they will continue to embrace change, work to continuously improve, and in so doing ensure that they remain a successful and profitable business. One within which people want to work; one with which clients and subcontractors want to work; and one which strives to clearly outperform the sector.

# Development and benefits of the quality system

Harris and Sutherland had started developing their quality system in 1991, publishing their first quality policy in May 1992. The precursor to this system was the design office manual: a set of sample engineering calculations and a set of sample engineering drawing details.

Once the formal quality system was implemented, however, the system ensured that procedures were actually being followed and a verification record was available to prove it. It was also found that documenting procedures provided support for the growing organization: new staff are more easily inducted into procedures when they are well documented.

Furthermore, since the advent of ISO 9000:2001 there has been a need for performance measurement to demonstrate continuous improvement; this has led to six monthly project audits. These audits, in turn, have become the basis for assessing the ongoing benefits derived from implementing the quality system, and identifying problems to target for further improvement.

At a time when all sectors of construction in Hong Kong have been in recession for some years, Babtie Asia has managed to grow the business, building employee numbers to above pre-recession levels. Growth opportunities have come through the development of new services and on the basis of innovation. The project management of the school refurbishments for the government has provided the company with stable work over a number of years. The demand for Independent Certifying Engineering (ICE) services has created an opportunity for growth in a new service area and a value engineering-based approach to working for contractors has paid off. Here, new value has been created by developing new and more efficient solutions to those already proposed. The firm's capacity for innovation has created business opportunities where fees are paid directly out of savings generated for the client.

# Performance measurement

Performance measurement has been developed to support quality improvement as a key part of the quality system. Some of the areas where it is being used include:

- Project quality audit scores are used to compare the quality outcomes between projects and on projects throughout construction. A comprehensive internal audit scheme is implemented with target scores in relation to compliance, client rating and complaint numbers. At regular intervals independent third party audits are conducted to review the performance of the quality system on individual projects and on the overall performance of Babtie Asia.

- Customer satisfaction with service is assessed on every project, and weaknesses are identified as targets for ongoing improvement.
- Every member of staff is assessed each year and counselled regarding his or her strengths, weaknesses and career development – annual increments reflect individual performance.
- A cost control system tracks budgets and cost performance as well individual level of work commitment.

# Client satisfaction

Babtie Asia has implemented a client satisfaction survey in which it asks customers to rate its performance against five dimensions:

- Programme/progress/target achievement.
- Quality (may include presentation standard) of work/deliverables.
- Resources, facilities, and competence and conduct of project staff.
- Project (may include subconsultant) management and co-ordination.
- Effectiveness in responding to queries and solving problems.

# Focus on people

## Investment in people

IIP (Investor in People)* and CAT (Capability Assessment Toolkit)† are two measurement tools which are increasingly being used to identify 'best in class' organizations and are essential to successful business development. In January 2004, Babtie announced that the entire organization had been accredited to the IIP standard.

## Graduate training scheme

Commitment to employing and developing the best and the brightest has made Babtie Asia one of the leading consultancy firms in Hong Kong. The Graduate Training Scheme has been designed to provide training and experience to help young graduates become qualified as a chartered engineer.

Babtie Asia places great emphasis in training both in the office and on site. Babtie engineers and building surveyors are encouraged to keep up to date with advances in technology through structured professional development programmes. Engineering graduates who join Babtie undergo a three-year training programme. Each graduate is allocated a supervising engineer and in most cases another engineer is delegated to give closer support.

---

* Investors in People UK was established in 1993 to provide national ownership of the Investors in People Standard and is responsible for its promotion and branding, quality assurance and development. Since 1991 tens of thousands of UK employers have become involved with the http://www.iipuk.co.uk/IIP/Internet/default.htm
† The Capability Assessment Tool (CAT) is a product and service from the UK Office of Government Commerce to support the capability of its suppliers to deliver successful IT-enabled business change programmes and projects.

The training scheme has been designed to achieve the following objectives:

- Training and personal development programmes, including approved schemes.
- Technical and managerial career advancement.
- Responsibility and a role in forward thinking teams.
- Challenging design and site work.
- The opportunity to work with a successful and expanding consultancy.

## Staff review process

Every member of staff is reviewed annually against three dimensions of performance: technical, personal and commercial competence. On the basis of this, each staff member is given a performance grade, which is reviewed in a face-to-face interview with his or her immediate supervisor and a director.

# Recent recognition for excellence

GEO Award – letter of commendation from residents for slope design and construction.

Seven times winner of the GEO Award for the best landscaped slope of the month.

Quality building award for Murray House, Ma hang Village Phase 3.

NCE International Consultant of the Year – reflects the group's strong international growth, supported by high levels of local manpower in its regional offices.

HKIA Award for Cathay Pacific Airways Headquarters.

GEO Award for achievement of programme milestones – LPM.

HKIA Award for Lok Fu Shopping Centre II.

HKIA Award for factory for Thomas de la Rue.

The company also has numerous letters of commendation from previous clients for excellent services provided.

# Review questions

1. Explain the role of client satisfaction played in Babie's total quality journey.
2. Evaluate the deployment of the quality system in Babtie, paying particular attention to the focus on people.
3. Discuss the approach to performance measurement and suggest areas for improvement.

# Acknowledgements

The authors gratefully acknowledge the assistance of David Knight, Director of Babtie Asia. Most of the content is taken from company information and from a face-to-face interview.

# Quality management at Contractor Woh Hup

# Company background

Woh Hup was founded as a building and civil engineering company in 1940 and to this day operates as a family company. It is one of the leading building and civil engineering companies in Singapore and has a well-established reputation for the quality of its work. The company has won numerous awards for its quality and safety management achievements.

Many of the projects undertaken by the company are well known landmarks in the region. From sixty-storey office buildings, luxury hotels and condominiums and hospitals to 65 000 m$^2$ warehousing and factory buildings.

The company's 1000 strong, skilled workforce consistently delivers projects at high quality standards. In Singapore the company is seen as one of the leaders in the CONQUAS scheme for quality workmanship.

# Introduction to quality management

In 1994, when ISO 9000 was mandated by the Singaporean government, Woh Hup was the first construction contractor to gain certification; however, they quickly realized that very little, if anything, had changed. Both clients and subcontractors continued to operate largely in the same way as they had before, even when these clients represented the Government.

Business as usual included all the well-known ills of the sector:

- ambiguous use of conditions of contract in relation to ISO 9000 qualifying criteria;
- low price, determined tender success, ignoring quality requirements;
- adversarial relationships and mistrust;
- bureaucratic methods of reviewing and approving work;
- construction teams' skills very limited in relation to quality systems;
- reluctance of people to take responsibility;
- poor specifications;
- quality plans seldom checked for compliance – ad hoc methods deployed;
- QA/QC seldom discussed at team and progress meetings;
- project teams reluctant to make decisions;
- bureaucratic paperwork preferred to practical systems;
- rework cost ignored as a measure of quality;
- lengthy text-based procedures – difficult for workers;
- clashes in trade sequence due to subcontractor control of works.

# Drivers for excellence

Out of a frustration that little had changed, senior management formulated a business strategy that focused on value generation for the firm, the clients and business partners. The initial quality management system had five components:

i. employee appraisal;
ii. internal audits;
iii. non-conformance reporting;
iv. corrective action;
v. management review.

These five components have formed the foundation of the Woh Hup QMS. The company's chairman and board of directors have firmly supported the quality management strategy, and company personnel have driven the redesign of the QMS tools. At Woh Hup each process owner and director is committed to pursuing value creation through the continuous development of the QMS.

Strategy for the further development of the system has involved modifying project management processes in a way that fully integrates client representation and subcontractor/suppliers activities. By 1997 this had created a corporate and project environment that offered greater efficiency for construction operations. It has also led to an integrated system for all project stakeholders based on enhanced process control with particular emphasis on architectural activities.

Collaborative teams from QA, administration, engineering and construction determined the weaknesses of the existing QMS, developed a new set of tools and then an analysis was conducted to identify reasons for poor value creation. This led to agreement that a core structure had to be created to act as the catalyst for the company's strategy to work. All agreed that a redesign of the project Quality Plans, and Inspection and Test Plans would provide the basic structure for the other elements of the system.

It was also agreed that emphasis should be placed on value creation at foreman and worker level. Therefore, to make it easier for the multi-national workforce to understand and use, pictorial and graphical representation of reference material took precedence in training at all levels.

The goals for the new system were defined in the following terms.

- To gain world-class recognition for quality service provided.
- To improve coordination of trade activities.
- To provide career development paths for advancement.
- To deliver excellent work and reduce defects.
- To establish a well-trained, informed and highly productive construction team.
- To increase competitiveness.
- To produce synergy within the project structure.

The system included a focus on training; the attributes being communicated to all staff through induction training and to external parties through training at the project quality planning stage. This has created an informed workforce as well as a high standard of supervision and accurate process monitoring with real time decision-making, resulting in timely and properly coordinated handovers between trade disciplines and systematic handovers of units/apartments.

## Key Performance Indicators (KPIs)

To monitor progress, the company developed the following six Key Performance Indicators in its Quality Policy to measure the effectiveness of the system and its tools.

- Safety    – reduction in the average frequency and severity rate of accidents.
- Quality    – reduction in the number of defects at handover.
- Cost contron    – reduction in the project construction costs.
- Planning and productivity – reduction in the project duration by increased productivity.
- Payment collections    – reduction in debtor days.
- Customer satisfaction    – increase in clients' satisfaction level to above 80% (based on *Customer Satisfaction Questionnaire*).

**Table C5** Management tool development – strategic

| Requirements | Description of management tools | Description of design/re-design |
|---|---|---|
| **Management Philosophy** | Management Review | Changed to focus on KPIs and efficiency improvements. |
| | Company Quality Philosophy | Changed to explore better methods of improving business processes through systems and company culture. |
| | Corrective Action | Changed to empower employees to recommend system improvement. |
| | Customer Satisfaction | The development of a weighting system for analyzing customer satisfaction, which is reviewed every 6 months. |
| **Training, Awareness and Competency** | Best Practice Training | Based on Inspection and Test Plans and guidelines, training modules including a completion test for all site staff. Successful candidates were given safety helmet stickers to identify their competency level. |
| | Multi-Skills Training | All workers at Woh Hup have undertaken a stringent training program to certify their skill levels; the focus is now on multi-skill training in the areas of steelfixing and formwork. |
| | | Supervisory staff undergo CONQUAS 21 training as well as in house training on the quality issues in the structural, M&E and architectural packages. |
| | | New staff member (including transferred staff) to a job site are given induction and specific familiarization training on the standard of work required of them and their subcontractors. |
| | | Subcontractors are also given training in the company's objectives. |
| **Resource Management** | Employee Appraisal | Appraisals are conducted by 2 parties and the information is used as a basis for performance rating, promotion, bonus quantum and salary review. |
| | Intranet Library | The company has taken the step to extend its office LAN System to all new projects through leased lines. This has allowed the company to design and publish an electronic library of management procedures standards, guides, etc. and minimize admin of manual system. |

The company, in addition to internal KPIs also gauges its performance against the assessment of external accredited certification auditors, and CONQUAS quality scores are regularly compared to those of other contractors.

The objectives of the QMS Policy are clearly stated in the company's revised quality policy, which is endorsed by all company directors. The policy sets out the company's drive to cultivate a project environment that delivers excellent quality with value creation for the client as well as Woh Hup.

**Table C6**  Management tool development – operational

| Requirements | Description of management tools | Description of design/re-design |
|---|---|---|
| Product Realization | Graphical and Pictorial representation of Management System Reference Materials | Designed new procedures using flowcharts, pictures and wallcharts. |
| | Inspection & Test Plans | By far the biggest impact on company processes was achieved by breaking down the work sequences for a typical project: comprehensive Inspection and Test Plans were prepared for each activity, leading to over 30 new plans for activities such as waterproofing, painting, plastering, etc. |
| | Company QC Guidelines | New company standards were developed using the new Inspection and Test plans, CONQUAS standards, best practices learnt from previous projects and an analysis of maintenance issues. These sources were used to establish a framework to guide the project team. |
| | Defects Management | The company recognized the deficiencies in trying to capture all defects at handover stage and designed an electronic system to capture defects early, notify teams and track progress. |
| | Subcontractor Violations and Incentives | Having clearly established subcontract performance requirements, the company acknowledges contractors who make a meaningful contribution to quality and cost efficiency by giving them preferred status and, hence, a greater opportunity for repeat business. At the same time, subcontractors who violate their contractual responsibilities are penalized. |
| Measurement, Analysis and Improvement | Internal Audits | The company restructured its audit program to progressively evaluate engineering and construction activities. The company employed IRCA registered auditors. |
| | Non-Conformance Reporting | All non-conformance reports are given to the relevant staff/subcontractor/supplier for their onward action/reference. |
| | Corrective Action | Changed to allow employees the opportunity to recommend system improvement. |
| | Second Party Audits | Introduced supplier and subcontractor audit programs for each project to follow in line with procurement schedule. |
| | Project QA/QC Appraisal | Woh Hup is the first company to set-up a project QA/QC Appraisal System that follows the CONQUAS System. This allows the company to know its standard of quality relative to CONQUAS for physical work on a bi-weekly and cumulative basis. The company purchased equipment and assigned QA/QC Engineers to this task. The results are used to make adjustments/ improvements during construction. |
| | Defects Tracking System | A defect tracking system was set up to identify the trend analysis of defects reported at Defects Liability Period. The results will form part of the awareness training of subsequent projects and best practice guides. |

# QMS System achievements

Woh Hup received a total of 57 non-conformances (NCs) in nine years of certification (only one was received in the last 4 years).

The company gathers data and analyses the KPI indicators at a 3-month frequency. The results of the analyses are reported and discussed at quarterly management review meetings, project management team meetings and subsequent supplier/subcontractor meetings.

A summary of the improvements achieved by using the QMS Management Tools relative to the company's KPIs is shown in the figures below.

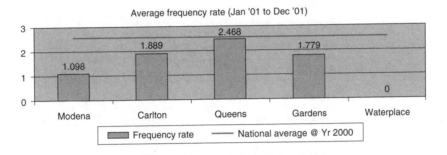

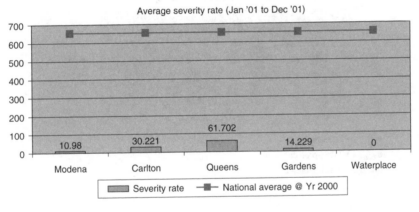

**Figure C31**
Measures of safety performance

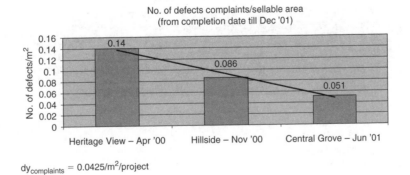

$dy_{complaints} = 0.0425/m^2/project$

**Figure C32**
Measures of quality

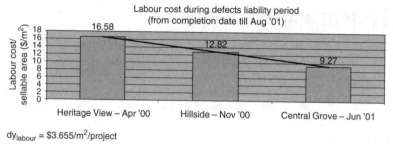

$dy_{labour}$ = \$3.655/m²/project

**Figure C32** (Continued)

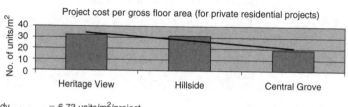

$dy_{project\ cost}$ = 6.73 units/m²/project

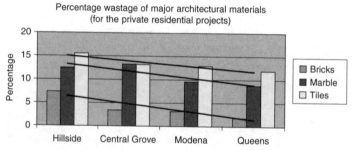

$dy_{bricks}$ = 1.67%/project
$dy_{tiles}$ = 1.117%/project
$dy_{marble}$ = 1.467%/project

**Figure C33**
Measures of cost

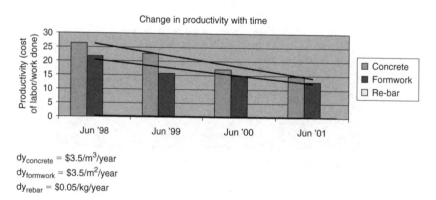

$dy_{concrete}$ = \$3.5/m³/year
$dy_{formwork}$ = \$3.5/m²/year
$dy_{rebar}$ = \$0.05/kg/year

**Figure C34**
Measures of planning and productivity

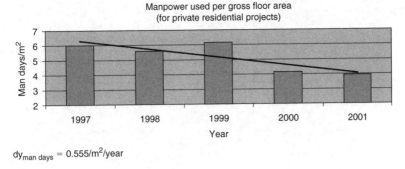

Manpower used per gross floor area
(for private residential projects)

$dy_{man\ days} = 0.555/m^2/year$

**Figure C34** (Continued)

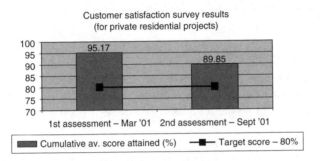

**Figure C35**
Customer satisfaction

# Benefits for the company

The implementation of the QMS has had a number of very important benefits for the company. For example, the control of occupational health and safety of the workers and the reduction in average frequency rate and severity rate of accidents has instilled confidence in the workforce and has provided motivation for improved productivity and quality.

Product realization through improved management processes, training and measuring tools have led to reductions in the number of defects as well as improved the reliability and maintainability of buildings and their components; thereby, reducing lifecycle costs (refer to example below).

For instance, the company identified plaster hollowness as a detrimental problem that required improvement. Allowing this defect to remain, particularly, on the external façade may affect building performance in a number of ways.

■ Collection of water behind the plaster will saturate the brickwork and lead to moisture/water penetration on internal surfaces or walls.
■ Plaster may spall-off and fall (serious risk to the general public at high levels).
■ Encourage fungus and algae to build-up, break brick bond and penetrate paintwork.

The company reviewed industry practice for plastering, traditional methods, new methods, materials, manpower, mixing and application to come up with a method that would reduce hollowness.

| Table C7 Areas of savings as a result of QMS system | | | | | |
|---|---|---|---|---|---|
| Impacted items | Indicative savings as a result of applying the QMS tools | | | | |
| | Honeycomb in concrete structure | Plaster-hollowness | Scratches on doors | Scratches on marble/granite | Scratches on timber flooring |
| Preliminaries | | | | | |
| Labour | | | | | |
| Material | | | | | |
| DSC | | | | | |
| Maintenance work (DLP) | | | | | |
| Others | | | | | |

Studies were then carried out on new projects that had adopted the method and showed that the hollowness had dropped from approx. 18% to <5%. The impact for the company has been considerable. The method dramatically reduced the high cost of remedial work, improved sequencing and handover of works and produced a more reliable building.

The plaster example is one of many exercises carried out by applying the QMS Tools either in combination or separately to reduce defective works. The control of defect and the effect on project cost elements are represented below in Table C7.

Through the introduction of QMS tools project costs have dropped, productivity has risen and quality improved. The high percentage of repeat business further demonstrates the effectiveness and impact of Woh Hup's QMS Tools and has enabled the company to focus even more closely on client satisfaction.

Overall, the impact on the company has been to enable a paradigm shift from the old style of construction – dirty, demanding and dangerous – to one that is best described as professional, productive and progressive. The major impact has been achieved in the following areas:

## Productivity

- Sequenced handovers have produced better coordination and control among trades.
- Smaller, more efficient construction teams have reduced on-site manpower.
- Common project system approach has created collaborative project teams.

## Quality

- In-process verification of satisfactorily completed works (in real time).
- Effective control has reduced remedial works and inherent defects.

## Knowledge sharing and training

- Better informed workforce through planning and awareness training.
- Multi-skilled workforce through training and site testing in lead practices.
- Demonstrable and systematic execution of construction management.

### Competitiveness

■ Clients are offered credible demonstration of management capability.
■ Efficient utilization of resources provides the edge to compete nationally and regionally.

### Dependability and reliability

■ The stability created by the QMS tools has increased the dependability and reliability of Woh Hup services for their clients.

# Pervasiveness and training

The QMS Tools are used company wide by all departments and construction teams. The company conducts inductions into the system for all staff, and department/project specific training to each functional group within each department/site to highlight the uses of the system and its applicability to particular job functions.

As part of the in-house best practice training, the company nurtures a sense of pride in the workforce through knowledge sharing; at the end of each training session participants are tested to determine how much information was understood. This information, together with results of audits and the Employee Appraisal System is used to identify the competency level for each member of the workforce.

The company has planned to enhance the current competency level of the workforce by introducing multi-lingual interactive training modules. These modules are designed to assist the workforce to further develop their skills and knowledge.

# Sustainability and the future

The Employee Appraisal System has been implemented since February 1994; however, the QMS Tools were only introduced in 1997. Since that time, the company has continuously measured and plotted its performance in terms of quality costs among other measures (see in Figure C36 below).

The Directors and Senior Managers evaluate company performance and the effectiveness of the system on a quarterly basis as part of the continuous improvement

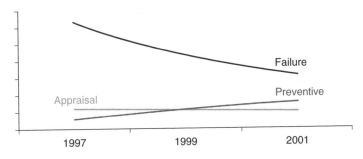

**Figure C36**
Plot of quality costs
Preventive costs – management review, preparation PQPs, provision of guidelines, auditing, training, etc.
Appraisal costs – site based inspections, laboratory testing
Failure costs – remedial works, repair, scrap, downtime, re-inspection

process. Their recommendations are incorporated into a quarterly executive report for the management review committee to discuss.

Up till now, Woh Hup is one of a few contractors in the Singaporean market that have employed a dedicated team of site based professionals, supported by the Head Office team to deliver a value driven management system.

The team has embarked on a rigorous program of training and appraisal. Once the construction team has learnt to apply the QMS tools effectively, the long-term view of the company is to reduce the QA/QC team to a smaller support unit. Given their current performance, this should be achieved within the next 3–5 years and will further reduce the company's quality prevention costs.

## Review questions

1. Write an award submission document for one of the so-called Excellence Awards, such as the EFQM or the Baldrige, based on the material presented here.
2. Discuss the links between the drivers for excellence in Woh Hup and its performance results.
3. What role did the Quality Management System play in helping Woh Hup achieve its results?

# Bibliography

# Understanding quality

Bank, J. (1992) *The Essence of Total Quality Management*, Prentice-Hall, Hemel Hempstead.

Beckford, J. (2002) *Quality*, 2nd edition, Routledge, London.

Cole, R. (2000) 'The early years of the quality movement.' *The quality movement and organization theory*, edited by R.E. Cole and W.R. Scott. London, Sage Publications Inc., pp. 67–88.

Crosby, P.B. (1979) *Quality is Free*, McGraw-Hill, New York.

Crosby, P.B. (1984) *Quality Without Tears*, McGraw-Hill, New York.

Dale, B.G. (ed) (2000) *Managing Quality*, 3rd edition, Philip Alan, Hemel Hempstead.

Deming, W.E. (1982) *Out of the Crisis*, MIT, Cambridge, Mass.

Deming, W.E. (1993) *The New Economies*, MIT, Cambridge, Mass.

Feigenbaum, A.V. (1991) *Total Quality Control*, 3rd edition, revised, McGraw-Hill, New York.

Garvin, D.A. (1988) *Managing Quality: the strategic competitive edge*, The Free Press (Macmillan), New York.

Ishikawa, K. (translated by D.J. Lu) (1985) *What is Total Quality Control? The Japanese Way*, Prentice-Hall, Englewood Cliffs, NJ.

Juran, J.M. (ed) (1988) *Quality Control Handbook*, McGraw-Hill, New York.

Kehoe, D.F. (1996) *The Fundamentals of Quality Management*, Chapman & Hall, London.

Macdonald, J. and Piggot, J. (1990) *Global Quality: the new management culture*, Mercury Books, London.

Murphy, J.A. (2000) *Quality in Practice*, 3rd edition, Gill and MacMillan, Dublin.

Price, F. (1990) *Right Every Time*, Gower, Aldershot.

Soin, S.S. (1992) *Total Quality Control Essentials – key elements, methodologies and managing for success*, McGraw-Hill, New York.

Stahl, M.J. (ed) (1999) *Perspectives in Total Quality*, Quality Press, Milwaukee.

Thomas, R., Marosszeky, M., Karim, K., Davis, S. and McGeorge, D. (2003) *Enhancing Project Completion*, Australian Centre for Construction Innovation, p. 28.

Wille, E. (1992) *Quality: achieving excellence*, Century Business, London.

# Models and frameworks for TQM

BQF (British Quality Foundation) (2000) *The Model in Practice*, London.

BQF (British Quality Foundation) (2002) *The Model in Practice 2*, London.

Brown, M.G. (2002) *Baldrige Award Winning Quality: how to interpret the Malcolm Baldrige Award criteria*, 11th edition, ASQ, Milwaukee.

EFQM (European Foundation for Quality Management) (2003) *The EFQM Excellence Model*, Brussels.

Hart, W.L. and Bogan, C.E. (1992) *The Baldrige: what it is, how it's won, how to use it to improve quality in your company*, McGraw-Hill, New York.

Mills Steeples, M. (1992) *The Corporate Guide to the Malcolm Baldrige National Quality Award*, ASQ, Milwaukee.

National Institute of Standard and Technology (2003) *USA Malcolm Baldrige National Quality Award, Criteria for Performance Excellence*, NIST, Gaithesburg.

# Leadership and commitment

Adair, J. (1987) *Not Bosses but Leaders: how to lead the successful way*, Talbot Adair Press, Guildford.

Adair, J. (1988) *The Action-Centred Leader*, Industrial Society, London.

Adair, J. (1988) *Effective Leadership*, 2nd edition, Pan Books, London.

Davidson, H. (2002) *The Committed Enterprise*, Butterworth-Heinemann, Oxford.

Juran, J.M. (1989) *Juran on Leadership for Quality: an executive handbook*, The Free Press (Macmillan), New York.

Labovitz, G., Chang, Y.S. and Rosansky, V. (1993) *Making Quality Work – a leadership guide for the results-driven manager*, Harper Business, London.

Levicki, C. (1998) *The Leadership Gene*, F.T. Management, London.

Townsend, P.L. and Gebhardt, J.E. (1992) *Quality in Action – 93 lessons in leadership, participation and measurement*, John Wiley Press, New York.

## Policy, strategy and goal deployment

Collins, J.C. (2001) *Good to Great*, Random House, New York.

Collins, J.C. and Porras, J.I. (1998) *Built to Last*, Random House, New York.

Hardaker, M. and Ward, B.K. (1987) 'Getting things done – how to make a team work', *Harvard Business Review*, Nov/Dec, pp. 112–119.

Hussey, D. (1998) *Strategic Management*, 4th edition, Butterworth-Heinemann, Oxford.

Johnson, G. and Scholes, K. (2002) *Exploring Corporate Strategy, Text and Cases*, 6th edition, Prentice-Hall, London.

Juran, J. (1988) *Juran on Planning for Quality*, Free Press, New York.

Kay, J. (1995) *The Foundations of Corporate Success*, Oxford University Press, Oxford.

## Partnerships and resources

Ansari, A. and Modarress, B. (1990) *Just-in-time Purchasing*, The Free Press (Macmillan), New York.

Bineno, J. (1991) *Implementing JIT*, IFS, Bedford.

Harrison, A. (1992) *Just-in-Time Manufacturing in Perspective*, Prentice-Hall, Englewood Cliffs, NJ.

Lysons, K. (2000) *Purchasing and Supply Chain Management*, Prentice-Hall, New York.

Muhlemann, A.P., Oakland, J.S. and Lockyer, K.G. (1992) *Production and Operations Management*, 6th edition, Pitman, London.

Voss, C.A. (ed) (1989) *Just-in-Time Manufacture*, IFS Publications, Bedford.

## Design for quality

Abdul-Rahman, H., Kwan, C.L. and Woods, P.C. (1999) 'Quality function deployment in construction design: application in low-cost housing design', *International Journal of Quality & Reliability Management*, Vol. 16 No. 6, pp. 591–605. MCB University Press, 0256-671X.

Albrecht, K. and Zenke, R. (1995) *Service America! – Doing business in the New Economy*, Dow Jones-Irwin, Homewood 111 (USA).

Austin, S., Baldwin, A., Li, B. and Waskett, P. (2000) 'Analytical Design Planning Technique (ADePT): A Dependency Structure Matrix Tool to Schedule the Building Design Process', *Construction Management and Economics*, Vol. 18, pp. 173–182.

Carlzon, J. (1987) *Moments of Truth*, Ballinger, Cambridge, Mass.

European Centre for Business excellence (the Research and Education Division of Oakland Consulting plc, 33 Park Square, Leeds LS1 2PF)/British Quality Foundation/Design Council (1998) *Designing Business Excellence*.

Fox, J. (1993) *Quality Through Design*, MGLR.

Juran, J.J. (1992) *Juran on Quality by Design*, Free Press, New York.

Khanzode, A., Fischer, M. and Reed, D. (2005) Case study of the implementation of the lean project delivery system using virtual building technologies on a larger healthcare project, *Proceedings 13th International Group for Lean Construction Conference*, (ed R. Kenley) Sydney, pp. 153–160. Available at: http://www.iglc.net/conferences/2005/papers/session04/18_036_Khanzode_Fischer_Reed.pdf

Marsh, S., Moran, J., Nakui, S. and Hoffherr, G.D. (1991) *Facilitating and Training in QFD*, ASQ, Milwaukee.

King Taylor, L. (1992) *Quality: total customer service* (a case study book), Century Business, London.

Mastenbrock, W. (ed) (1991) *Managing for Quality in the Service Sector*, Basil Blackwell, Oxford.

Parasuraman, A., Zeithaml, V.A. and Berry, L.L. (1988) 'SERVQUAL: a multiple-item scale for consumer perceptions of service quality', *Journal of Retailing*, Vol. 64, No. 1, pp. 12–40.

Parasuraman, A., Zeithaml, V.A. and Berry, L.L. (1991) 'Refinement and reassessment of the SERVQUAL scale', *Journal of Retailing*, Vol. 67, No. 4, pp. 420–450.

Parasuraman, A., Zeithaml, V.A. and Berry, L.L. (1994) 'Alternative scales for measuring service quality: a comparative assessment based on psychometric and diagnostic criteria', *Journal of Retailing*, Vol. 70, No. 3, pp. 201–230.

Steward, D.V. (1965) 'Partitioning and Tearing Systems of Equations', *SIAM Journal On Numerical Analysis*, Vol. 2, No. 2, pp. 345–365.

Zeithaml, V.A., Parasuraman, A. and Berry, L.L. (1990) *Delivering Quality Service: balancing customer perceptions and expectations*, The Free Press (Macmillan), New York.

# Performance measurement frameworks

British Standards Institute (1992) *BS6143-1: 1992, Guide to the economics of quality, Process cost model*, BSI, London.

British Standards Institute (1992) *British Standard BS6143-2: 1990, Guide to the economics of quality, Prevention, appraisal and failure model*, BSI, London.

Cameron, K. and Quinn, R.E. (1999) *Diagnosing and Changing Organizational Culture: Based on the competing values framework*, Addison Wesley, New York.

Dale, B.G. and Plunkett, J.J. (1991) *Quality Costing*, Chapman and Hall, London.

Dixon, J.R., Nanni, A. and Vollmann, T.E. (1990) *The New Performance Challenge – measuring operations for world class competition*, Business One Irwin, Homewood.

Hall, R.W., Johnson, J.Y. and Turney, P.B.B. (1991) *Measuring Up – charting pathways to manufacturing excellence*, Business One Irwin, Homewood.

Kaplan, R.W. (ed) (1990) *Measures for Manufacturing Excellence*, Harvard Business School Press, Boston, MA.

Kaplan, R.S. and Norton, P. (1996) *The Balanced Scorecard*, Harvard Business School Press, Boston, MA.

Karim, K., Davis, S., Naik, N. and Marosszeky, M. (2004) 'Designing an Effective Framework for Performance Measurement', *CIOB Special Publication – Quality Management*, January.

Marosszeky, M., Karim, K., Davis, S. and Naik, N. (2004). Lessons Learnt in Developing Effective Performance Measures for Construction Safety Management, *Proceedings 12th Conference of the International Group for Lean Construction (Ed. Carlos T. Formoso)*, Denmark.

Marosszeky, M. and Karim, K. (2002) 'Enterprise process monitoring using key performance indicators', In: *Building in Value*, eds R. Best and G. de Valence, Butterworth-Heinemann, Oxford.

Marosszeky, M., Karim, K., Perera, P. and Davis, S. (2005) 'Improving Work Flow Reliability Through Quality Control Mechanisms', *Proceedings 13th Conference of the International Group for Lean Construction* (Ed: Russel Kenley), Sydney, Australia.

Neely, A. (2002) *Measuring Business Performance*, 2nd edition, Economist Books, London.

Porter, L.J. and Rayner, P. (1992) 'Quality costing for TQM', *International Journal of Production Economics*, 27, pp. 69–81.

Parasuraman, A., Zeithaml, V.A. and Berry, L.L. (1994) 'Alternative scales for measuring service quality: a comparative assessment based on psychometric and diagnostic criteria', *Journal of Retailing*, Vol. 70, No. 3, pp. 201–230.

Talley, D.J. (1991) *Total Quality Management: performance and cost measures*, ASQ, Milwaukee.

Thomas, R., Marosszeky, M., Karim, K., Davis, S. and McGeorge, D. (2003) *Enhancing Project Completion*, Australian Centre for Construction Innovation, p. 28.

Zairi, M. (1992) *TQM-Based Performance Measurement*, TQM Practitioner Series, Technical Communication (Publishing), Letchworth.

Zairi, M. (1994) *Measuring Performance for Business Results*, Chapman and Hall, London.

# Self-assessment, audits and reviews

Arter, D.R. (2000) *Quality audits for improved performance*, 2nd edition, Quality Press, Milwaukee.

BQF (1999) *The X-Factor, winning performance through business excellence*, European Centre for Business Excellence/British Quality Foundation, London.

BQF (2000) *The Model in Practice*, British Quality Foundation, London (Prepared by the European Centre for Business Excellence).

BQF (2002) *The Model in Practice 2*, British Quality Foundation, London (Prepared by the European Centre for Business Excellence).

EFQM (2003) *Assessing for Excellence – a practical guide for self-assessment*, EFQM, Brussels.

International Standards Organisation (2003) *ISO 19011: Guidelines on quality and/or environmental management systems auditing*, ISO.

JUSE (Union of Japanese Scientists and Engineers) (2003) *Deming Prize Criteria*, JUSE, Tokyo.

Keeney, K.A. (2002) *The ISO 9001:2000 Auditor's Companion*, Quality Press, Milwaukee.

NIST (National Institute of Standards and Technology) (2003) *The Malcolm Baldrige National Quality Award Criteria*, NIST, Gaithersburg.

Porter, L.J. and Tanner, S.J. (2003) *Assessing Business Excellence*, 2nd edition, Butterworth-Heinemann, Oxford.

Porter, L.J., Oakland, J.S. and Gadd, K.W. (1998) *Evaluating the European Quality Award Model for Self-Assessment*, CIMA, London.

Pronovost, D. (2000) *Internal Quality Auditing*, Quality Press, Milwaukee.

Tricker, R. (2001) *ISO 9001:2000 Audit Procedures*, Butterworth-Heinemann, Oxford.

# Benchmarking

Bendell, T. (1993) *Benchmarking for competitive advantage*, Longman, London.

Camp, R.C. (1995) *Business Process Benchmarking: finding and implementing best practice*, ASQ Quality Press, Milwaukee.

Camp, R.C. (1998) *Global cases in benchmarking: best practices from organisations around the world*, Quality Press, Milwaukee.

Macdonald, J. and Tanner, S.J. (1996) *Understanding Benchmarking in a Week*, Institute of Management, London.

Marosszeky, M. and Karim, K. (2002) 'Enterprise process monitoring using key performance indicators', In: *Building in Value*, eds R. Best and G. de Valence, Butterworth-Heinemann, Oxford.

Morling, P. and Tanner, S.J. (2000) 'Benchmarking a public service business management system', *Total Quality Management*, Vol. 11, No. 4, 5 & 6, pp. 417–426.

Spendolini, M.J. (1992) *The Benchmarking Book*, ASQ, Milwaukee.

Zairi, M. (1996) *Benchmarking for Best Practice*, Butterworth-Heinemann, Oxford.

Zairi, M. (1998) *Effective Management of Benchmarking Projects*, Butterworth-Heinemann, Oxford.

# Process management

Besterfield, D. (2000) *Quality Control*, 6th edition, Prentice-Hall, Englewood Cliffs, NJ.

BQF (1999) *The X-Factor, winning performance through business excellence*, European Centre for Business Excellence/British Quality Foundation, London.

BQF (2000) *The Model in Practice*, British Quality Foundation, London (Prepared by the European Centre for Business Excellence).

BQF (2002) *The Model in Practice 2*, British Quality Foundation, London (Prepared by the European Centre for Business Excellence).

Dimaxcescu, D. (1992) *The Seamless Enterprise – making cross-functional management work*, Harper Business, New York.

Francis, D. (1990) *Unblocking the Organizational Communication*, Gower, Aldershot.

Harrington, H.J. (1995) *Total Improvement Management*, McGraw-Hill, New York.

Rummler, G.A. and Brache, A.P. (1998) *Improving Performance: how to manage the white space on the organization chart*, 2nd edition, Jossey-Bass Publishing, San Francisco, CA.

Senge, P.M. (1990) *The Fifth Discipline*, Century Business, London.

Senge, P.M., Roberts, C., Ross, R.B., Smith, B.J. and Kleiner A. (1994) *The Fifth Discipline Fieldbook – Strategies and tools for building a learning organization*, Nicholas Brearley, London.

Oakland J.S. (1999) *Total Organisational Excellence*, Butterworth-Heinemann, Oxford.

Warboys, B.C., Kawalek, P., Robertson, I. and Greenwood, R.M. (1999) *Business Information Systems: A Process Approach*, McGraw-Hill, New York.

# Process re-design

Arbulu, R.J. and Tommelein, I.D. (2002) Alternative Supply-Chain Configurations for Engineered or Catalogued Made-to-Order Components: case study on pipe supports used in power plants, *Proceedings IGLC-10*, August, Gramado, Brazil. Available at: http://www.cpgec.ufrgs.br/norie/iglc10/

Braganza, A. and Myers, A. (1997) *Business Process Redesign – a view from the inside*, International Thomson Business Press, London.

Davis, S. and Naik, N. (2004) Lessons Learnt in Developing Effective Performance Measures for Construction Safety Management, A Case Study, *Proceedings IGLC-12*, August, Ellsinore, Denmark. Available at: http://www.iglc2004.dk/13728

Elfving, J., Tommelein, I.D. and Ballard, G. (2003) An International Comparison of the Delivery Process of Power Distribution Equipment, *Proceedings IGLC-11*, August, Blacksburg, USA. Available at: http://strobos.cee.vt.edu/IGLC11/

Hammer, M. and Champy, J. (1993) *Re-engineering the corporation*, Nicholas Brearley, London.

Hammer, M. and Stanton, S.A. (1995) *The Re-engineering Revolution – the handbook*, BCA, Glasgow.

Jacobson, I. (1993) *The Objective Advantage – Business Process Re-engineering with Object Technology*, John Wiley, Chichester.

Johansson, H.J., McHugh, P., Pendlebury, A.J. and Wheeler, W.A. (1993) *Business Process Re-engineering*, John Wiley, London.

Koskela, L. and Howell, G. (2002) The Theory of Project Management: explanation to novel methods, *Proceedings IGLC-10*, August, Gramado, Brazil. Available at: http://www.cpgec.ufrgs.br/norie/iglc10/

Marosszeky, M., Sauer, C., Johnson, K., Karim, K. and Yetton, P. (2000) Information Technology in the Building and Construction Industry: The Australian Experience, *INCITE 2000 – Implementing IT to Obtain a Competitive Advantage in the 21st Century*, The Hong Kong Polytechnic University, Hong Kong.

Marosszeky, M., Thomas, R., Karim, K., Davis, S. and McGeorge, D. (2002) Quality Management Tools for Lean Production: moving from enforcement to empowerment, *Proceedings IGLC-10*, August, Gramado, Brazil, p. 87. Available at: http://www.cpgec.ufrgs.br/norie/iglc10/frame_proceedings.htm

Marosszeky, M., Karim, K., Davis, S. and Naik, N. (2004) Lessons Learnt in Developing Effective Performance Measures for Construction Safety Management, *Proceedings 12th Conference of the International Group for Lean Construction (Ed: Carlos T. Formoso)*, Denmark.

Saurin, T., Formoso, C., Guimarães, L.M. and Soares, A. (2002) Safety and Production: an integrated planning and control model, *Proceedings IGLC-10*, August, Gramado, Brazil (p. 61). Available at: http://www.cpgec.ufrgs.br/norie/iglc10/frame_proceedings.htm

Tommelein, I.D., Akel, N.G. and Boyers, J.C. (2004) Application of Lean Supply Chain Concepts to a Vertically-Integrated Company: A Case Study, *Proceedings IGLC-12*, August, Ellsinore, Denmark. Available at: http://www.iglc2004.dk/13729

Womack, J.D., Jones, D.T. and Roos, D. (1991) *The machine that changed the world*, Harper Collins.

# Quality management systems

Born, G. (1994) *Process Management to Quality Improvement*, John Wiley, Chichester.

British Standards Institute (BSI) (2000) *BS EN ISO 9001:2000, Quality Management Systems*, BSI, London.

Cianfrani, C.A., Tsiakals, J.J. and West, J.E. (2001) *ISO 9001:2000 explained*, 2nd edition, Quality Press, Milwaukee.

Cianfrani, C.A., Tsiakals, J.J. and West, J.E. (2001) *The ASQ ISO 9000:2000 Handbook*, Quality Press, Milwaukee.

Dissanayaka, S.M., Kumaraswamy, M.M., Karim, K. and Marosszeky, M. (2000) 'Evaluating Outcomes from IS0 9000 Certified Quality Systems of Hong Kong Constructors', *Total Quality Management Journal*, Vol. 12, No. 1.

Federal Information Processing Standards (FIPS) (1993) *Publications 183 and 194*, National Institute of Standards and Technology (NIST), Gaithesburg.

Hall, T.J. (1992) *The Quality Manual – the application of BS5750 ISO 9001 EN29001*, John Wiley, Chichester.

Hill, N. (ed) (2002) *Customer Satisfaction Measurement for ISO 9000:2000*, Butterworth-Heinemann, Oxford.

Hoyle, D. (2001) *Integrated Management Systems*, Butterworth-Heinemann, Oxford.

Hoyle, D. (2001) *ISO 9000 Quality Systems Handbook*, 4th edition, Butterworth-Heinemann, Oxford.

International Standards Organisation:

ISO9000-1: 1994, *Quality management and quality assurance standards – Part 1: Guidelines for selection and use*

ISO 9001:2000, *Quality management systems – Requirements*

ISO 9004:2000, *Quality management systems – Guidelines for performance improvements*

ISO 10006:1997, *Quality management – Guidelines to quality in project management*

ISO 10012:2003, *Quality assurance requirements for measuring equipment*

ISO 10013:1995, *Guidelines for developing quality manuals*

ISO/TR 10017, *Guidance on statistical techniques for ISO 9001:1994*

ISO 10241, *International terminology standards – Preparation and layout*

ISO/TR 13425, *Guide for the selection of statistical methods in standardization and specification*

ISO 14001:1996, *Environmental management systems – Specification with guidance for use*

# Continuous improvement

Bauer, J.E., Duffy, G.Z. and Westcott, R.T. (eds) (2002) *The quality improvement handbook*, Quality Press, Milwaukee.

Bendell, T., Wilson, G. and Millar, R.M.G. (1990) *Taguchi Methodology with Total Quality*, IFS, Bedford.

Bhote, K.R. (1999) *World Class Quality – using design of experiments to make it happen*, 2nd edition, AMACOM, New York.

Carlzon, J. (1987) *Moments of Truth*, Ballinger, Cambridge, Mass.

Checkland, P. (1990) *Soft systems methodology in action*, Wiley.

Clark, T.J. (1999) *Success through quality: support guide for the journey to continuous improvement*, Quality Press, Milwaukee.

Harry, M. and Schroeder, R. (2000) *Six-Sigma – the breakthrough management strategy revolutionising the world's top corporations*, Doubleday, New York.

Joiner, B. (1994) *Fourth Generation Management*, McGraw-Hill, New York.

Kinlaw, D.C. (1992) *Continuous Improvement and Measurement for Total Quality – a team-based approach*, Pfieffer & Business One.

Logothetis, N. (1992) *Managing for Total Quality*, Prentice-Hall Intl., London.

Neave, H. (1990) *The Deming Dimension*, SPC Press, Knoxville.

Oakland, J.S. (2003) *Statistical Process Control: a practical guide*, 5th edition, Butterworth-Heinemann, Oxford.

Pande, P.S., Neumann, R.P. and Cavanagh, R.R.(2000) *The Six-Sigma Way – how GE, Motorola and other top companies are honing their performance*, McGraw-Hill, New York.

Porter, L., Six sigma excellence (2002) *Quality world*, April, pp. 12–15.

Price, F. (1985) *Right First Time*, Gower, London.

Ranjit, R. (1990) *A Primer on the Taguchi Method*, Van Nostrand Reinhold, New York.

Ryuka Fukuda (1990) *CEDAC – a tool for continuous systematic improvement*, Productivity Press, Cambridge, Mass.

Scherkenbach, W.W. (1991) *Deming's Road to Continual Improvement*, SPC Press, Knoxville.

Shingo, S. (1986) *Zero Quality Control: source inspection and the Poka-yoke system*, Productivity Press, Stamford, Conn.

Wheeler, D.J. (1992) *Understanding Statistical Process Control*, 2nd edition, SPC Press, Knoxville.

Wheeler, D.J. (1993) *Understanding Variation*, SPC Press, Knoxville.

White, A. (1996) *Continuous Quality Improvement*, Piatkus, London.

Tennor, A.R. and De Toro, I.J. (1992) *Total Quality Manager – three steps to continuous improvement*, Addison-Wesley, Reading, MA.

# Human resource management

Blanchard, K. and Herrsey, P. (1982) *Management of Organizational Behaviour: Utilizing Human Resources*, 4th edition, Prentice-Hall, Englewood Cliffs, NJ.

Boyatsis, R. (1982) *The Competent Manager: A Model for Effective Performance*, Wiley, New York.

Choppin, J. (1997) *Quality Through People: a blueprint for proactive total quality management*, Rushmere Wynne, Bedford.

Dale, B.G., Cooper, C. and Wilkinson, A. (1992) *Total Quality and Human Resources – a guide to continuous improvement*, Blackwell, Oxford.

Imai, M. (1986) *Kaizen*, McGraw-Hill, New York.

Imai, M. (1997) *Gemba Kaizen: A common sense, low cost approach to management*, Quality Press, Milwaukee.

Kotter, J.P. and Heskett, J.L. (1992) *Corporate Culture and Performance*, The Free Press, New York.

Maguire, S. (1995) 'Learning to change', *European Quality*, Vol. 2, No. 6, p. 8.

Marchington, M. and Wilkinson, A. (1997) *Core Personnel and Development*, Institute of Personnel and Development, London.

Oakland, S. and Oakland, J.S. (2001) 'Current people management activities in world-class organisations', *Total Quality Management*, Vol. 12, No. 6, pp. 25–31.

Ouchi, W. (1985) *Theory Z: How American Business can meet the Japanese Challenge*, Addison Wesley, New York.

Stone, R.J. (2002) *Human Resource Management*, 4th edition, Wiley, London.

# Culture change through teamwork

Adair, J. (1987) *Effective Teambuilding*, 2nd edition, Pan Books, London.

Atkinson, P. (1990) *Creating culture change: The key to successful Total Quality Management*, IFS, Bedford.

Blanchard, K. and Herrsey, P. (1982) *Management of Organizational Behaviour: Utilizing Human Resources*, 4th edition, Prentice-Hall, Englewood Cliffs, NJ.

Katzenbach, J.R. and Smith, D.K. (1994) *The Wisdom of Teams – creating the high performance organisation*, McGraw-Hill/Harvard Business School, Boston, MA.

Kormanski, C. (1985) 'A situational leadership approach to groups using the Tuckman Model of Group Development', *The 1985 Annual: Developing Human Resources*, University Associates, San Diego, CA.

Kormanski, C. and Mozenter, A. (1987) 'A new model of team building: a technology for today and tomorrow', *The 1987 Annual: Developing Human Resources*, University Associates, San Diego, CA.

Bibliography

Krebs Hirsh, S. (1992) *MBTI Team Building Program, Team Member's Guide*, Consulting Psychologists Press, Palo Alto, CA.

Krebs Hirsh, S. and Kummerow, J.M. (1987) *Introduction to Type in Organizational Settings*, Consulting Psychologists Press, Palo Alto, CA.

McCaulley, M.H. (1975) 'How individual differences affect health care teams', *Health Team News*, Vol. 1, No. 8, pp. 1–4.

Myers-Briggs, I. (1987) *Introduction to Type: a description of the theory and applications of the Myers-Briggs Type Indicator*, Consulting Psychologists Press, Palo Alto.

Scholtes, P.R. (1990) *The Team Handbook, Joiner Associates*, Madison, NY.

Schutz, W. (1958) *FIRO: A Three-Dimensional Theory of Interpersonal Behaviour*, Mill Valley WSA, CA.

Schutz, W. (1978) *FIRO: Awareness Scales Manual*, Consulting Psychologists Press, Palo Alto, CA.

Schutz, W. (1994) *The Human Element – Productivity, Self-esteem and the Bottom Line*, Jossey-Bass, San Francisco, CA.

Tannenbaum, R. and Schmidt, W.H. (1973) 'How to choose a leadership pattern', *Harvard Business Review*, May-June.

Tuckman, B.W. and Jensen, M.A. (1977) 'States of small group development revisited', *Group and Organizational Studies*, Vol. 2, No. 4, pp. 419–427, New York.

Webb, J. and Cleary, D. (1994) *Organisational change and the management of expertise*, Routledge, London.

Wellins, R.S., Byham, W.C. and Wilson, J.M. (1991) *Empowered Teams*, Jossey Bass, San Francisco, CA.

Wilkinson, A. and Willmott, H. (eds) (1995) *Making Quality Critical – new perspectives on organisational change*, Routledge, London.

# Communication, innovation and learning

Adair, J. (1990) *The challenge of innovation*, Talbot Adair Press, Guildford.

Dawson, R. (2000) *Developing knowledge-based client relationships*, Butterworth-Heinemann, Oxford.

Ettlie, J.E. (2000) *Managing Technological Innovation*, Wiley, London.

Francis, D. (1990) *Unblocking the Organisational Communication*, Gower, Aldershot.

Larkin, T.J. and Larkin, S. (1994) *Communicating Change – winning employee support for new business goals*, McGraw-Hill, New York.

Purdie, M. (1994) *Communicating for Total Quality*, British Gas, London.

Tidd, J., Bessant, J. and Pavitt, K. (2001) *Managing Innovation*, Wiley, London.

Zairi, M. (1999) *Best practice process innovation management*, Butterworth-Heinemann, Oxford.

# Implementing TQM

Albin, J.M. (1992) *Quality Improvement in Employment and other Human Services – managing for quality through change*, Paul Brookes Publications.

Antony, J. and Preece, D. (2002) *Understanding, managing and implementing quality*, Routledge, London.

Ciampa, D. (1992) *Total Quality – a user's guide for implementation*, Addison-Wesley, Reading, Mass.

Cook, S. (1992) *Customer Care – implementing total quality in today's service driven organisation*, Kogan Page, London.

Economist Intelligence Unit (1992) *Making Quality Work: Lessons from Europe's Leading Companies*, EIU, London.

Fox, R. (1994) *Six Steps to Total Quality Management*, McGraw-Hill, Sydney.

Hiam, A. (1992) *Closing the Quality Gap – lessons from America's leading companies*, Prentice-Hall, Englewood Cliffs, NJ.

Kanji, G.P. and Asher, M. (1996) *100 Methods for Total Quality Management*, Sage, London.

Morgan, C. and Murgatroyd, S. (1994) *Total Quality Management in the Public Sector*, Open University Press, Milton Keynes.

Munro-Faure, L. and Munro-Faure, M. (1992) *Implementing Total Quality Management*, Pitman, London.

Oakland, J.S. (2001) *Total Organizational Excellence*, Butterworth-Heinemann, Oxford.

Senge, P.M., Kleiner, A., Roberts, C., Ross, R., Roth, G. and Smith, B. (2000) *A Fifth Discipline Resource – the dance of change*, Nicholas Brearley, London.

Strickland, F. (1998) *The Dynamics of Change*, Routledge, London.

Taguchi, G. (1987) *Systems of experimental design*, Vols 1&2, Unipub/Kraus Int., New York.

Tunks, R. (1992) *Fast Track to Quality*, McGraw-Hill, New York.

Whitford, B. and Bird, R. (1996) *The Pursuit of Quality*, Beaumont, Herts.

# Index